STRUCTURE & FUNCTION OF THE BODY

STRUCTURE & FUNCTION
OF THE BODY

Gary A. Thibodeau, Ph.D.

Chancellor and Professor of Biology,
University of Wisconsin–River Falls,
River Falls, Wisconsin

Catherine Parker Anthony,
R.N., B.A., M.S.

Formerly Assistant Professor of Nursing, Science Department,
and Assistant Instructor of Anatomy and Physiology,
Frances Payne Bolton School of Nursing,
Case Western Reserve University, Cleveland, Ohio;
formerly Instructor of Anatomy and Physiology, Lutheran
Hospital and St. Luke's Hospital, Cleveland, Ohio

with 330 illustrations, including 287 in color

Illustration program by

Ernest W. Beck
Lake Forest, Illinois

Joan M. Beck, Medical Illustrator,
Minneapolis, Minnesota

James M. Krom
Spring Valley, Minnesota

Eighth Edition

TIMES MIRROR/MOSBY COLLEGE PUBLISHING

ST. LOUIS · TORONTO · SANTA CLARA   1988

Editor: David Kendric Brake
Developmental editor: Kathy Sedovic
Project manager: Carlotta Seely
Production editors: Radhika Rao Gupta, Linda Kocher, Jeanne Gulledge
Design: Rey Umali
Cover illustration: Bob Lapsley

EIGHTH EDITION

Copyright © 1988 by Times Mirror/Mosby College Publishing
A division of The C.V. Mosby Company
11830 Westline Industrial Drive
St. Louis, Missouri 63146

Previous editions copyrighted 1960, 1964, 1968, 1972, 1976, 1980, 1984

Printed in the United States of America

Library of Congress Cataloging-in-Publication Data

Thibodeau, Gary A., 1938-
    Structure & function of the body/Gary A. Thibodeau, Catherine
Parker Anthony.
        p.   cm.
    Anthony's name appears first on earlier editions.
    Bibliography: p.
    Includes index.
    1. Human physiology. 2. Anatomy, Human.   I. Anthony, Catherine
Parker, 1907-   . II. Title. III. Title: Structure and function of
the body.
    [DNLM: 1. Anatomy. 2. Physiology. QS 4 T427s]
QP34.5.T5   1988
612—dc19   87-25552
                                                        CIP

ISBN 0-8016-5047-X (Paperback)   03/C/315
ISBN 0-8016-5036-4 (Hardbound) (SE)   03/B/350

GW/VH/VH 9 8 7 6 5 4 3

Preface

In the 1980s the study of anatomy and physiology presents new and demanding challenges for students and teachers alike. The subject matter in both areas represents a body of knowledge that is large, complex, and ever-growing. Proper presentation in an introductory course is difficult indeed. The eighth edition of *Structure and Function of the Body* is a unique new text designed specifically to meet the changing needs of today's students and teachers. It is unique because it incorporates the successful elements and insights gained from previous editions with the most current information in both disciplines. Teachers familiar with previous editions will note the depth and extent of the revision. Students facing the challenge of learning about human anatomy and physiology for the first time will appreciate and welcome a contemporary new text that is relevant and readable.

The writing style, depth of coverage, and selection of material in this edition of the text are based on the longtime classroom experience of both authors and the suggestions of hundreds of teachers, especially health professionals, who have worked with students at the introductory level. It remains an introductory textbook—a teaching book and not a reference text. Texts that attempt to provide information about all aspects of anatomy and physiology often discourage learning. This is a book that students will read—and learn from. Growing numbers of introductory anatomy and physiology texts on the market today confuse rather than assist students by failing to focus on fundamental principles and separate them from the mass of available material. This text presents the core material of anatomy and physiology in such a way that students can understand it and put it to use. Emphasis is on material required for entry into more advanced courses, completion of professional licensing examinations, and successful application of information in a practical work-related environment.

Two major themes dominate this book: the complementarity of normal structure and function and homeostasis. In every chapter of the book the student is shown how organized anatomical structures of a particular size, shape, form, or placement are intended to serve unique and specialized functions. Repeated emphasis of this principle encourages students to integrate otherwise isolated factual information into a cohesive and understandable whole. As a result, anatomy and physiology emerge as living and dynamic topics of personal interest and importance to the student. The integrating principle of homeostasis is used to show how the "normal" interaction of structure and function is achieved and maintained by dynamic counterbalancing forces. Numerous and highly selective clinical and pathological examples are used throughout the book to help students understand that the disease process is a disruption in homeostasis and a breakdown of the normal integration of form and function.

Instructors looking for a contemporary presentation of anatomy and physiology that covers essentials without wordiness, places emphasis on concepts rather than on descriptions, and correlates structure with function will welcome this text. Concise yet comprehensive, the content is enhanced by numerous pedagogical aids, outstanding new artwork, and many full-color micrographs and illustrations that will reinforce learning and make the teaching and learning of this fascinating subject less difficult and more enjoyable.

ORGANIZATION

The twenty chapters of *Structure and Function of the Body* are grouped into seven organizational units. Sequencing of material follows the pattern most commonly used by instructors of anatomy and physiology at the undergraduate level. Each unit or group of chapters develops one major theme, a theme clearly identified by the unit's title. The sequence of units or of chapters within units may be modified by the instructor, depending upon the needs of the course and the students. Flexibility in the presentation of course material is possible because each unit following Unit One is designed and written to stand alone.

Unit One: The Body as a Whole (Chapters 1 to 4) has been extensively revised. In Chapter 1 (An Introduction to the Structure and Function of the Body) the text's two major themes—homeostasis and complementarity of structure and function—are presented and explained. The basic concept of organization in living organisms is introduced and the importance of learning and using correct terminology is stressed. Chapter 2 (Cells and Tissues) contains expanded material about physical and chemical phenomena that are so important to understanding the life process, such as the role played by permease systems in membrane transport mechanisms. Extensive new material on the structure and function of cellular organelles and basic tissue types is complemented by an outstanding series of new full-color artwork and photomicrographs. Chapter 3 presents an overview of the structure and function of the body's organ systems and is intended to provide a basic "road map" to help the student anticipate and prepare for the more detailed information in subsequent chapters of the text.

Unit Two: Systems That Form the Framework of the Body and Move It (Chapters 5 to 6) introduces the organ system level of organization, using the integumentary system as an example. Chapter 4 (The Integumentary System and Body Membranes) is new to this edition. Chapter 5 (The Skeletal System) discusses the key information about the anatomy and physiology of skeletal tissue and classifies joints or articulations. Chapter 6 (The Muscular System) contains expanded coverage of muscle physiology and a listing of the important muscles of the body and their functions.

Unit Three: Systems That Provide Communication and Control (Chapters 7 to 9) covers the anatomy and physiology of the nervous system and sense organs (Chapter 7), the elements of the autonomic nervous systems (Chapter 8), and the endocrine system (Chapter 9).

Chapter 7 also includes expanded information on neurotransmitters, impulse propagation, synaptic transmission, and sense organ physiology. Extensive new material on endocrine diseases and regulation of hormone activity and secretion has been added to Chapter 9.

Unit Four: Systems That Provide Transportation and Immunity (Chapters 10 to 12) has been extensively revised and reillustrated with full-color diagrams and photomicrographs. New information in Chapter 10 (Blood) includes the anemias and other blood diseases and an expanded discussion of both red and white blood cell functions. Chapter 11 (The Circulatory System: Cardiovascular and Lymphatic Systems) has been extensively rewritten. Additional information includes expanded coverage of the heart chambers, pericardium, valves, heart sounds, coronary circulation, the conduction system, and the normal electrocardiogram. In addition, material on the relationship of the systemic and pulmonary circulations has been expanded and new material on the hepatic portal and the fetal circulations has been added. Chapter 12 (The Immune System) now includes new information on opsonization and on AIDS.

Unit Five: Systems That Process and Distribute Foods and Eliminate Wastes (Chapters 13 to 15) details the anatomy and physiology of three important areas: the digestive, respiratory, and urinary systems. Information presented in Chapter 13 (The Digestive System) has been expanded in several areas and many new full color illustrations have been added. New or more developed material includes coverage of the anatomy of the wall of the gastrointestinal tract, functions of the teeth and saliva, cause of ulcers, and functions of the large intestine. In Chapter 14 (The Respiratory System) discussion of the system's structural plan has been expanded and correlated with the pathological condition called Infant Respiratory Distress Syndrome or IRDS. New information on the nose, paranasal sinuses, and tonsils has also been added. The section on gas exchange in the lungs and tissues has also been completely rewritten.

Unit Six: The Cycle of Life (Chapters 16 to 18) discusses the male and female reproductive systems and the biological process of growth and development. These chapters have been revised to reflect changes in nomenclature used to describe both anatomical structures and functional activity. Numerous illustrations in all three chapters have been modified or replaced with more functional, accurate, and visually attractive full-color artwork.

Unit Seven: Fluid, Electrolyte, and Acid-Base Balance (Chapters 19 to 20) are "stand alone" chapters that provide key information about fluid, electrolyte, and acid-base balance.

SPECIAL FEATURES

CONTENT

As noted earlier, *Structure and Function of the Body* does not attempt to cover every detail of the subject; rather it discusses the basic facts about human anatomy and physiology. It is intended to provide a conceptual framework and the necessary information upon which to build. In each chapter of the text appropriate physiological content balances the anatomical information that is presented. At every level of organization, care has been taken to couple structural information with the important functional concepts. As a result, the student has a more integrated understanding of human structure and function. Throughout the text examples that stress the "complementarity of structure and function" have been consciously selected to emphasize the importance of homeostasis as a unifying concept. Supportive information has been integrated by use of carefully selected clinical examples and other aids. The style of presentation of material, its readability, accuracy, and level of coverage have been carefully developed to meet the needs of undergraduate students

taking an introductory course in anatomy and physiology. Although it is particularly well suited for nursing courses and for use in other allied health areas, we feel *Structure and Function of the Body* will be a useful source of information for students, whatever their particular fields of interest and study.

PEDAGOGY

Structure and Function of the Body is a student-oriented text. Written in a very readable style, it has numerous pedagogical aids that maintain interest and motivation. Every chapter contains the following elements that facilitate learning and the retention of information in the most effective manner:

Chapter Outline: An overview outline introduces each chapter and enables the student to preview the content and direction of the chapter at the major concept level prior to the detailed reading.

Chapter Objectives: This is a *new feature* in this edition. Each chapter opening page contains approximately five measurable objectives for the student. Each objective clearly identifies for the student, before he or she reads the chapter, what the key goals should be and what information should be mastered.

Key Terms and Pronunciation Guide: Key terms, when introduced and defined in the text body, are now identified in **boldface** to highlight their importance. Another feature *new to this edition*, a pronunciation guide, follows each new term that students may find difficult to pronounce correctly.

Boxed Inserts: Numerous brief boxed inserts appear in every chapter. These inserts include information ranging from sidelights on recent research to clinical applications or additional information on issues of importance in anatomy and physiology.

Boxed Essays: Throughout the text there are longer boxed essays on topics of special interest to students. Pathological conditions are sometimes explained in essay format to help students better understand the relationship between normal structure and function. Examples include ulcers, shingles, and acne.

Outline Summaries: Extensive and detailed end-of-chapter summaries in outline format provide excellent guides for students as they review the text materials when preparing for examinations. Many students also find such detailed guides useful as a chapter preview in conjunction with the chapter outline.

Chapter Tests: New objective-type Chapter Test questions are included at the end of each chapter. They serve as quick checks for the recall and mastery of important subject matter. They are also designed as aids to increase the retention of information. Answers to all Chapter Test questions are provided at the end of the text.

Review Questions: Subjective review questions at the end of each chapter allow students to use a narrative format to discuss concepts and synthesize important chapter information for review by the instructor. The answers to these review questions are available in the Instructor's Manual that accompanies the text.

Additional learning and study aids at the end of the text include **Common Medical Abbreviations, Prefixes, and Suffixes;** an extensive **Glossary** of terms to assist students in mastering the vocabulary of anatomy and physiology; and a detailed **Index** that serves as a ready reference for locating information.

Illustrations: A major strength of *Structure and Function of the Body* is the exceptional quality, accuracy, and beauty of the illustration program. The truest test of any illustration is how effectively it can complement and strengthen written information found in the text and how successfully it can be used by the student as a learning tool. Extensive use has been made of full-color illustrations, micrographs, and dissection photographs throughout the text. Illustrations proven pedagogically effective in previous editions of *Structure and Function of the Body* have been retained and supplemented by many additional figures to provide accurate information and visual appeal. Each illustration is carefully referred to in the text and is designed to support the text discussion. The illustrations are an in-

tegral part of the learning process and should be carefully studied by the student.

SUPPLEMENTS

The supplements package has been carefully planned and developed to assist instructors and to enhance their use of the text. Each supplement, including the test items and study guide, has been thoroughly reviewed by many of the same instructors who reviewed the text.

INSTRUCTOR'S MANUAL AND TEST BANK

The Instructor's Manual and Test Bank, prepared by Barbara Felen of Community College of Allegheny County, provides text adopters with substantial support in teaching from the text. The following features are included in every chapter:

Brief chapter overview highlighting key points and concepts in the chapter.

Chapter outline that includes page numbers referred to in the text.

Chapter objectives that correspond with the objectives in the text and provide explanations of answers.

Suggestions to the lecturer contains helpful hints to the instructor for presenting information in the chapter, including suggested review activities, films and filmstrips, and classroom demonstrations.

New words list with pronunciation guide to identify key terms.

TRANSPARENCY ACETATES

A set of 75 transparency acetates in both two and four color is available free to adopters of the text for use as teaching aids.

STUDY GUIDE

The Study Guide, written by Linda Swisher of Sarasota County Vocational Technical Center, provides students with additional self-study aids, including chapter overviews, topic reviews, review questions keyed to specific pages in the text, and application and labeling exercises. These learning aids have been specially designed to help prepare students for class discussion and exams.

A WORD OF THANKS

Many people have contributed to the development and success of *Structure and Function of the Body*. We extend our thanks and deep appreciation to the various students and classroom instructors who have provided us with helpful suggestions following their use of earlier editions of this text.

A specific "thank you" to the following instructors who critiqued in detail the most recent edition of this text. Their invaluable comments were instrumental in the development of this new edition.

Benja Allen
El Centro College

Judith Carpenter
Columbus Technical Institute

Barbara Felen
*Community College of
Allegheny County*

Ruth McFarland
Mt. Hood Community College

Linda Strause
*University of California
at San Diego*

Linda Swisher
*Sarasota County Vocational
Technical Center*

Sandra Uyeshiro
Modesto Junior College

Shirley Yeargin
Rend Lake College

Marshall Yokell
Middlesex Community College

For their contributions to the illustration program our thanks go to Ernest W. Beck, Joan M. Beck, and James M. Krom. Professor Robert Calentine, University of Wisconsin–River Falls, produced the outstanding color photomicrographs. A special acknowledgment is also due Dr. Branislav Vidić, M.D., and his late colleague, Dr. Faustino R. Suarez, M.D., for the use of illustrations from their excellent text, *Photographic Atlas of the Human Body*.

At Times Mirror/Mosby College Publishing thanks are due all who have worked with us in bringing this new edition to completion. We wish especially to acknowledge the support and effort of our editor, David Brake; developmental editor, Kathy Sedovic; project manager, Carlotta Seely; and production editors, Radhika Rao Gupta and Linda Kocher, all of whom were instrumental in bringing this edition to successful completion.

Gary A. Thibodeau
Catherine Parker Anthony

Contents

UNIT ONE

The Body as a Whole

1 An Introduction to the Structure and Function of the Body

Human anatomy and physiology
Structural levels of organization
Some words used in describing body structures
Planes or body sections
Anatomical position
Body regions
Some basic facts about body functions

2 Cells and Tissues

Cells
Movement of substances through cell membranes
Cell reproduction
Tissues

3 Organ Systems of the Body

Integumentary
Skeletal
Muscular
Nervous
Endocrine
Circulatory
Urinary
Digestive
Respiratory
Reproductive

4 The Integumentary System and Body Membranes

The skin
Types of body membranes

CHAPTER

1

An Introduction to the Structure and Function of the Body

OBJECTIVES

After you have completed this chapter, you should be able to:

1 Define the terms *anatomy* and *physiology*.

2 List and discuss in order of increasing complexity the levels of organization of the body.

3 Define the *anatomical position*.

4 Discuss and contrast the *axial* and the *appendicular* subdivisions of the body. Identify a number of specific anatomical regions in each area.

5 List the nine abdominal regions and the four abdominal quadrants.

6 List and define the principal directional terms and body sections (planes) used in describing the body and the relationship of body parts to one another.

7 List the major cavities of the body and the subdivisions found in each.

8 Explain the meaning of the term *homeostasis* and give an example of a typical homeostatic mechanism.

Wonders are many in our world today, but none is more wondrous than the human body. This is a textbook about that incomparable structure. Chapter 1 relates a few basic facts about the body's structure and its functions. It also defines some of the words scientists use in talking about the body.

HUMAN ANATOMY AND PHYSIOLOGY

This text deals with two very distinct and yet interrelated sciences: **anatomy** and **physiology.** As a science anatomy is often defined as the study of the structure of an organism and the relationships of its parts. The word *anatomy* is derived from two Greek words that mean "a cutting up." Anatomists learn about the structure of the human body literally by cutting it apart. This process, called **dissection,** is still the principal technique used to isolate and study the structural components or parts of the human body. Physiology is the study of the functions of living organisms and their parts. It is a dynamic science that requires active experimentation. In the chapters that follow, you will see again and again that anatomical structures seem "designed" to perform specific functions. Each has a particular size, shape, form, or position in the body because it is intended to perform a unique and specialized activity.

STRUCTURAL LEVELS OF ORGANIZATION

Before you begin the study of the structure and function of the human body and its many parts, it is important to think about how those parts are organized and how they might logically fit together into a functioning whole. Examine Figure 1-1. It illustrates the differing levels of organization that influence body structure and function. Note that the levels of organization illustrated in Figure 1-1 progress from the least complex (chemical level) to the most complex (body as a whole).

Organization is one of the most important characteristics of body structure. Even the word *organism,* used to denote a living thing, implies organization.

Although the body is a single structure, it is made up of billions of smaller structures. Atoms and molecules are often referred to as the **chemical level** of organization. The very existence of life itself depends on the proper levels and proportions of many chemical substances in the cytoplasm of cells. Many of the physical and chemical phenomena that play important roles in the life process will be reviewed in the next chapter. Such information provides an understanding of the physical basis for life and for the study of the next levels of organization so important in the study of anatomy and physiology—cells, tissues, organs, and systems.

Cells are considered to be the smallest "living" units of structure and function in our body. Although long recognized as the simplest units of living matter, cells are far from simple. They are extremely complex, a fact you will discover in Chapter 2.

Tissues are somewhat more complex units than cells. By definition a tissue is an organization of a great many similar cells that act together to perform a common function. Cells are held together and surrounded by varying amounts and varieties of gluelike nonliving intercellular substances.

Organs are more complex units than tissues. An organ is a group of several different kinds of tissues arranged so that together they can perform a special function. For instance, the lungs shown in Figure 1-1 are an example of organization at the organ level.

Systems are the most complex units that make up the body. A system is an organization of varying numbers and kinds of organs arranged so that together these organs can perform complex functions for the body. The organs of the respiratory system shown in Figure 1-1 permit air to enter the body and travel to the lungs, where the eventual exchange of oxygen and carbon dioxide can occur. Organs of the respiratory system include the nose, the windpipe or trachea, and the complex series of bronchial tubes that permit passage of air into the substance of the lungs.

Figure 1-1   Structural levels of organization in the body.

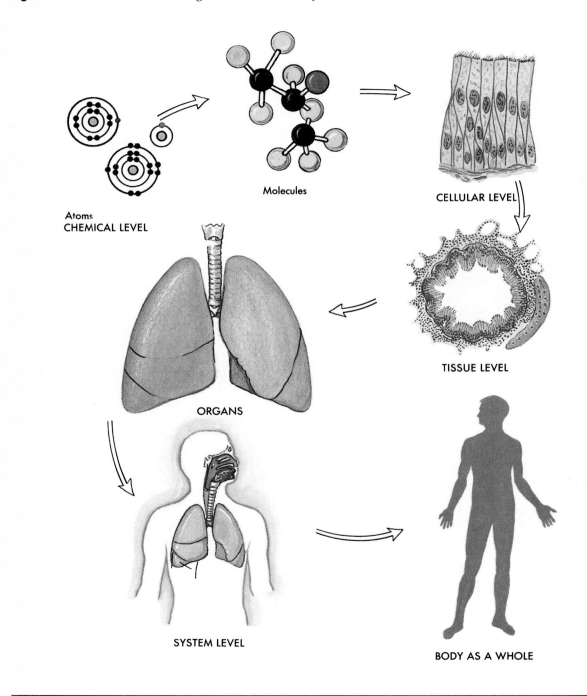

Contrary to its external appearance, the body is not a solid structure. It is made up of open spaces or cavities that in turn contain compact, well-ordered arrangements of internal organs. The two major body cavities are called the **ventral** and **dorsal body cavities.** The location and outlines of the body cavities are illustrated in Figure 1-2. The **thoracic cavity** is the space that you perhaps think of as your chest cavity. Its midportion is a subdivision of the thoracic cavity called the **mediastinum,** and its other subdivisions are called the right and left **pleural cavities.** Figure 1-2 also identifies an **abdominal cavity** and a **pelvic cavity.** Actually they form only one cavity—the **abdominopelvic cavity**—since no physical partition of any kind separates them. In the figure a dotted line is used to show the approximate point of separation between the abdominal and pelvic subdivisions. Notice, however, that an actual physical partition, represented in the figure as a white band, does separate the thoracic cavity from the abdominal cavity. This muscular partition is called the **diaphragm.** It is dome-shaped and is the most important muscle we have for breathing.

The space inside the skull contains the brain and is called the **cranial cavity.** The space inside the spinal column is called the **spinal cavity;** it contains the spinal cord. The cranial and spinal cavities are **dorsal cavities,** whereas the thoracic and abdominopelvic cavities are called **ventral cavities.** Dorsal (or posterior in humans) means "back." Ventral (or anterior in humans) means "front", that is, the abdominal side.

Some of the organs contained in the largest body cavities are visible in Figure 1-3. For example, it shows the trachea, aorta, and heart in the mediastinal portion of the thoracic cavity and the lungs in the pleural portions. Observe the many organs shown in the abdominal cavity: liver, gallbladder, stomach, spleen, pancreas, and parts of the small intestine (cecum and ascending, transverse, and descending colon). The sigmoid colon, rectum, and urinary bladder lie in the pelvic portion of the abdominopelvic cavity. Although there is no specific anatomical structure that separates the abdominal and pelvic portions of the abdominopelvic cavity, the area above the hipbones is considered to be abdominal, and the area below the hipbones is considered pelvic. Find each body cavity in a model of the human body if you have access to one. Try to identify the organs in each cavity, and try to visualize their locations in your own body. Study Figures 1-2 and 1-3.

The structure of the body changes in many ways and at varying rates during a lifetime. Before young adulthood it develops and grows; after young adulthood it gradually undergoes various degenerative changes. With advancing age there is a generalized decrease in size or a wasting away of many body organs and tissues that affects the structure and function of many body areas. This degenerative process is called **atrophy.** Nearly every chapter of this book will refer to a few of these changes.

SOME WORDS USED IN DESCRIBING BODY STRUCTURES

1 Superior and **inferior**—as you can see in Figure 1-4, superior means "toward the head," and inferior means "toward the feet." Superior also means "upper" or "above," and inferior means "lower" or "below." Examples: The lungs are located superior to the diaphragm muscle, whereas the stomach is located inferior to it.

2 Anterior and **posterior**—anterior means "front" or "in front of"; posterior means "back" or "in back of." Examples: The knees are located on the anterior surface of the body, and the shoulder blades are located on its posterior surface. The liver lies anterior to the gallbladder, and the gallbladder lies posterior to the liver. The terms *anterior* and *posterior* are used in reference to humans. In reference to animals, ventral and dorsal are substituted for anterior and posterior.

3 Medial and **lateral** (Figure 1-4)—medial means "toward the midline of the body"; lateral means "toward the side of the body or away from its midline." Examples: The great toe is located at the medial side of the foot, and the little toe is located at its lateral side. The heart lies medial to the lungs, and the lungs lie lateral to the heart.

Figure 1-2   Location and subdivisions of the dorsal and ventral body cavities as viewed from the front (anterior) and from the side (lateral).

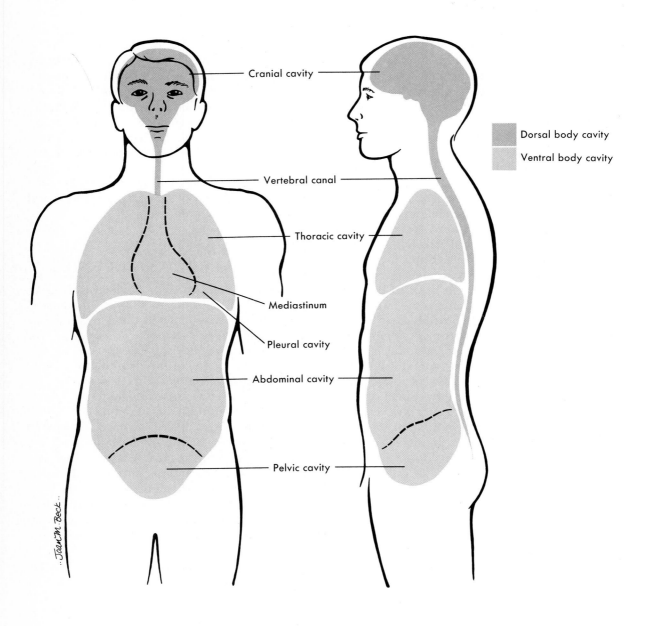

Figure 1-3 Organs of the thoracic and abdominal cavities viewed from the front.

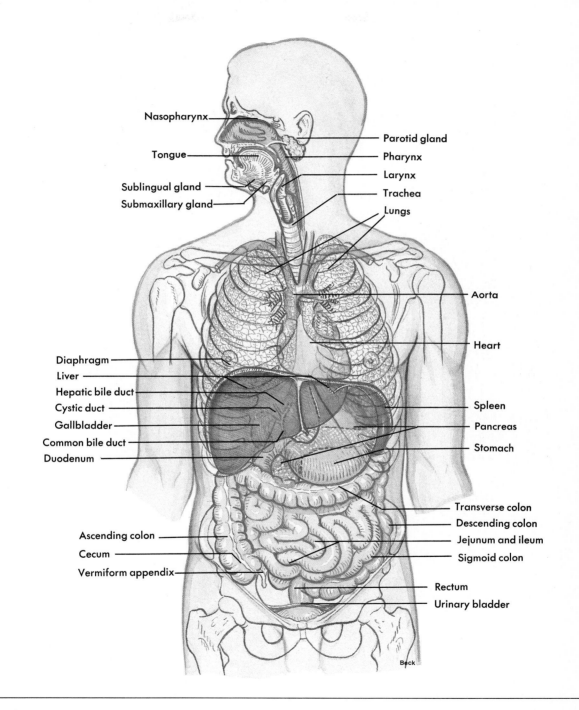

Figure 1-4 Directions and planes of the body.

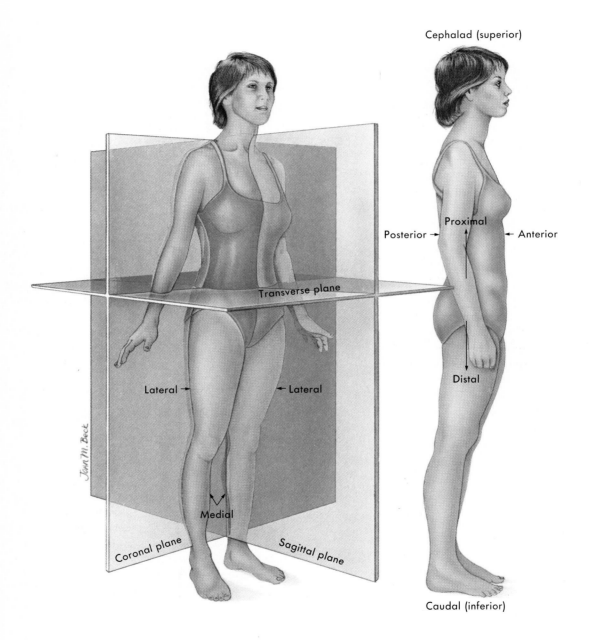

Figure 1-5   The nine regions of the abdomen showing the most superficial organs.

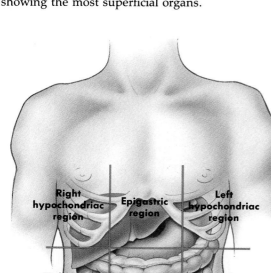

4 Proximal and **distal**—proximal means "toward or nearest the trunk of the body, or nearest the point of origin of one of its parts"; distal means "away from or farthest from the trunk or the point of origin of a body part." Examples: The elbow lies at the proximal end of the lower arm, whereas the hand lies at its distal end.

5 Superficial and **deep**—superficial means nearer the surface; deep means farther away from the body surface. Examples: The skin of the arm is superficial to the muscles below it, and the humerus of the arm is deep to the muscles that surround and cover it.

6 Abdominal regions—to make it easier to locate abdominal organs, anatomists have divided the abdomen into the nine regions shown in Figure 1-5 and defined them as follows:

a *Upper abdominal regions*—right hypochondriac, epigastric, and left hypochondriac regions; these lie above an imaginary line across the abdomen at the level of the ninth rib cartilages

b *Middle regions*—right lumbar, umbilical, and left lumbar regions; lie below an imaginary line across the abdomen at the level of the ninth rib cartilages and above an imaginary line across the abdomen at the top of the hipbones

c *Lower regions*—right iliac, hypogastric, and left iliac regions; lie below an imaginary line across the abdomen at the level of the top of the hipbones (see Figure 1-5)

Another, and perhaps easier, way to divide the abdomen is shown in Figure 1-6. This method is frequently used by health professionals and is useful for locating pain or describing the location of a skin lesion or abdominal tumor. As you can see in Figure 1-6, midsagittal and transverse planes, which both pass through the umbilicus, will divide the abdomen into **four quadrants:** right upper or superior, right lower or inferior, left upper or superior, and left lower or inferior.

PLANES OR BODY SECTIONS

To facilitate the study of individual organs or the body as a whole, it is often useful to subdivide or "cut" it into smaller segments. In order to do this, body planes or sections have been identified by special names. Read the following definitions and identify each term in Figure 1-4.

1 Sagittal—a sagittal cut or section is a lengthwise plane running from front to back. It divides the body or any of its parts into right and left sides. The sagittal plane shown in Figure 1-4 divides the body into two *equal halves.* This is a unique type of sagittal plane and is called a **midsagittal plane.**

Figure 1-6 Division of the abdomen into four quadrants. **A,** Photo showing surface outlines of *1*, right upper quadrant; *2*, left upper quadrant; *3*, right lower quadrant; *4*, left lower quadrant. **B,** Diagram showing relationship of internal organs to the four abdominal quadrants.

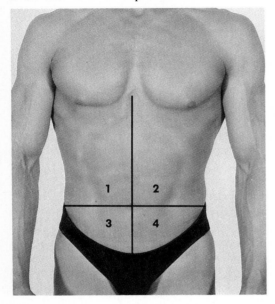

A

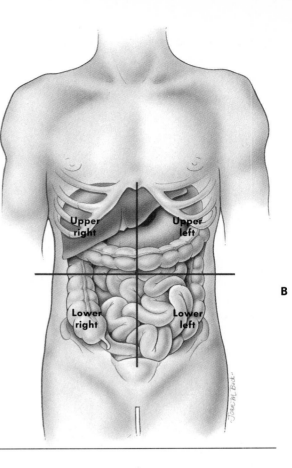

B

2 Coronal—a coronal plane is a lengthwise plane running from side to side. As you can see in Figure 1-4, a coronal plane divides the body or any of its parts into anterior and posterior (front and back) portions.

3 Transverse—a transverse plane is a horizontal or crosswise plane. Such a plane divides the body or any of its parts into upper and lower portions.

ANATOMICAL POSITION

Discussions about the body, how it moves, its posture, or the relationship of one area to another assume that the body as a whole is in a specific position called the **anatomical position.** In this reference position (Figure 1-7) the body is in an erect or standing posture with the arms at the sides and palms turned forward. The head and feet are also pointing forward. The anatomical position is a reference position that gives meaning to the directional terms used to describe the body parts and regions.

BODY REGIONS

In order to recognize an object, you usually first notice its overall or generalized structure and form. For example, a car is recognized as a car before the specific details of its tires, grill, or wheel covers are noted. Recognition of the human form also occurs as you first identify overall shape and basic outline. However, in order for more specific identification to occur, details of size, shape, and appearance of individual body areas must be described. Individuals differ in

Figure 1-7   Anatomical position: the body is in an erect or standing posture with the arms at the sides and the palms forward. The head and feet are also pointing forward.

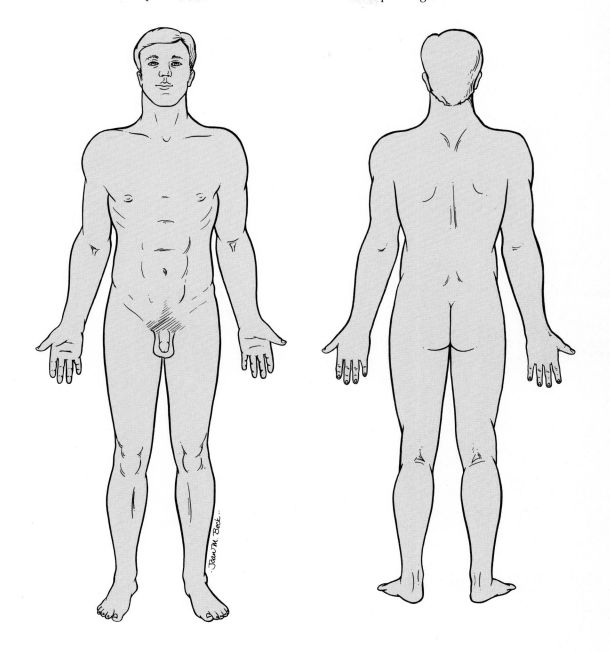

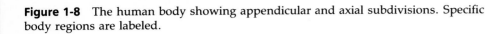

Figure 1-8 The human body showing appendicular and axial subdivisions. Specific body regions are labeled.

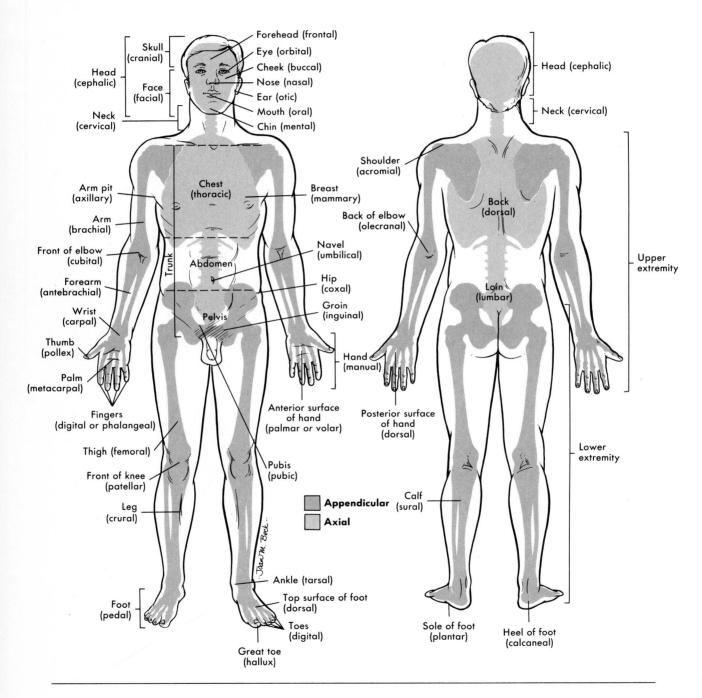

overall appearance because specific body areas such as the face or torso have unique identifying characteristics. Detailed descriptions of the human form require that specific regions be identified and appropriate terms be used to describe them.

The ability to identify and correctly describe specific body areas is particularly important in the health sciences. For a patient to complain of pain in the head is not as specific, and therefore not as useful to a physician or nurse, as a more specific and localized description. Saying that the pain is facial provides additional information and helps to identify the area of pain more specifically. By using correct anatomical terms such as forehead, cheek, or chin to describe the area of pain, attention can be focused even more

quickly on the specific anatomic area that may need attention. Familiarize yourself with the more common terms used to describe specific body regions identified in Figure 1-8 and listed in Table 1-1.

The body as a whole can be subdivided into two major portions or components: **axial** and **appendicular.** The axial portion of the body consists of the head, neck, and torso or trunk; the appendicular portion consists of the upper and lower extremities. Each major area is subdivided as shown in Figure 1-8. Note, for example, that the torso is composed of thoracic, abdominal, and pelvic areas, and the upper extremity is divided into arm, forearm, wrist, and hand components. Although most terms used to describe gross body regions are well understood, misuse

Table 1-1 Body regions

Body region	Area or example
Abdominal (ab-DOM-in-al)	Anterior torso below diaphragm
Antebrachium (an-te-BRA-ke-um)	Forearm
Antecubital (an-te-KU-bital)	Depressed area just in front of elbow
Arm	Upper extremity between shoulder and elbow
Axillary (AK-si-ler-e)	Armpit
Brachial (BRA-ke-al)	Arm
Buccal (BUK-al)	Mouth
Carpal (CAR-pal)	Wrist
Cephalic (se-FAL-ik)	Head
Cervical (SER-vi-kal)	Neck
Costal (KOS-tal)	Ribs
Cranial (CRA-ne-al)	Skull
Crural (KROOR-al)	Leg
Cubital (KU-bi-tal)	Elbow
Cutaneous (ku-TANE-e-us)	Skin (or body surface)
Digital (DIJ-i-tal)	Fingers or toes

Continued.

Table 1-1 Body regions—cont'd

Body region	Area or example
Dorsum (DOR-sum)	Back
Epigastric (ip-i-GAS-trik)	Upper middle area of abdomen
Facial (FAY-shal)	Face
Frontal (FRON-tal)	Forehead
Oral (OR-al)	Mouth
Orbital or **ophthalmic** (OR-bi-tal or op-THAL-mik)	Eyes
Nasal (NA-sal)	Nose
Zygomatic (zi-go-MAT-ik)	Cheek
Femoral (FEM-or-al)	Thigh
Forearm	Upper extremity between elbow and wrist
Gluteal (GLOO-te-al)	Buttock
Groin	Root of thigh between lower extremity and abdomen
Inguinal (ING-gwi-nal)	Groin
Leg	Lower extremity between knee and ankle
Lumbar (LUM-bar)	Lower back between ribs and pelvis
Mammary (MAM-er-e)	Breast
Mastoid (MAST-oid)	Area of skull just below and behind the ear
Occipital (ok-SIP-i-tal)	Back of lower skull
Palmar (PAHL-mar)	Palm of hand
Pectoral (PEK-tor-al)	Chest
Pedal (PED-al)	Foot
Pelvic	Lower portion of torso
Perineal (per-i-NE-al)	Area (perineum) between anus and genitals
Precordial (pre-COR-di-al)	Chest area over heart
Plantar (PLAN-tar)	Sole of foot
Popliteal (pop-li-TEA-al)	Behind knee
Supraclavicular (supra-cla-VIC-u-lar)	Above clavicle
Tarsal (TAR-sal)	Ankle
Temporal	Side of skull
Thoracic (tho-RAS-ik)	Chest
Umbilical (um-BIL-i-cal)	Area around umbilicus
Volar (VO-lar)	Palm or sole

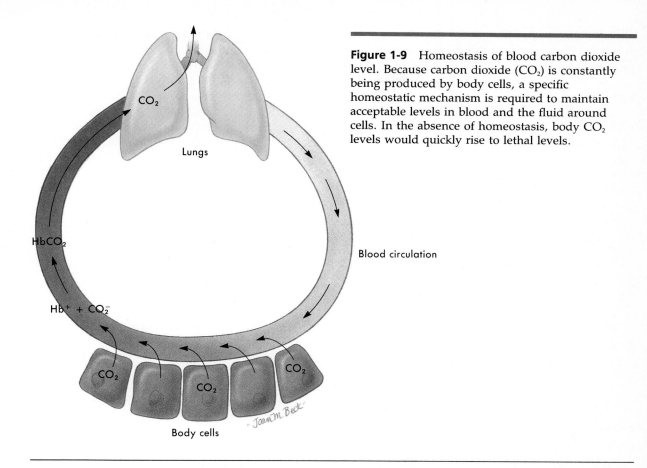

Figure 1-9 Homeostasis of blood carbon dioxide level. Because carbon dioxide (CO_2) is constantly being produced by body cells, a specific homeostatic mechanism is required to maintain acceptable levels in blood and the fluid around cells. In the absence of homeostasis, body CO_2 levels would quickly rise to lethal levels.

is common. The word *leg* is a good example: it refers to the area of the lower extremity between the knee and ankle and *not* to the entire lower extremity.

SOME BASIC FACTS ABOUT BODY FUNCTIONS

1 Survival is the body's most important business—survival of itself and survival of the human species. Although the body carries on a great many different functions, each one contributes in some way to the survival of the individual or of humankind.

2 Survival depends on the body's maintaining or restoring homeostasis. **Homeostasis** is the relative constancy of the internal environment. More specifically, homeostasis means that the chemical composition, the volume, and certain other characteristics of blood and interstitial fluid (fluid around cells) remain constant within narrow limits.

3 Homeostasis depends on the body's ceaselessly carrying on many activities. It must continually respond to changes in its environment, exchange materials between its environment and its cells, metabolize food, and control all of its diverse activities.

A typical homeostatic mechanism is shown in Figure 1-9. At the bottom of the illustration the waste gas carbon dioxide (CO_2) is shown leaving the tissue cells, where it is produced, and being carried by the blood to the lungs, where it exits from the body in the expired air. The carbon dioxide combines with a specialized pigment called hemoglobin (Hb) to form a carrier substance ($HbCO_2$) that serves to transport CO_2 to the lungs. The constant circulation of the blood allows for continuous removal of the CO_2 produced by body cells. In the absence of this homeostatic mechanism, body CO_2 levels would rapidly rise to toxic levels and death would result.

4 All body functions are ultimately cell functions.

5 Body functions are related to age. During childhood, body functions gradually become more and more efficient and effective. They operate with maximum efficiency and effectiveness during young adulthood. During late adulthood and old age, they gradually become less and less efficient and effective. Changes and functions that occur during the early years are called developmental processes; those that occur after young adulthood are called aging processes. In general, developmental processes improve functions; aging processes usually diminish them. Future chapters will refer to many functional age changes.

OUTLINE SUMMARY

Structural Levels of Organization
A Organization is an outstanding characteristic of body structure
B The body is a unit constructed of the following smaller units—cells, tissues, organs, systems
  1 Cells—the smallest structural units; organizations of various chemicals
  2 Tissues—organizations of similar cells
  3 Organs—organizations of different kinds of tissues
  4 Systems—organizations of many different kinds of organs
C The presence of the following cavities is a prominent feature of body structure
  1 Ventral cavities
    a Thoracic cavity
      (1) Mediastinum—midportion of thoracic cavity; heart, trachea, esophagus, and thymus gland located in mediastinum
      (2) Pleural cavities—right lung located in right pleural cavity, left lung in left pleural cavity
    b Abdominopelvic cavity
      (1) Abdominal cavity contains stomach, intestines, liver, gallbladder, pancreas, and spleen
      (2) Pelvic cavity contains reproductive organs, urinary bladder, lowest part of intestine
  2 Dorsal cavities
    a Cranial cavity contains brain
    b Spinal cavity contains spinal cord

Some Words Used in Describing Body Structures
A Superior—toward the head, upper, above
  Inferior—toward the feet, lower, below
B Anterior—front, in front of
  Posterior—back, in back of
C Medial—toward midline of a structure
  Lateral—away from midline or toward side of a structure
D Proximal—toward or nearest trunk, or nearest point of origin of a structure
  Distal—away from or farthest from trunk, or farthest from a structure's point of origin
E Superficial—nearer body surface
  Deep—farther away from body surface
F Abdominal regions (see Figures 1-5 and 1-6)
  1 Nine regions of abdomen
  2 Four quadrants of abdomen

Planes or Body Sections
A Sagittal plane—lengthwise plane that divides a structure into right and left sections
B Midsagittal—sagittal plane that divides body into two equal halves
C Coronal plane—lengthwise plane that divides a structure into anterior and posterior sections

D Transverse plane—horizontal plane that divides a structure into upper and lower sections

Anatomical Position (Figure 1-7)
Standing erect with arms at sides and palms turned forward

Body Regions (Figure 1-8)
A Axial region—head, neck, and torso or trunk
B Appendicular region—upper and lower extremities

Some Basic Facts about Body Functions
A Survival is the body's most important business—survival of the individual and of the human species
B Survival depends on the maintenance or restoration of homeostasis (relative constancy of the internal environment; see Figure 1-9)
C Homeostasis depends on never-ceasing activities: response to changes in the external and internal environment, exchange of materials between the environment and cells, metabolism, and control
D All body functions are ultimately cell functions
E Body functions are related to age; peak efficiency during young adulthood, diminishing efficiency after young adulthood

NEW WORDS

abdominal quadrants (4)
abdominal regions (9)
anatomical position
atrophy
cavities
  abdominal
  cranial
  mediastinal
  pelvic
  pleural
  spinal
  thoracic

directional terms
  superior
  inferior
  anterior
  posterior
  medial
  lateral
  proximal
  distal
  superficial
  deep

homeostasis
organization
  (structural levels)
  chemical
  cellular
  tissue
  organ
  system

planes of section
  sagittal
  midsagittal
  coronal
  transverse

CHAPTER TEST

1. The study of the structure of an organism and the relationship of its parts is called _____ ; the study of the functions of that organism is called _____ .

2. An organization of many similar cells that together perform a common function is called a _____ .

3. The cranial and spinal cavities are called _____ cavities, whereas the thoracic and abdominopelvic cavities are called _____ cavities.

4. The term _____ means "toward the side of the body."

5. A coronal plane divides the body or any of its parts into _____ and _____ portions.

6. The most complex organizational units that make up the body are called _____ .

7. An individual standing erect with arms at the sides and palms turned forward is said to be in the _____ position.

8. The body as a whole can be subdivided into two major portions: _____ and _____ .

9. The relative constancy of the body's internal environment is described by the term _____ .

10. The area of the lower extremity between the knee and ankle is called the _____ .

Match the body area in column B with the area or example in column A. (Only one answer is correct.)

Column A	Column B
11. ____ Skull	a. Mammary
12. ____ Groin	b. Thoracic
13. ____ Breast	c. Digital
14. ____ Sole of foot	d. Carpal
15. ____ Chest	e. Inguinal
16. ____ Wrist	f. Precordial
17. ____ Fingers or toes	g. Gluteal
18. ____ Chest area over heart	h. Cranial
19. ____ Area between anus and genitals	i. Perineal
20. ____ Buttock	j. Plantar

21. The mediastinum is a subdivision of the:
    a. Thoracic cavity
    b. Pleural cavity
    c. Abdominal cavity
    d. Pelvic cavity

22. Which of the following is an example of a lower abdominal region?
    a. Epigastric region
    b. Umbilical region
    c. Right hypochondriac region
    d. Hypogastric region
23. The body can be divided or "cut" into right and left portions by which of the following planes?
    a. Coronal plane
    b. Sagittal plane
    c. Transverse plane
    d. Horizontal plane
24. Which of the following represents the least complex of the structural levels of organization?
    a. System
    b. Tissue
    c. Cell
    d. Organ
25. The diaphragm separates the:
    a. Right and left pleural cavities
    b. Abdominal and pelvic cavities
    c. Thoracic and abdominal cavities
    d. Mediastinum and pleural cavities

REVIEW QUESTIONS

1 Name the four kinds of structural units of the body. Define each briefly.
2 In what cavity could you find each of the following?

appendix	liver
brain	lungs
esophagus	pancreas
gallbladder	spinal cord
heart	spleen
intestines	urinary bladder

3 In one word, what is the one dominant function of the body or of any living thing?
4 Explain briefly what the term *homeostasis* means.
5 What do the terms *proximal* and *distal* mean?
6 Besides the maintenance of homeostasis and the carrying on of metabolism, name another major function the body must perform in order to survive.
7 On what surface of the body are the toenails located?
8 What structures lie lateral to the bridge of the nose?
9 Which joint—hip or knee—lies at the distal end of the thigh?
10 Define the term *anatomical position*.
11 List the major subdivisions of the axial and appendicular areas of the body.
12 List and define the four major planes of section.

CHAPTER
2   Cells and Tissues

CHAPTER OUTLINE

Cells
Size and shape
Composition
Structural parts
Cell functions

Movement of substances through cell membranes
Passive transport processes
Active transport processes

Cell reproduction
Stages of mitosis

Tissues
Epithelial tissue
Connective tissue
Muscle tissue
Nervous tissue

BOXED ESSAY

Tonicity

OBJECTIVES

After you have completed this chapter, you should be able to:

1   Identify and discuss the basic structure and function of the three major components of a cell.

2   Compare the major passive and active transport processes that act to move substances through cell membranes.

3   Discuss the stages of mitosis and explain the importance of cellular reproduction.

4   Explain how epithelial tissue can be grouped according to shape and arrangement of cells.

5   List and briefly discuss the major types of connective and muscle tissue.

6   List the three structural components of a neuron.

About 300 years ago Robert Hooke looked through his microscope—one of the very early, somewhat primitive ones—at some plant material. What he saw must have surprised him. Instead of a single magnified piece of plant material, he saw many small pieces. Since they reminded him of miniature prison cells, that is what he called them—cells. Since Hooke's time, thousands of individuals have examined thousands of plant and animal specimens and found them all, without exception, to be composed of cells. This fact, that cells are the smallest structural units of living things, has become the foundation stone of modern biology. Many living things are so simple that they consist of just one cell. The human body, however, is so complex that it consists not of a few thousand or millions or even billions of cells, but of many trillions of them. This chapter discusses cells first and then tissues.

CELLS

SIZE AND SHAPE

Human cells are microscopic in size; that is, they can be seen only when magnified by a microscope. However, they vary considerably in size. An ovum (female sex cell), for example, has a diameter of a little less than 1,000 micrometers* (about ¹⁄₂₅ of an inch), whereas red blood cells have a diameter of only 7.5 micrometers. Cells differ even more notably in shape than in size. Some are flat, some are brick shaped, some are threadlike, and some have irregular shapes.

---

*A micrometer is one millionth of a meter. (Micron is another name for micrometer.) In the metric system the units of length are as follows:

$$
\begin{aligned}
1 \text{ meter (m)} &= 39.37 \text{ inches} \\
1 \text{ centimeter (cm)} &= 1/100 \text{ meter} \\
1 \text{ millimeter (mm)} &= 1/1,000 \text{ meter} \\
1 \text{ micrometer } (\mu m) \text{ or micron } (\mu) &= 1/1,000,000 \text{ meter} \\
1 \text{ nanometer (nm)} &= 1/1,000,000,000 \text{ meter} \\
1 \text{ Angstrom } (\text{Å}) &= 1/10,000,000,000 \text{ meter}
\end{aligned}
$$

Approximately equal to 1 inch:

    2.5 cm
    25 mm
    25,000 μm
    25,000,000 nm
    250,000,000 Å

COMPOSITION

Cells contain **cytoplasm** (SI-to-plazm), or "living matter," a substance that exists only in cells. Each cell in the body is surrounded by a thin membrane—the **plasma** or **cytoplasmic membrane.** It is this membrane that separates the cell contents from the dilute salt water solution called **tissue fluid** that bathes every cell in the body. Contained within the cytoplasm of each cell is a small circular body called the **nucleus** (NU-kle-us) and numerous specialized structures called **organelles** (or-gan-ELS), which will be described below.

Cytoplasm is a good example of why organization is important in living things. It is composed of a number of ordinary substances. For instance, it might surprise you to learn that cytoplasm consists mostly of water. It is the *organization* of a relatively few chemical substances in water that makes life possible.

The word **element** is used to identify a substance that cannot be broken down into two or more different substances. In all of nature there are only about 100 known chemical elements. Four of these elements compose over 98% of our body. They are **oxygen, carbon, hydrogen,** and **nitrogen** (Table 2-1).

There are about 20 other elements that are called **trace elements.** Although present in the body only in very small quantities, trace elements are necessary for many important body functions. Examples of trace elements include calcium and phosphorus for bone growth, iron for healthy blood, and elements such as sodium and calcium that are essential for the smooth functioning of the nervous and muscular systems. Iodine is necessary for the prevention of a disease condition called **goiter** (GOI-ter), which will be discussed further in Chapter 9. Many foods and nutritional supplements are now "fortified" by the addition of one or more of the trace elements (Table 2-1). Iodized salt and "enriched" bread are good examples of such foods.

A **compound** is a substance composed of two or more elements. Elements are represented by the symbols shown in Table 2-1. Important

Table 2-1 Important elements found in the human body

Name	Symbol
Major elements*	
Oxygen	O
Carbon	C
Hydrogen	H
Nitrogen	N
Trace elements†	
Calcium	Ca
Phosphorus	P
Sodium	Na (Latin—natrium)
Chlorine	Cl
Iron	Fe (Latin—ferrum)
Iodine	I

*These four elements compose more than 98% of body weight.
†Examples of the more than 20 trace elements found in the body.

chemical compounds found in the body are represented by combinations of these symbols. Water, for example, is H_2O (two parts hydrogen and one part oxygen), and common table salt, or sodium chloride, is NaCl (one part sodium and one part chlorine). The four major elements in cytoplasm combine to form three compounds that constitute the bulk of our body substance. These compounds are called (1) **carbohydrates,** (2) **proteins,** and (3) **fats.**

Carbohydrates and fats are composed of carbon, hydrogen, and oxygen put together in different ways. Protein compounds also have nitrogen. Carbohydrates are used as an energy source for the body. Fats serve to store food energy and insulate our body by accumulating in a layer beneath the skin. Proteins transport substances in the blood and communicate information to cells. Hair and nails are examples of body structures that are composed largely of protein.

STRUCTURAL PARTS

The three main parts of a cell are the following:
1 Cytoplasmic (plasma) membrane
2 Cytoplasm
3 Nucleus

Cytoplasmic Membrane

As the name suggests, the **cytoplasmic membrane** is the membrane that encloses the cytoplasm and forms the outer boundary of the cell. It is an incredibly delicate structure—only about 3/10,000,000 of an inch thick! Yet it has a precise, orderly structure (Figure 2-1). According to a widely held concept, two layers of phosphate-containing fat molecules called **phospholipids** form a fluid framework for the cytoplasmic membrane. Note in Figure 2-1 that protein molecules lie at both outer and inner surfaces of this framework, and many extend all the way through it.

Despite its seeming fragility, the cytoplasmic membrane is strong enough to keep the cell whole and intact. It also performs other life-preserving functions for the cell. It serves as a well-guarded gateway between the fluid inside the cell and the fluid around it. It allows certain substances to move through it but bars the passage of others. It was recently discovered that the cytoplasmic membrane even functions as a communication device. How? Some of the protein molecules on the membrane's outer surface serve as receptors for certain other molecules when they contact them. In other words, certain molecules bind to certain receptor proteins. For example, some hormones (chemicals secreted into blood from ductless glands) bind to membrane receptors, and a change in cell functions follows. We might therefore think of such hormones as chemical messages, communicated to cells by binding to their cytoplasmic membrane receptors. A majority of hormones are themselves composed of proteins.

Cytoplasm

Cytoplasm is the specialized living material sometimes called *protoplasm* (PRO-to-plazm). It lies between the cytoplasmic membrane and the

Figure 2-1 Structure of the cytoplasmic or plasma membrane. Note that protein molecules may penetrate completely through the two layers of phospholipid molecules.

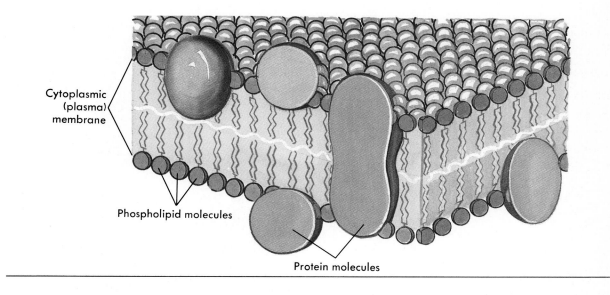

Cytoplasmic (plasma) membrane

Phospholipid molecules

Protein molecules

nucleus, which can be seen in Figure 2-2 as a round or spherical structure in the center of the cell. Numerous small structures are located in the cytoplasm. As a group they are called **organelles,** that is, little organs, an appropriate name since they function for the cell much as organs function for the body.

Look again at Figure 2-2. Notice how many different kinds of structures you can see in the cytoplasm of this cell. A generation ago almost all of these organelles were unknown. They are so small that, even when magnified 1,000 times by a light microscope, they are invisible. Electron microscopes brought them into view by magnifying them many thousands of times. We shall briefly discuss the following organelles, which are found in cytoplasm:

1 Endoplasmic reticulum
2 Ribosomes
3 Mitochondria
4 Lysosomes
5 Golgi apparatus
6 Centrioles

Endoplasmic reticulum The **endoplasmic reticulum** (en-do-PLAS-mik re-TIK-u-lum) **(ER)** is a network of connecting sacs and canals that wind tortuously through a cell's cytoplasm, all the way from its cytoplasmic membrane to its nucleus. The tubular passageways or canals in the ER carry proteins and other substances through the cytoplasm of the cell from one area to another. There are two types of ER: *smooth* and *rough.* Smooth ER is found in cells that handle or manufacture fatty substances, and rough ER is found in cells that manufacture proteins. In both cases the ER functions as a miniature circulatory system for the cell.

Ribosomes The **ribosomes** (RI-bo-somes), shown as red dots in Figure 2-2, are attached to and are part of the rough ER. Although these vitally important organelles are usually attached to ER, they may also be free in the cytoplasm. Ribosomes perform a very complex function, that of making enzymes and other protein compounds. Their nickname, ''protein factories,'' indicates this function. Since the cells of the pan-

Figure 2-2 Artist's interpretation of cell structure as seen under an electron microscope. Note the many mitochondria, popularly known as the "power plants of the cell." Note, too, the innumerable dots bordering the endoplasmic reticulum. These are ribosomes, the cell's "protein factories."

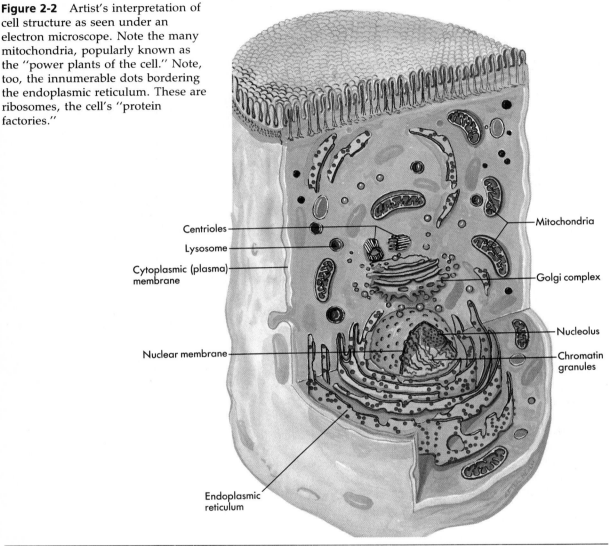

Centrioles

Lysosome

Cytoplasmic (plasma) membrane

Nuclear membrane

Endoplasmic reticulum

Mitochondria

Golgi complex

Nucleolus

Chromatin granules

creas, which secrete digestive enzymes, are rich in ribosomes, they are the cell type that has been researched the most to discover how ribosomes work.

Mitochondria The **mitochondria** (mi-toe-KON-dre-ah) are another kind of organelle present in all cells. Observe their appearance in Figure 2-2, where they are magnified thousands of times. Mitochondria are so tiny that a lineup of 15,000 or more of them would fill a space

only 1 inch long. Two membranous sacs, one inside the other, compose a mitochondrion. The inner membrane forms folds that look like miniature incomplete partitions. Within a mitochondrion's fragile walls, complex, energy-releasing chemical reactions go on continuously. Because these reactions supply most of the power for doing cellular work, mitochondria have been nicknamed the cell's "power plants." The survival of cells and therefore of the body depends on mitochondrial chemical reactions. Without these reactions, the cells would soon have no energy for doing the work that keeps them alive.

Lysosomes The **lysosomes** (LI-so-soms) are membranous-walled organelles that in their active stage look like small sacs, often with tiny particles in them (Figure 2-2). Because lysosomes contain chemicals (enzymes) that can digest food compounds, one of their nicknames is "digestive bags." Lysosomal enzymes can also digest substances other than foods. For example, they can digest and thereby destroy microbes that manage to invade the cell. Thus lysosomes can protect cells against destruction by microbes. Yet, paradoxically, lysosomes sometimes kill cells instead of protecting them. If their powerful enzymes escape from the lysosome sacs into the cytoplasm, they kill the cell by digesting it. This fact has earned lysosomes their other nickname, which is "suicide bags."

Golgi apparatus The **Golgi** (GOL-je) **apparatus,** long a mystery organelle, consists of tiny sacs stacked one upon the other near the nucleus. It is now known to make certain carbohydrate compounds, combine them with certain protein molecules, and package the product in neat little globules. Then these globules move slowly outward to and through the cell membrane. Once outside the cell they break open and their contents spill out. An example of a Golgi apparatus product is the slippery substance called mucus. If we wanted to nickname the Golgi apparatus, we might call it the cell's "carbohydrate-producing and -packaging factory."

Centrioles The **centrioles** (SEN-tri-ols) are paired organelles. Two of these rod-shaped structures are present in every cell. They are arranged so that they lie at right angles to each other (Figure 2-2). Each centriole is composed of fine tubules that play an important role during the process of cell division.

Nucleus

Viewed under a light microscope, the **nucleus** of a cell looks like a very simple structure indeed—just a small sphere in the central portion of the cell. However, its simple appearance be-

lies the complex and critical role it plays in cell function. It is the nucleus that ultimately controls every organelle in the cytoplasm. It also controls the complex process of cell reproduction. In other words, the nucleus must function properly for a cell to accomplish its normal activities and be able to duplicate itself.

Note that the cell nucleus in Figure 2-2 is surrounded by a **nuclear membrane.** The membrane serves to enclose a special type of protoplasm in the nucleus called **nucleoplasm.** Like cytoplasm, nucleoplasm contains a number of specialized structures, two of the most important of which are shown in Figure 2-2. They are the **nucleolus** (nu-KLE-o-lus) and the **chromatin** (KRO-muh-tin) **granules.**

Nucleolus and chromatin granules The nucleolus is of critical importance in protein formation because it "programs" the formation of ribosomes in the nucleus. The ribosomes then migrate through the nuclear membrane into the cytoplasm of the cell and function to produce proteins. Chromatin granules in the nucleus are in reality threadlike structures made up of proteins and the hereditary material called **DNA** or **deoxyribonucleic** (de-ok-se-ri-bo-NU-kle-IK) **acid.** DNA is the genetic material often described as the chemical "blueprint" of the body. It determines everything from sex to body build and hair color in every human being. The importance and function of DNA will be explained in greater detail in the section on cell reproduction later in this chapter.

CELL FUNCTIONS

Every human cell performs certain functions, some that maintain its own survival and others that help maintain the body's survival. Many cells, but not all, perform another kind of function that maintains the survival of their species—they reproduce themselves. Of all the functions that maintain a cell's own life, in this chapter we shall discuss only the major processes by which various substances move through the cell membranes. Before beginning the study of tissues, we shall also describe cell reproduction.

MOVEMENT OF SUBSTANCES THROUGH CELL MEMBRANES

The cytoplasmic membrane in every healthy cell serves to separate the contents of the cell from the tissue fluid that surrounds it. At the same time the membrane must permit certain substances to enter the cell and allow others to leave. Heavy traffic moves continuously in both directions through cell membranes. Streaming in and out of all cells in endless procession go molecules of water, foods, gases, wastes, and many other substances. A number of processes allow for this mass movement of substances into and out of cells. These transport processes are classified under two general headings:

1 Passive transport processes
2 Active transport processes

The difference between the two categories is based on whether or not energy is required to effect the movement of something through the cell membrane. As implied by their name, active transport processes require the expenditure of energy by the cell and passive transport processes do not. The energy required for active transport processes is obtained from a very important chemical substance called **adenosine triphosphate** (ah-DEN-o-sen tri-FOS-fate) or **ATP** for short. ATP is produced in the cell from nutrients and is capable of releasing energy that in turn enables the cell to do work. The breakdown of ATP and use of the energy that is released is required for active transport processes to occur.

The details of both active and passive transport of substances across cell membranes will be much easier to understand if you keep in mind the following two key facts: (1) in passive transport processes no cellular energy is required to move substances from a high concentration to a low concentration; and (2) in active transport processes cellular energy is required to move substances from a low concentration to a high concentration.

PASSIVE TRANSPORT PROCESSES

The primary **passive transport** processes that move substances through the cell membranes include the following:

1 Diffusion
   a Osmosis
   b Dialysis
2 Facilitated diffusion
3 Filtration

Scientists describe the movement of substances in passive systems as going "down a concentration gradient." This simply means that substances in these passive systems will move from a region of high concentration to a region of low concentration until they reach equal proportions on both sides of the membrane.

Diffusion

Diffusion is a good example of a passive transport process. By definition diffusion simply is the process by which substances scatter themselves evenly throughout an available space. The system does not require any additional energy for this movement to occur. To demonstrate diffusion of particles throughout a fluid, perform this simple experiment the next time you pour yourself a cup of coffee or tea. Place a cube of sugar on a teaspoon and lower it gently to the bottom of the cup. Let it stand for 2 or 3 minutes, and then, holding the cup steady, take a sip off the top. It will taste sweet. Why? Because some of the sugar molecules will have diffused from the area of high concentration near the sugar cube at the bottom of the cup to the area of low concentration at the top of the cup.

The process of diffusion is shown in Figure 2-3. In this illustration a membrane that is permeable to both salt, or sodium chloride (NaCl), and water separates a 10% NaCl solution from a 20% NaCl solution. The container on the left shows the two solutions separated by the membrane at the start of the diffusion process. The container on the right shows the result of diffusion on the system after time.

Note that both substances diffuse rapidly through the membrane in both directions. However, as indicated by the red arrows, more sodium chloride moves out of the 20% solution, where the concentration is higher, into the 10% solution, where the concentration is lower, than in the opposite direction. This is an example of

Figure 2-3 Diffusion. See text for discussion.

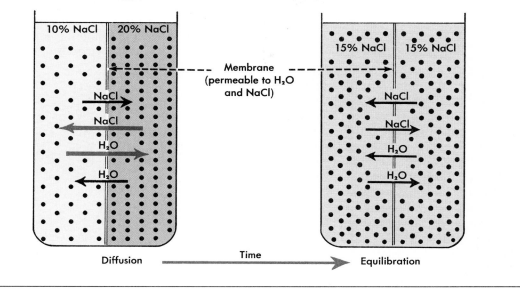

movement down a concentration gradient. Simultaneously, more water moves from the 10% solution, where there are more water molecules, into the 20% solution, where there are fewer water molecules. This is also an example of movement down a concentration gradient. Water moves from high to low concentration. Result? Equilibration of the concentrations of the two solutions after an interval of time. From then on, equal amounts of salt will diffuse in both directions, as will equal amounts of water.

Osmosis and dialysis Osmosis (oz-MOE-sis) and **dialysis** (di-AL-i-sis) are specialized examples of diffusion. In both cases the diffusion occurs across what is called a **selectively permeable membrane.** The cytoplasmic membrane of a cell is said to be selectively permeable because it will permit the passage of certain substances and not of others. This is a necessary property if the cell is to permit some substances, such as nutrients, to gain entrance to the cell while excluding others. Osmosis is defined as the diffusion of *water* across a selectively permeable membrane. However, in the case of dialysis, substances called **solutes,** which are dissolved

particles in water, move across a selectively permeable membrane by the process of diffusion.

Facilitated Diffusion

Like ordinary diffusion, **facilitated diffusion** is a passive process that moves a substance down its own concentration gradient, that is, from an area where it is in high concentration to an area of lower concentration. This movement is "facilitated" or assisted by a special process that involves protein molecules that extend all the way through the cytoplasmic membrane. Examine Figure 2-1 and note again that certain protein molecules do indeed extend completely through the membrane. Some of these are called "carrier molecules." In the process of facilitated diffusion these protein carrier molecules bind to the substance that is to be diffused across the membrane. According to one theory, the carrier molecules then rotate in the membrane and thereby carry the substance that is being diffused rapidly from one side of the membrane to the other. There the carrier molecule dissociates itself from the substance moving across the cell membrane and releases it into the cytoplasm of the cell. It has fulfilled its function of facilitating (accelerating) the substance's diffusion through the membrane.

Filtration

Filtration is the movement of both water and solutes through a membrane because of a greater pushing force on one side of the membrane than on the other side. The force is called *hydrostatic pressure*, which is simply the force or weight of a fluid pushing against some surface. A principle about filtration that is of great physiological importance is that it always occurs *down* a hydrostatic pressure gradient. This means that when two fluids have unequal hydrostatic pressures and are separated by a membrane, water and diffusible solutes or particles (those to which the membrane is permeable) filter out of the solution that has the higher hydrostatic pressure into the solution that has the lower hydrostatic pressure. Filtration is the process responsible for urine formation in the kidney: wastes are filtered out of the blood into the kidney tubules because of a difference in hydrostatic pressure.

ACTIVE TRANSPORT PROCESSES

Active transport is the uphill movement of a substance through a living cell membrane. "Uphill" means "up a concentration gradient," that is, from a lower to a higher concentration. The energy that is required for moving a substance up its concentration gradient is obtained from ATP. Because the formation and breakdown of ATP requires complex cellular activity, active transport mechanisms can take place only through living membranes.

Permease System

A specialized cellular component called the **permease** (PER-me-aze) **system** makes a number of active transport mechanisms possible. Permease is a protein complex in the cell membrane that uses energy from ATP to actively move many substances across cell membranes *against* their concentration gradients. Many permease-driven transport systems are called **pumps,** an appropriate term, since it suggests that active

Figure 2-4   Phagocytosis of bacteria by a white blood cell. Phagocytosis is an active transport mechanism that requires expenditure of energy. Note how an extension of cytoplasm envelops the bacteria, which are drawn through the cell membrane and into the cytoplasm.

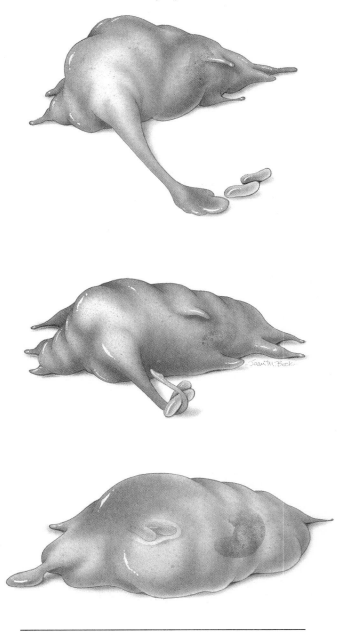

transport moves a substance in an uphill direction just as a water pump, for example, moves water uphill. Permeases are very specific, and different permease pumps are required to move different substances. The sodium pump, for example, moves sodium (and only sodium) through the cytoplasmic membrane from the inside of the cell, where sodium concentration is low, to the outside of the cell, where sodium concentration is high. In addition to pumping sodium, other specific permease pumps transport certain sugars and important amino acids into cells.

Phagocytosis and Pinocytosis

Phagocytosis (fag-o-si-TOE-sis) is another example of how a cell can use its active transport mechanism to move an object or substance through the plasma membrane and into the cytoplasm. The term *phagocytosis* comes from a

Tonicity

A salt (NaCl) solution is said to be **isotonic** if it contains the same concentration of salt normally found in a living red blood cell. A 0.9% NaCl solution is isotonic; that is, it contains the same level of NaCl as found in red cells. A solution that contains a higher level of salt (above 0.9%) is said to be **hypertonic,** and one containing less (below 0.9%) is **hypotonic.** With what you now know about filtration, diffusion, and osmosis, can you predict what would occur if red blood cells were placed in isotonic, hypotonic, and hypertonic solutions?

Examine the figure below. Note that red blood cells placed in isotonic solution remain unchanged. The movement of water into and out of the cells is about equal. This is not the case with red cells placed in hypertonic salt solution: they immediately begin to lose water from their cytoplasm into the surrounding solution and they shrink. This process is called **crenation.**

The opposite occurs if red cells are placed in a hypotonic solution; they will swell as water enters the cell from the surrounding solution. Eventually the cells will break or **lyse,** and the hemoglobin they contain will be released into the solution.

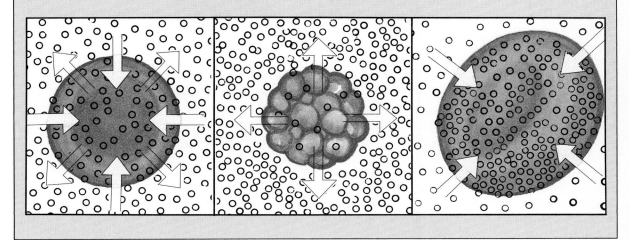

ISOTONIC   HYPERTONIC   HYPOTONIC

Greek word meaning "to eat." The word is appropriate since the process permits a cell to engulf and literally "eat" some foreign material (Figure 2-4). Certain white blood cells function to destroy bacteria in the body, and they do so by phagocytosis. In this process the cell membrane forms a pocket around the material to be moved into the cell and, by expenditure of energy from ATP, the object is moved to the interior of the cell. Once inside the cytoplasm, the bacterium (shown in Figure 2-4) will fuse with a lysosome and be destroyed. **Pinocytosis** (pi-no-si-TOE-sis) is an active transport mechanism used to incorporate fluids or dissolved substances into cells. Once again, the term is appropriate because it comes from the Greek word meaning "drink."

CELL REPRODUCTION

All human cells, other than sex cells, reproduce by a process called **mitosis** (mi-TO-sis). In this process a cell divides in order to multiply: one cell divides to form two cells. But before a cell divides, the chromosomes in its nucleus undergo certain changes. Chromosomes are composed largely of a compound named deoxyribonucleic acid but almost always called DNA. DNA is probably the most important compound in the world. Such an extravagant claim can be justified by stating DNA's nickname: the "heredity molecule." DNA makes heredity possible. Its molecules pass on the capabilities for developing the same characteristics as the parent cells from one generation of cells to the next. By so doing, DNA molecules transmit from each generation of parents to their children all the traits, both physical and mental, that they inherit from their ancestors.

Stating this function is a simple matter, but explaining it is not. It requires the telling of a long, complicated, and still unfinished story. We shall attempt only a brief synopsis.

The main theme of the DNA story revolves around the strange and unique structure of the DNA molecule. To try to visualize the shape of the DNA molecule, picture first an extremely long, narrow ladder made of a pliable material. Now see it twisting round and round on its axis

and taking on the shape of the DNA molecule—a double helix (Greek word for spiral).

Structurally, the DNA molecule is made up of many smaller units, namely, a sugar, bases, and phosphate units. As you can see in Figure 2-5, the name of the sugar is deoxyribose. The bases are adenine, thymine, guanine, and cytosine. Observe that the sides of the DNA ladder consist of deoxyribose alternating with phosphate units. Look next at the ladder's steps. Notice that each one consists of a pair of bases. Only two combinations of bases occur. The same two bases invariably pair off with each other in a DNA molecule. Adenine and thymine always go together, as do guanine and cytosine. This characteristic of DNA structure is called **complementary base pairing.** It is an important fact that we shall refer to again when we describe DNA's function.

Another important fact about DNA structure is that the sequence of its base pairs is not the same in all DNA molecules, although the base pairs are the same. This fact has tremendous functional importance because it is the sequence of base pairs that determines heredity. In fact, that is what a *gene* is—a specific sequence of an average of about 1,000 base pairs. Approximately 100,000 genes, according to one estimate, compose the DNA molecule of just one human chromosome. If we do some simple multiplication, we arrive at some staggering figures:

1 gene consists of a sequence of 1,000 base pairs.

1 chromosome consists of 100,000 genes.

1 chromosome consists of $(100,000 \times 1,000)$ or 100 million base pairs.

1 human cell contains 46 chromosomes.

1 human cell contains $(46 \times 100,000)$ or over 4½ million genes.

Is it any wonder, then, with more than 4½ million genes or "heredity-bearers" in each of our cells, that no two of us inherit exactly the same traits?

How do genes bring about heredity? There is, of course, no short and easy answer to that question. In general, genes tell ribosomes what enzymes and other proteins they are to make. According to a current theory, each gene—or,

Figure 2-5 Structure of DNA molecule.

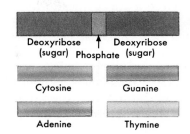

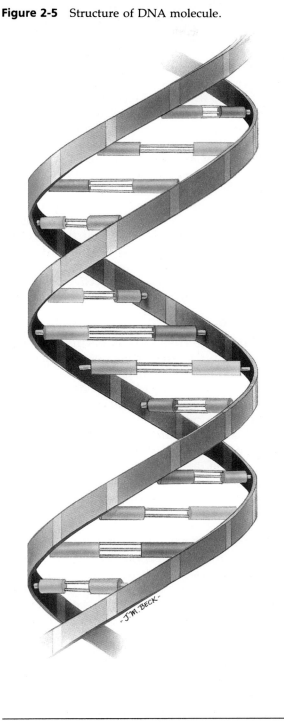

in other words, each sequence in a DNA molecule of a thousand or so base pairs—is a code indicating the structure of a polypeptide, that is, a part of a protein molecule. This information is relayed to the ribosomes, and these cellular protein factories then make the enzymes and other proteins. Enzymes are vital substances. Their job is to keep innumerable chemical reactions going on at a fast enough pace to keep cells and therefore the body alive. In summary, genes control enzyme production, enzymes facilitate cellular chemical reactions, and cellular chemical reactions determine both cell structure and function and therefore heredity.

Ribosomes contain a chemical named ribonucleic acid (abbreviation RNA). RNA resembles DNA in that both are composed of four bases, a sugar, and phosphate. RNA, however, differs from DNA in three ways: RNA molecules contain the base uracil instead of thymine; RNA molecules contain the sugar ribose instead of deoxyribose; RNA molecules are smaller than DNA molecules.

DNA molecules possess a unique ability that no other molecules in the world have. They can make copies of themselves, a process called **DNA replication.** Before a cell divides to form two new cells, each DNA molecule in its nucleus forms another DNA molecule just like itself. When a DNA molecule is not replicating, it has the shape of a tightly coiled double helix. As it begins the process of replication, short segments of the DNA molecule uncoil and the two strands of the molecule pull apart between their base pairs. The separated strands therefore contain unpaired bases. Each unpaired base in each of the two separated strands of the DNA molecule attracts its complementary base (present in the nuclear fluid) and binds to it. Specifically, each adenine attracts and binds to a thymine and each cytosine attracts and binds to a guanine. These steps repeat themselves over and over again throughout the length of the DNA molecule. Thus each half of a DNA molecule becomes a whole DNA molecule identical to the original DNA molecule.

STAGES OF MITOSIS

A cell is ready to reproduce itself by mitosis after the DNA molecules have duplicated themselves. The process of mitosis involves the division of both the nucleus and the cytoplasm. After the process is complete, two daughter cells result—both with the same genetic material as the cell that preceded them. As you can see in Figure 2-6, the specific and visible stages of cell division are preceded by a period called **interphase** (IN-ter-faze). During interphase the cell is said to be "resting." However, it is resting only from the standpoint of reproductive activity. In all other aspects it is exceedingly active. During interphase and just before mitosis begins, the DNA of each chromosome replicates itself.

The stages of mitosis are listed below with a brief description of the changes that occur during each stage.

Prophase

Look at Figure 2-6 and note the changes that identify the first stage of mitosis, prophase (PRO-faze). The chromatin granules in the nucleus have become "organized." Chromosomes in the nucleus have formed two strands called **chromatids** (KRO-mah-tids). Note that the two chromatids are held together at their centers by a beadlike structure called the **centromere** (SEN-tro-mer). In the cytoplasm the centrioles can be seen moving away from each other while forming a network of tubules called **spindle fibers.** These spindle fibers serve as "guidewires" and will assist the chromosomes to move toward opposite ends of the cell later in mitosis.

Metaphase

By the time metaphase (MET-ah-faze) begins, nuclear membrane and nucleolus have disappeared. Note in Figure 2-6 how the chromosomes have aligned themselves across the center of the cell. Also, the centrioles have now migrated to opposite ends of the cell, and spindle fibers can be seen attached to each chromatid.

Anaphase

As anaphase (AN-ah-faze) begins, the beadlike centromeres, which were holding the paired chromatids together, break apart. As a result, the individual chromatids, now identified once again as chromosomes, move away from the center of the cell. Movement of chromosomes occurs along spindle fibers attached to centrioles. Note in Figure 2-6 how chromosomes are being pulled to opposite ends of the cell. A **cleavage furrow** that begins to divide the cell into two daughter cells can be seen for the first time at the end of anaphase.

Telophase

During telophase (TEL-o-faze) the process of cell division is completed. Two nuclei appear and chromosomes become less distinct and appear to break up. As the nuclear membrane forms around the chromatin granules, the cleavage furrow completely divides the cell into two parts. Before division is complete, each nucleus is surrounded by cytoplasm in which organelles have been equally distributed. By the end of telophase, two separate daughter cells, each having identical genetic characteristics, are

Figure 2-6 Diagram of mitosis. For simplicity only four chromosomes are shown.

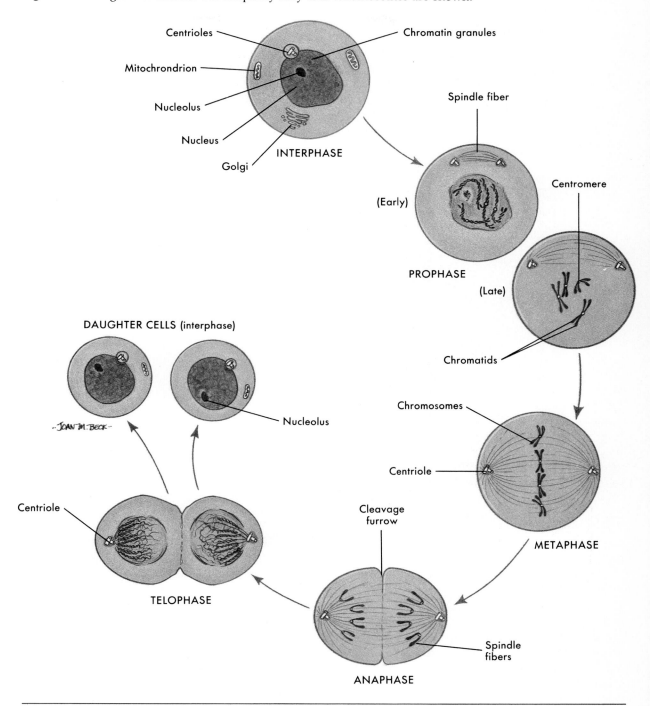

Centrioles

Chromatin granules

Mitochrondrion

Nucleolus

Nucleus

Golgi

INTERPHASE

Spindle fiber

(Early)

PROPHASE

Centromere

(Late)

Chromatids

DAUGHTER CELLS (interphase)

Nucleolus

JOAN M. BECK

Chromosomes

Centriole

METAPHASE

Centriole

Cleavage furrow

TELOPHASE

Spindle fibers

ANAPHASE

Figure 2-7 Classification of epithelial tissues according to the shape and arrangement of the cells.

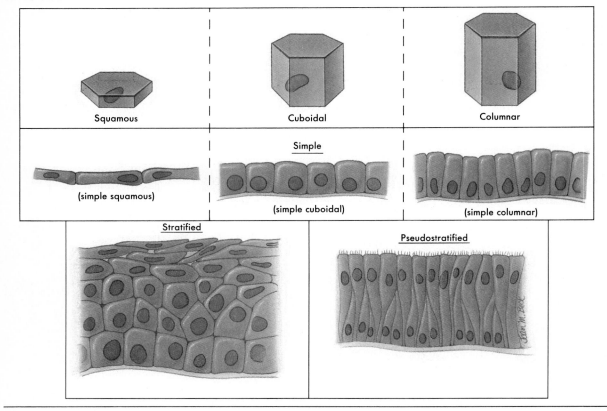

formed. Each cell is fully functional and will itself undergo mitosis in the future.

TISSUES

The four main kinds of tissues that compose the body's many organs are:

1 Epithelial tissue
2 Connective tissue
3 Muscle tissue
4 Nervous tissue

Tissues differ from each other in the size and shape of their cells, in the amount and kind of material present between the cells, and in the special functions they perform to help maintain the body's survival. In Table 2-2 you will find a listing of the four major tissues and the various subtypes of each. The table also includes examples of the location of the tissues and a primary function of each tissue type.

EPITHELIAL TISSUE

Epithelial (ep-i-THEE-le-al) **tissue** covers the body and many of its parts. It also lines various parts of the body. Because epithelial cells are packed close together with little or no intercellular material between them, they form continuous sheets that contain no blood vessels. Examine Figure 2-7. It illustrates how this large group of tissues can be subdivided according to the **shape** and **arrangement** of the cells found in each type.

Table 2-2 Tissues

Tissue	Location	Function
Epithelial		
Simple squamous	Alveoli of lungs	Absorption by diffusion of respiratory gases between alveolar air and blood
	Lining of blood and lymphatic vessels	Absorption by diffusion, filtration, and osmosis
Stratified squamous	Surface of lining of mouth and esophagus	Protection
	Surface of skin (epidermis)	Protection
Simple columnar	Surface layer of lining of stomach, intestines, and parts of respiratory tract	Protection; secretion; absorption
Stratified transitional	Urinary bladder	Protection
Connective (most widely distributed of all tissues)		
Areolar	Between other tissues and organs	Connection
Adipose (fat)	Under skin	Protection
	Padding at various points	Insulation; support; reserve food
Dense fibrous	Tendons; ligaments	Flexible but strong connection
Bone	Skeleton	Support; protection
Cartilage	Part of nasal septum; covering articular surfaces of bones; larynx; rings in trachea and bronchi	Firm but flexible support
	Disks between vertebrae	
	External ear	
Blood	Blood vessels	Transportation
Muscle		
Skeletal (striated voluntary)	Muscles that attach to bones	Movement of bones
	Eyeball muscles	Eye movements
	Upper third of esophagus	First part of swallowing
Cardiac (striated involuntary)	Wall of heart	Contraction of heart
Visceral (nonstriated involuntary or smooth)	In walls of tubular viscera of digestive, respiratory, and genitourinary tracts	Movement of substances along respective tracts
	In walls of blood vessels and large lymphatic vessels	Changing of diameter of blood vessels
	In ducts of glands	Movement of substances along ducts
	Intrinsic eye muscles (iris and ciliary body)	Changing of diameter of pupils and shape of lens
	Arrector muscles of hairs	Erection of hairs (gooseflesh)
Nervous		
	Brain; spinal cord; nerves	Irritability; conduction

Figure 2-8 Photomicrograph of lung tissue showing the thin simple squamous epithelium lining the alveolar air sacs. (×140.)

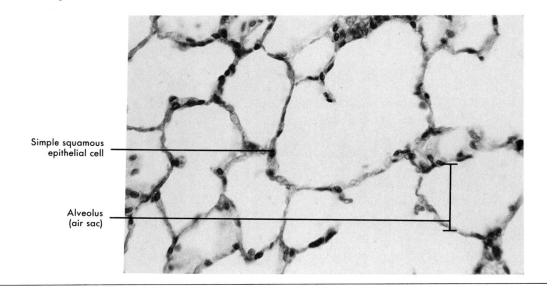

Simple squamous
epithelial cell

Alveolus
(air sac)

Shape of Cells

If classified according to *shape,* epithelial cells are:
   1 Squamous (flat and scalelike)
   2 Cuboidal (cube shaped) or
   3 Columnar (higher than they are wide)

Arrangement of Cells

If categorized according to *arrangement* of cells, epithelial tissue can be classified as:
   1 Simple (a single layer of cells of the same shape) or
   2 Stratified (many layers of cells of the same shape) or
   3 Transitional (several layers of cells of differing shapes)
Several types of epithelium are listed below. Each type is described in the paragraphs that follow and illustrated in Figures 2-8 to 2-11.

Simple Squamous Epithelium

Simple squamous epithelium consists of a single layer of scalelike cells. Because of its struc-

ture, substances can readily pass through simple squamous epithelial tissue, making absorption its special function. Absorption of oxygen into the blood, for example, takes place through the simple squamous epithelium that forms the tiny air sacs in the lungs (Figure 2-8).

Stratified Squamous Epithelium

Stratified squamous epithelium (Figure 2-9) consists of several layers of closely packed cells, an arrangement that makes this tissue a specialist at protection. For instance, stratified squamous epithelial tissue protects the body against invasion by microorganisms. Most microbes cannot work their way through a barrier of stratified squamous tissue such as that which composes the surface of skin and of mucous membranes.

One way of preventing infections, therefore, is to take good care of your skin. Don't let it become cracked from chapping, and guard against cuts and scratches.

Figure 2-9   **A,** Photomicrograph of stratified squamous epithelium. (×140.) **B,** Sketch of photomicrograph. Note the many layers of epithelial cells that have been stained yellow.

A

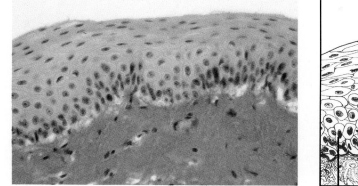

B

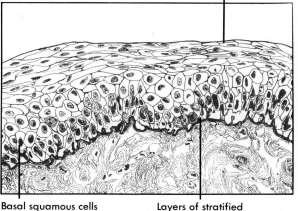

Superficial squamous cells

Basal squamous cells

Layers of stratified squamous epithelial cells

Simple Columnar Epithelium

Simple columnar epithelium can be found lining the inner surface of the stomach, intestines, and some areas of the respiratory and reproductive tracts. Note in Figure 2-10 that the simple columnar cells are arranged in a single layer lining the inner surface of the colon or large intestine. These epithelial cells are higher than they are wide, and the nuclei are located toward the bottom of each cell. What appear to be open spaces between the cells are actually specialized **goblet cells** that produce mucus. The regular columnar-shaped cells specialize in absorption.

Stratified Transitional Epithelium

Stratified transitional epithelium is typically found in body areas that are subjected to stress and must be able to stretch, such as the wall of the urinary bladder. In many instances up to 10 layers of cuboidal-shaped cells of varying sizes are present in the absence of stretching. When stretching occurs, the epithelial sheet expands, the number of cell layers decreases, and cell shape changes from cuboidal to squamous (flat) in appearance. The fact that transitional epithelium has this ability keeps the bladder wall from tearing under the pressures of stretching. Stratified transitional epithelium is shown in Figure 2-11.

Glandular Epithelium

Glandular epithelium differs from the membranous and sheetlike arrangement of epithelial cells just described. Instead of occurring in protective coverings or linings, glandular epithelial cells are specialized for secretory activity. These specialized cells may function singly or in clusters or groups of secretory cells commonly called **glands.** Glandular secretions produced by these epithelial cells may be discharged into ducts, directly into the blood, or onto the body surface. Examples of glandular secretions include saliva produced by the salivary glands, digestive juices, sweat or perspiration, and numerous hormones such as those secreted by the pituitary or thyroid glands.

Figure 2-10 **A,** Sketch of photomicrograph. **B,** Photomicrograph of simple columnar epithelium lining the colon. (×140.) Some goblet or mucus-producing cells are present.

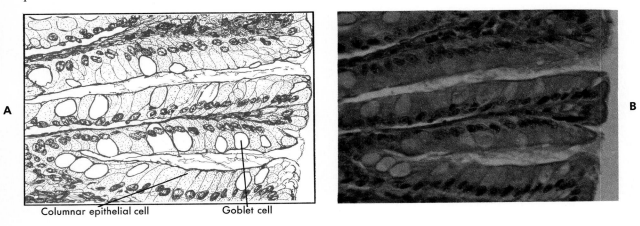

A   B

Columnar epithelial cell   Goblet cell

Figure 2-11 **A,** Sketch of photomicrograph. Note the many layers of various shaped epithelial cells. **B,** Photomicrograph of stratified transitional epithelium of urinary bladder wall. (×140.)

Binucleate cell   Stratified transitional
epithelial cells

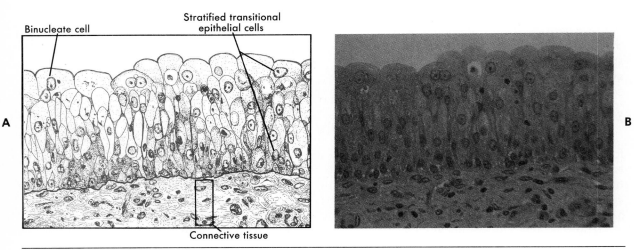

A   B

Connective tissue

CONNECTIVE TISSUE

Connective tissue is the most abundant and widely distributed type of tissue in the body. It also exists in more varied forms than any of the other tissue types. It is found in skin, membranes, muscles, bones, nerves, and all internal organs. Connective tissue exists as delicate, paper-thin webs that hold internal organs together and give them shape. It also exists as strong and tough cords, rigid bones, and even in the form of a fluid: blood.

The functions of connective tissue are as varied as its structure and appearance. It connects tissues to each other and forms a supporting framework for the body as a whole and for its individual organs. As blood it transports substances throughout the body. There are several other kinds of connective tissue that function to defend us against microbes and other invaders.

Connective tissue differs from epithelial tissue in the arrangement and variety of its cells and in the amount and kinds of intercellular material, called **matrix,** found between its cells. Besides the relatively few cells that are embedded in the matrix of most types of connective tissue, varying numbers and kinds of fibers are also present. It is the structural quality and appearance of the matrix and fibers that determine the qualities of each type of connective tissue. The matrix of blood, for example, is a liquid, but other types of connective tissue, such as cartilage, have the consistency of firm rubber. The matrix of bone is hard and rigid, although the matrix of connective tissues such as tendons and ligaments is strong and flexible.

The following list identifies a number of the major types of connective tissue in the body. Photomicrographs and sketches of several are shown below.

1 Areolar connective tissue
2 Adipose or fat tissue
3 Fibrous connective tissue
4 Bone
5 Cartilage
6 Blood

Areolar and Adipose Connective Tissue

Areolar (ah-RE-o-lar) **connective tissue** is the most widely distributed of all connective tissue types. It is the "glue" that gives form to the internal organs. It consists of delicate webs of fibers and of a variety of cells embedded in a loose matrix of soft, sticky gel.

Figure 2-12 Adipose (fat) tissue. **A,** Photomicrograph of human adipose tissue. (×140.) **B,** Sketch of the photomicrograph. Note the large storage spaces for fat inside the adipose tissue cells.

A

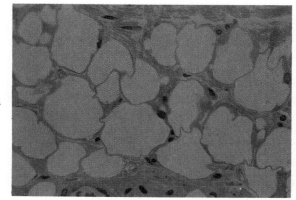

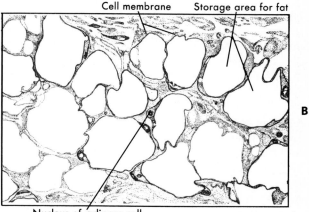

B

Cell membrane   Storage area for fat

Nucleus of adipose cell

Figure 2-13 Tendon. **A,** Sketch of photomicrograph. Note the multiple layers of flattened collagenous fibers arranged in parallel rows. **B,** Photomicrograph of dense fibrous connective tissue of tendon. (×35.)

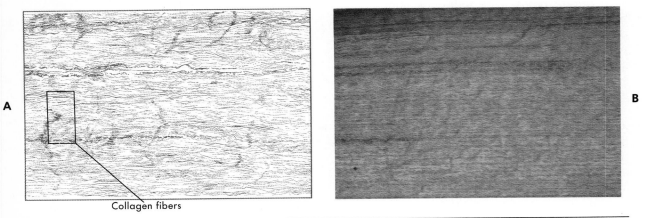

A

B

Collagen fibers

Adipose (AD-i-pose) or **fat tissue** is specialized to store lipids. Note in Figure 2-12 that numerous spaces have formed in the tissue so that large quantities of fat can be accumulated inside the cells.

Fibrous Connective Tissue

Fibrous connective tissue (Figure 2-13) consists mainly of bundles of strong white **collagenous** (kol-LAJ-e-nus) fibers arranged in parallel rows. This is the type of connective tissue that composes tendons. It provides for great strength and nonstretchability—desirable characteristics for these structures that anchor our muscles to our bones.

Bone and Cartilage

Bone is one of the most highly specialized forms of connective tissue. The matrix of bone is hard and calcified. It forms numerous structural building blocks called **Haversian** (ha-VER-shan) **systems.** When bone is viewed under a microscope, we can see these circular arrangements of calcified matrix and cells that give bone its characteristic appearance (Figure 2-14). Bones serve as a storage area for calcium and provide support and protection for the body.

Cartilage differs from bone in that its matrix is the consistency of a firm plastic or gristlelike gel. Cartilage cells, called **chondrocytes** (KON-dro-sits), are located in tiny spaces that are distributed throughout the matrix (Figure 2-15).

Blood and Hemopoietic Tissue

Because its matrix is liquid, blood is perhaps the most unusual form of connective tissue. It serves a transportative and protective function in the body. Red and white blood cells are the cell types common to blood (Figure 2-16).

Hemopoietic (he-mo-poi-ET-ik) **tissue** is the connective tissue found in the marrow cavities of bones and in such organs as the spleen, tonsils, and lymph nodes. This specialized connective tissue is responsible for the formation of blood cells and lymphatic system cells important in our defense against disease.

MUSCLE TISSUE

There are three kinds of muscle tissue: **skeletal, cardiac,** and **visceral.**

Figure 2-14   **A,** Photomicrograph of dried, ground bone. (×35.) **B,** Sketch of photomicrograph. Many wheel-like structural units of bone known as Haversian systems are apparent in this section.

A

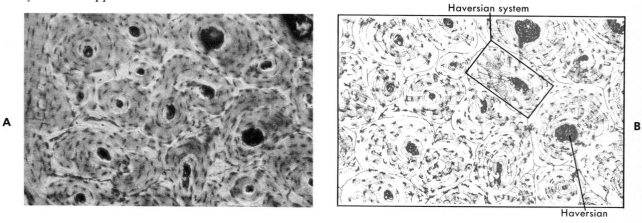

B

Figure 2-15   **A,** Photomicrograph of cartilage. (×140.) **B,** Sketch of photomicrograph. Note the chondrocytes distributed throughout the gel-like matrix.

A

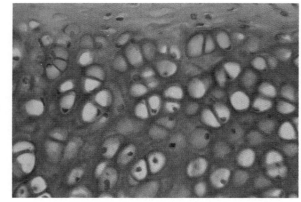

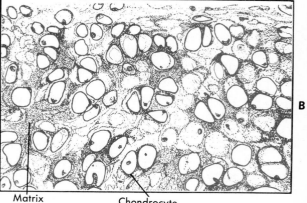

B

Figure 2-16 **A,** Sketch of photomicrograph. This smear shows two white blood cells surrounded by hundreds of smaller red blood cells. **B,** Photomicrograph of blood smear, human. (×350.)

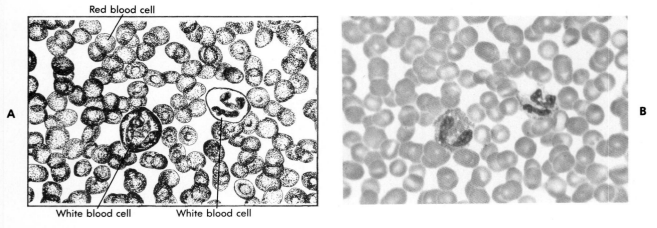

A

B

1 Skeletal or striated voluntary muscle—attaches to bones; microscopic view shows cells have cross striations (Figure 2-17); contractions are controlled voluntarily.

2 Cardiac or striated involuntary muscle—composes wall of heart; cells have cross striations (Figure 2-18); contractions ordinarily cannot be controlled.

3 Visceral or nonstriated (smooth) involuntary muscle—helps form walls of blood vessels, intestines, and various other tube-shaped structures; cells appear smooth, that is, without cross striations (Figure 2-19); contractions ordinarily cannot be controlled. In recent years many individuals have learned some voluntary control of smooth muscle contractions by using biofeedback devices.

All three types of muscle tissue specialize in contraction, a function that produces many kinds of movements in and of the body.

NERVOUS TISSUE

The basic function of **nervous tissue** is to make rapid communication between body structures and control of body functions possible. Actual nerve tissue consists of two basic kinds of cells: nerve cells, or **neurons** (NU-rons), which are the functional or conducting units of the system, and special connecting and supporting cells called **neuroglia** (nu-ROG-le-ah).

All neurons are characterized by a **cell body** and two types of processes: one **axon,** which transmits a nerve impulse away from the cell body, and one or more **dendrites** (DEN-drits), which carry impulses toward the cell body. Both neurons in Figure 2-20 have multiple dendrites extending from the cell body.

Figure 2-17 **A,** Photomicrograph of skeletal muscle. (×140.) **B,** Sketch of photomicrograph. This section of skeletal muscle shows bundles of cell fibers cut in both cross section *(top)* and longitudinally *(bottom)*.

A

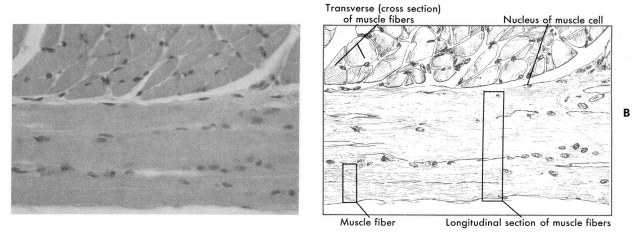

B

Figure 2-18 **A,** Photomicrograph of cardiac muscle. (×140.) **B,** Sketch of photomicrograph. The dark bands called intercalated discs, which are characteristic of cardiac muscle, are easily identified in this tissue section.

A

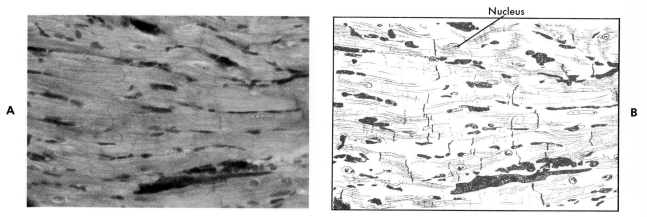

B

Figure 2-19   **A,** Sketch of photomicrograph. Note the central placement of nuclei in the spindle-shaped smooth muscle fibers. **B,** Photomicrograph of smooth muscle, longitudinal aspect. (×140.)

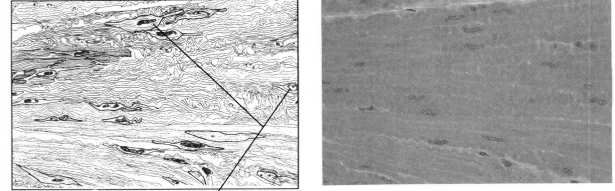

A   B

Smooth muscle cell

Figure 2-20   **A,** Sketch of photomicrograph. Both neurons in this slide show characteristic cell bodies and multiple cell processes. **B,** Photomicrograph of neurons in smear of spinal cord. (×35.)

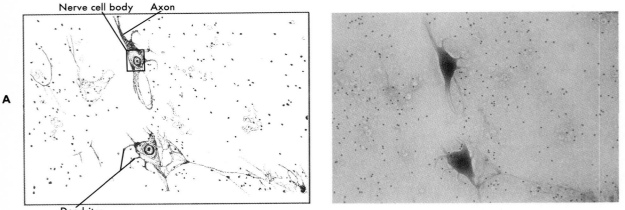

Nerve cell body   Axon

A   B

Dendrites

OUTLINE SUMMARY

Cells

A Size and shape
  1 Human cells vary considerably in size
  2 All are microscopic
  3 Cells differ notably in shape
B Composition
  1 Cytoplasm containing specialized organelles surrounded by a plasma membrane
  2 Organization of cytoplasmic substances important for life
  3 Four major elements—oxygen, carbon, hydrogen, and nitrogen make up 98% of body weight (Table 2-1)
  4 Over 20 trace elements important to life
C Structural parts
  1 Cytoplasmic membrane (Figure 2-1)
    a Forms outer boundary of cell
    b Thin two-layered membrane of phospholipids containing proteins
    c Is selectively permeable
  2 Cytoplasm (Figure 2-2)
    a Organelles
      (1) Endoplasmic reticulum (ER)
        (a) Network of connecting sacs and canals
        (b) Carry substances through cytoplasm
        (c) Types are rough and smooth
        (d) Smooth ER handle fatty substances
        (e) Rough ER handle proteins
      (2) Ribosomes
        (a) May attach to rough ER or lie free in cytoplasm
        (b) Manufacture proteins
        (c) Often called "protein factories"
      (3) Mitochondria
        (a) Composed of inner and outer membranes
        (b) Involved with energy-releasing chemical reactions
        (c) Often called "power plants" of the cell

      (4) Lysosomes
        (a) Membranous-walled organelles
        (b) Contain digestive enzymes
        (c) Have protective function
        (d) Often called "suicide bags"
      (5) Golgi apparatus
        (a) Collection of small sacs near nucleus
        (b) Manufacture carbohydrate secretion products
        (c) Called "carbohydrate-producing and -packaging factory"
      (6) Centrioles
        (a) Paired organelles
        (b) Lie at right angles to each other near nucleus
        (c) Function in cell reproduction
  3 Nucleus
    a Controls cytoplasmic organelles
    b Controls all steps in cell reproduction
    c Component structures include nuclear membrane, nucleoplasm, nucleolus, and chromatin granules
    d Chromatin granules contain DNA
D Cell functions
  1 Regulation of life processes
  2 Survival of species via reproduction of the individual

Movement of Substances Through Cell Membranes

A Passive transport processes—do not require added energy and result in movement "down a concentration gradient"
  1 Diffusion (Figure 2-3)
    a Substances scatter themselves evenly throughout an available space
    b It is unnecessary to add energy to the system
    c Movement is from high concentration to low concentration

    d Osmosis and dialysis are specialized examples of diffusion across a selectively permeable membrane

    e Osmosis is diffusion of water

    f Dialysis is diffusion of solutes

  2 Facilitated diffusion

    a Passive movement down a concentration gradient

    b Involves "carrier molecules" (proteins) in cytoplasmic membrane

    c Accelerates diffusion

  3 Filtration

    a Movement of both water and solutes caused by hydrostatic pressure on one side of membrane

    b Responsible for urine formation

B Active transport processes—occur only in living cells; movement of substances is "up the concentration gradient"; requires energy from ATP

  1 Permease systems

    a Permease is protein complex in cell membrane

    b Permease systems use energy from ATP to move substances across cell membranes against their concentration gradients

    c Permease-driven transport systems often called "pumps"—examples include sodium pump and movement of sugars and amino acids

  2 Phagocytosis and pinocytosis (Figure 2-4)

    a Both are active transport mechanisms

    b Phagocytosis is a protective mechanism often used to destroy bacteria

    c Pinocytosis is used to incorporate fluids or dissolved substances into cells

Tonicity

A Isotonic—contains the same concentration of salts as is found in living RBCs and will not cause cell injury

B Hypertonic—contains higher level of salts than found in RBCs. Will shrink cells by loss of water (crenation) if mixed with living cells

C Hypotonic—contains less salt than in living cells and will cause lysis or breakage of cells

Cell Reproduction

A DNA structure—large molecule shaped like a spiral staircase; sugar, deoxyribose, and phosphate units compose sides of the molecule; base pairs (adenine-thymine or guanine-cytosine) compose "steps"; base pairs always the same but sequence of base pairs differs in different DNA molecules; a gene is a specific sequence of about 1,000 base pairs; genes dictate formation of enzymes and other proteins by ribosomes, thereby indirectly determining cell's structure and functions; in short, genes are heredity determinants (Figure 2-5)

B DNA replication—process by which each half of a DNA molecule becomes a whole molecule identical to the original DNA molecule; precedes mitosis

C Mitosis—process of cell division that distributes identical chromosomes (DNA molecules) to each of new cells formed when the original cell divides; mitosis enables cells to reproduce their own kind; it makes heredity possible

D Stages of mitosis (Figure 2-6)

  1 Prophase–first stage

    a Chromatin granules become organized

    b Chromatids appear

    c Centrioles move away from nucleus

    d Spindle fibers appear

  2 Metaphase–second stage

    a Nuclear membrane disappears

    b Nucleolus disappears

    c Chromosomes align across center of cell

    d Centrioles move to opposite ends of cell

    e Spindle fibers attach themselves to each chromatid

  3 Anaphase–third stage

    a Centromeres break apart

    b Chromatids once again become chromosomes

    c Chromosomes are pulled to opposite ends of cell

    d Cleavage furrow develops

  4 Telophase–fourth stage

    a Cell division is completed during this phase

    b Nuclei appear in daughter cells

    c Nuclear membrane and nucleoli appear
    d Cytoplasm is divided
    e Daughter cells become fully functional

Tissues
A Epithelial tissue
  1 Covers body and lines body cavities
  2 Cells packed close together with little matrix
  3 Subdivided by shape and arrangement of cells
  4 Classified by shape of cells (Figure 2-7)
    a Squamous
    b Cuboidal
    c Columnar
  5 Classification by arrangement of cells
    a Simple
    b Stratified
    c Transitional
  6 Simple squamous epithelium (Figure 2-8)
    a Single layer of scalelike cells
    b Function in absorption
  7 Stratified squamous epithelium (Figure 2-9)
    a Several layers of closely packed cells
    b Protection is primary function
  8 Simple columnar epithelium (Figure 2-10)
    a Columnar cells arranged in a single layer
    b Line stomach and intestines
    c Contain mucus-producing goblet cells
    d Specialized for absorption
  9 Stratified transitional epithelium (Figure 2-11)
    a Found in body areas, such as urinary bladder, that are subjected to stretch
    b Up to 10 layers of cuboidal-shaped cells
  10 Glandular epithelium
    a Specialized for secretory activity
    b Cells grouped into glands
    c Secrete into ducts, directly into blood, and on body surface
    d Examples include saliva, digestive juice, and hormones
B Connective tissue
  1 Most abundant tissue in body
  2 Most widely distributed tissue in body
  3 Multiple types, appearances, and functions
  4 Relatively few cells
  5 Types
    a Areolar—glue that holds organs together
    b Adipose (fat)—lipid storage is primary function (Figure 2-12)
    c Fibrous—strong fibers; example is tendon (Figure 2-13)
    d Bone—matrix is calcified; function in support/protection (Figure 2-14)
    e Cartilage—chondrocyte is cell type (Figure 2-15)
    f Blood—matrix is fluid; function is transportation (Figure 2-16)
C Muscle tissue (Figures 2-17 to 2-19)
  1 Types
    a Skeletal—attaches to bones; also called striated or voluntary; control is voluntary; striations apparent when viewed under a microscope (Figure 2-17)
    b Cardiac—also called striated involuntary; composes heart wall; ordinarily cannot control contractions (Figure 2-18)
    c Visceral—also called nonstriated (smooth) or involuntary; no cross striations; found in blood vessels and other tube-shaped organs (Figure 2-19)
D Nervous tissue (Figure 2-20)
  1 Cell types
    a Neurons—conducting cells
    b Neuroglia—supportive and connecting cells
  2 Neurons
    a Cell components
      (1) Cell body
      (2) Axon (one) carry nerve impulse away from cell body
      (3) Dendrites (one or more) carry nerve impulse toward cell body
  3 Function—rapid communication between body structures and control of body functions

NEW WORDS

adenosine triphosphate
 (ATP)
adipose
areolar
axon
centriole
centromere
chondrocyte
chromatid
chromatin
collagen
columnar
compound
crenation

cuboidal
cytoplasm
dendrite
deoxyribonucleic acid
 (DNA)
dialysis
diffusion
element
endoplasmic reticulum
 (ER)
facilitated diffusion
filtration
goblet cell
Golgi apparatus

Haversian system
hemopoietic
hypertonic
hypotonic
interphase
lyse
lysosome
matrix
metaphase
mitochondria
mitosis
neuroglia
neuron
nucleolus

nucleoplasm
nucleus
organelle
osmosis
phagocytosis
pinocytosis
prophase
ribonucleic acid
 (RNA)
ribosome
spindle fiber
squamous
telophase

CHAPTER TEST

Which of the following groups of elements compose over 98% of our body?
a. Oxygen, nitrogen, iron, calcium
b. Oxygen, carbon, sodium, nitrogen
c. Oxygen, nitrogen, carbon, hydrogen
d. Oxygen, hydrogen, phosphorus, calcium

2. Goiter is a disease condition associated with which one of the following elements?
a. Sodium
b. Calcium
c. Chlorine
d. Iodine

3. Which of the following groups represents the three principal parts of a cell?
a. Cytoplasmic membrane, nucleus, cytoplasm
b. Ribosomes, cytoplasmic membrane, mitochondria
c. Lisosomes, centrioles, cytoplasmic membrane
d. Nucleus, Golgi apparatus, mitochondria

4. Which of the following is considered an active transport process?
a. Osmosis
b. Permease system
c. Filtration
d. Dialysis

5. Red blood cells placed in a hypertonic solution will:
   a. Undergo crenation
   b. Lyse
   c. Remain unchanged
   d. Undergo mitosis
6. Deoxyribonucleic acid or DNA:
   a. Is exactly the same as RNA
   b. Contains the compound uracil
   c. Is sometimes called the "heredity molecule"
   d. Is not involved with genes or chromosomes
7. The stage of mitosis in which the chromosomes align themselves across the center of the cell is called:
   a. Prophase
   b. Anaphase
   c. Telophase
   d. Metaphase
8. Which of the following terms describes flat, scalelike cells that permit substances to readily pass through them?
   a. Squamous
   b. Stratified
   c. Transitional
   d. Columnar
9. Haversian systems are associated with which of the following?
   a. Adipose or fat tissue
   b. Bone
   c. Fibrous connective tissue
   d. Cartilage
10. Which of the following types of muscle is considered "voluntary"?
   a. Cardiac
   b. Smooth
   c. Visceral
   d. Skeletal

Select the most appropriate answer in column B for each item in Column A. (Only one answer is correct.)

Column A	**Column B**

Column A

11. _____ Trace element
12. _____ Compound containing nitrogen
13. _____ Cellular organelle
14. _____ Example of diffusion
15. _____ Active transport mechanism
16. _____ Characteristic of DNA
17. _____ Stage of mitosis
18. _____ Type of connective tissue
19. _____ Composed of epithelial tissue
20. _____ Component of a nerve cell

Column B

a. Osmosis
b. Blood
c. Mitochondria
d. Anaphase
e. Sodium
f. Complementary base pairing
g. Pinocytosis
h. Dendrite
i. Proteins
j. Glands

21. A substance that cannot be broken down into two or more different substances is called an _____ .
22. The nickname "protein factories" is often used to describe _____ .
23. Two specialized examples of diffusion include _____ and _____ .
24. Substances move "up a concentration gradient" in _____ transport processes.
25. Red blood cells placed in a hypotonic solution will _____ .
26. The most important "heredity molecule" is _____ .
27. The first stage of mitosis is called _____ .
28. Absorption, protection, and secretion are important functions of _____ tissue.
29. Dendrites, a cell body, and an axon are characteristic of _____ .
30. No blood vessels are found in _____ tissue.

REVIEW QUESTIONS

1 One inch is equal to approximately how many centimeters? How many millimeters? How many micrometers?

2 Identify the three main parts of a cell.

3 Describe the structure and functions of the cytoplasmic membrane. Give another name for this structure.

4 What and where are the following? What functions do they perform?

nucleolus   lysosomes
endoplasmic reticulum   mitochondria
ribosomes   centrioles
chromatin   Golgi apparatus

5 Explain the difference between an element and a compound. Identify the four major elements in the body.

6 Explain what is meant by the term *trace element*.

7 Discuss the importance of the following three chemical substances: DNA, RNA, and ATP. Where in the cell would you expect to find each of these substances in the largest quantity?

8 What is a gene? How does a gene differ from a chromosome?

9 Discuss and contrast active and passive transport systems. List by name the major active and passive transport processes.

10 How does facilitated diffusion differ from a permease system?

11 What term would you use to describe a salt (NaCl) solution that caused a red blood cell to lyse? To crenate?

12 What actually moves during osmosis? During dialysis? What is required for both osmosis and dialysis to occur that is not required for diffusion?

13 How is ATP involved in active transport processes?

14 List the stages of mitosis and briefly describe what occurs during each period.

15 When does DNA replication occur with respect to mitosis?

16 Identify the stage of mitosis that could be described as "prophase in reverse."

17 Explain how epithelial tissue can be classified according to the shape and arrangement of the cells.

18 Explain how the structure of stratified transitional epithelium is related to its function.

19 Compare the matrix found in bone, areolar connective tissue, and blood.

20 What tissue is the most widely distributed in the body? What tissue exists in more varied forms than any other tissue type? In what type of tissue is appearance most determined by the nature of the matrix?

21 Identify and compare the three major types of muscle tissue.

22 Identify the two basic types of cells found in nervous tissue. What is the difference between an axon and a dendrite?

CHAPTER
3
Organ Systems of the Body

CHAPTER OUTLINE

Organ systems of the body
  Integumentary system
  Skeletal system
  Muscular system
  Nervous system
  Endocrine system
  Circulatory system
  Urinary system
  Digestive system
  Respiratory system
  Reproductive system

OBJECTIVES

After you have completed this chapter, you should be able to:

1 Define and contrast the terms *organ* and *organ system*.

2 List the 10 major organ systems of the body.

3 Identify and locate the major organs of each major organ system.

4 Briefly describe the major functions of each major organ system.

5 Identify and discuss the major subdivisions of the circulatory and reproductive organ systems.

The words **organ** and **system** were discussed in Chapter 1 as having special meanings when applied to the body. An **organ** is a structure made up of two or more kinds of tissues organized in such a way that together these tissues can perform a more complex function than can any one tissue alone. A **system** is a group of organs arranged in such a way that together they can perform a more complex function than can any one organ alone. This chapter gives an overview of the 10 major organ systems of the body.

In the chapters that follow, the presentation of information on individual organs and an explanation of how they work together to accomplish complex body functions will form the basis for the discussion of each organ system. For example, a detailed description of the skin as the primary organ of the integumentary system will be covered in Chapter 4, and information on the bones of the body as organs of the skeletal system will be presented in Chapter 5. A knowledge of individual organs and how they are organized into groups makes much more meaningful the understanding of how a particular organ system functions as a unit in the body.

When you have completed your study of the major organ systems in the chapters that follow, it will be possible to view the body not as an assemblage of individual parts but as an integrated and functioning whole. This chapter names the systems of the body and the major organs that compose them, and it briefly describes the functions of each system. It is intended to provide a basic "road map" to help you anticipate and prepare for the more detailed information that follows in the remainder of the text.

ORGAN SYSTEMS OF THE BODY

In contrast to cells, which are the smallest structural units of the body, organ systems are its largest and most complex structural units. The 10 major organ systems that compose the human body are listed below.

1 Integumentary
2 Skeletal
3 Muscular
4 Nervous
5 Endocrine
6 Circulatory
  a Cardiovascular subdivision
  b Lymphatic subdivision
7 Urinary
8 Digestive
9 Respiratory
10 Reproductive
  a Male subdivision
  b Female subdivision

Examine Figure 3-1 to find a diagrammatic listing of the body systems and the major organs in each. In addition to the information contained in Figure 3-1, each of the body organ systems is presented in visual form in Figure 3-2. Visual presentation of material is often useful in understanding the interrelationships that are so important in anatomy and physiology.

INTEGUMENTARY SYSTEM

Note in Figure 3-2 that the skin itself is the largest and most important organ in the **integumentary** (in-teg-u-MEN-tar-e) **system.** Its weight in most adults is 20 pounds or more, accounting for about 15% of total body weight and making it the body's heaviest organ. The integumentary system includes both the skin proper and its **appendages,** which include the hair, nails, and specialized sweat- and oil-producing glands. In addition, a number of microscopic and highly specialized sense organs are found embedded in the skin. They permit the body to respond to pain, pressure, touch, and changes in temperature.

The integumentary system is crucial to survival itself. Its primary function is **protection.** The skin protects underlying tissue against invasion by hordes of harmful bacteria, bars entry of most chemicals, and minimizes the chances of mechanical injury to underlying structures. In addition, the skin serves to regulate body temperature by sweating, synthesizes important chemicals and hormones, and functions as a sophisticated sense organ.

Text continued on p. 61.

Figure 3-1 Body systems and the organs comprising them.

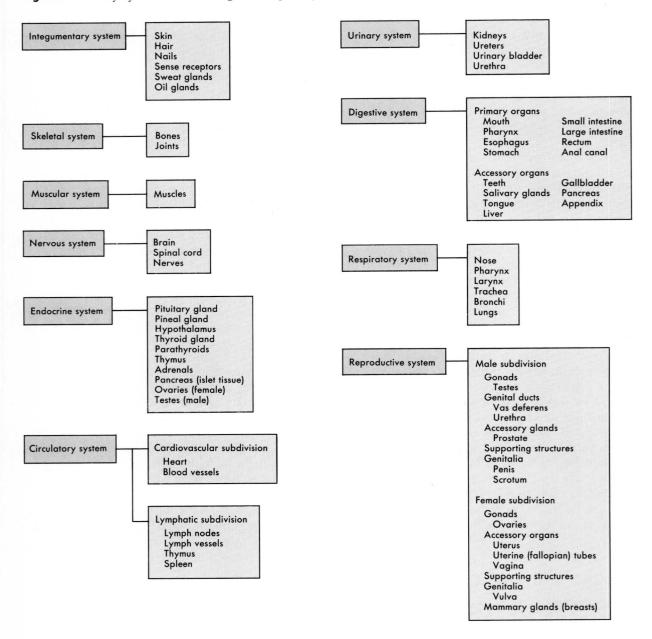

Figure 3-2 Organ systems of the human body.

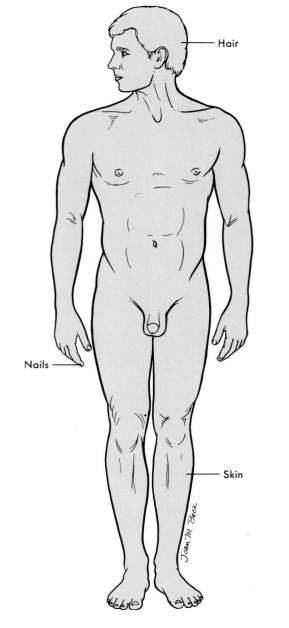

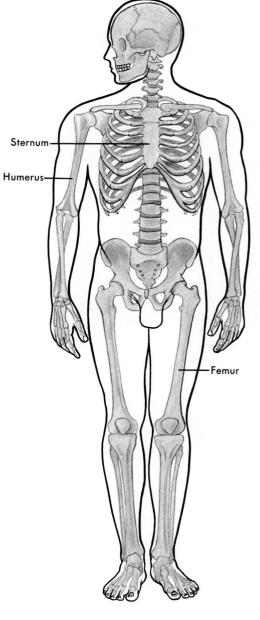

INTEGUMENTARY SYSTEM **SKELETAL SYSTEM**

Figure 3-2, cont'd

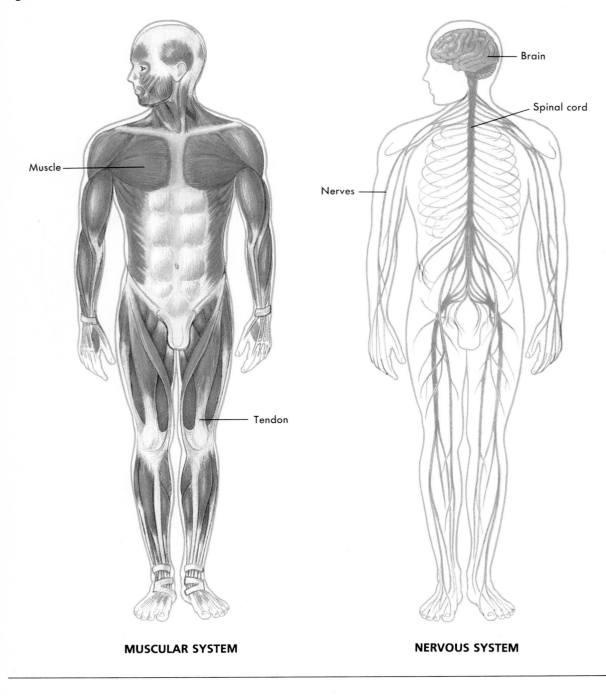

MUSCULAR SYSTEM   **NERVOUS SYSTEM**

Figure 3-2, cont'd

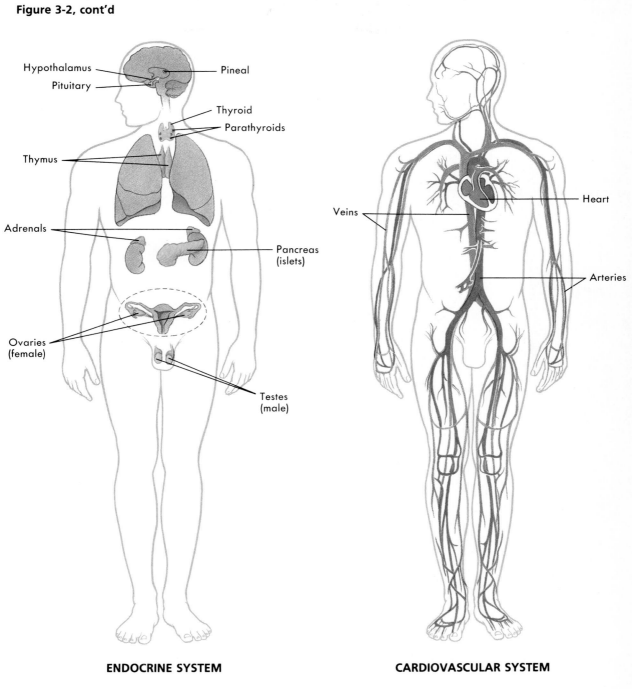

ENDOCRINE SYSTEM   **CARDIOVASCULAR SYSTEM**

Figure 3-2, cont'd

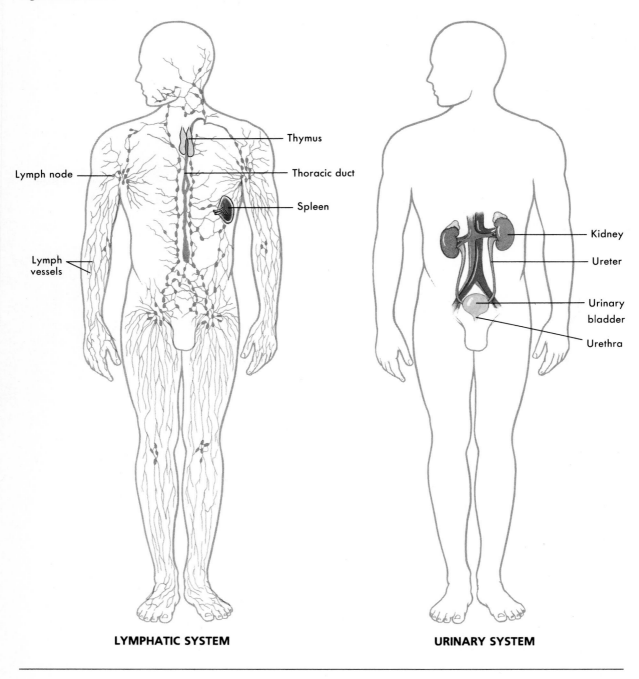

LYMPHATIC SYSTEM

URINARY SYSTEM

Figure 3-2, cont'd

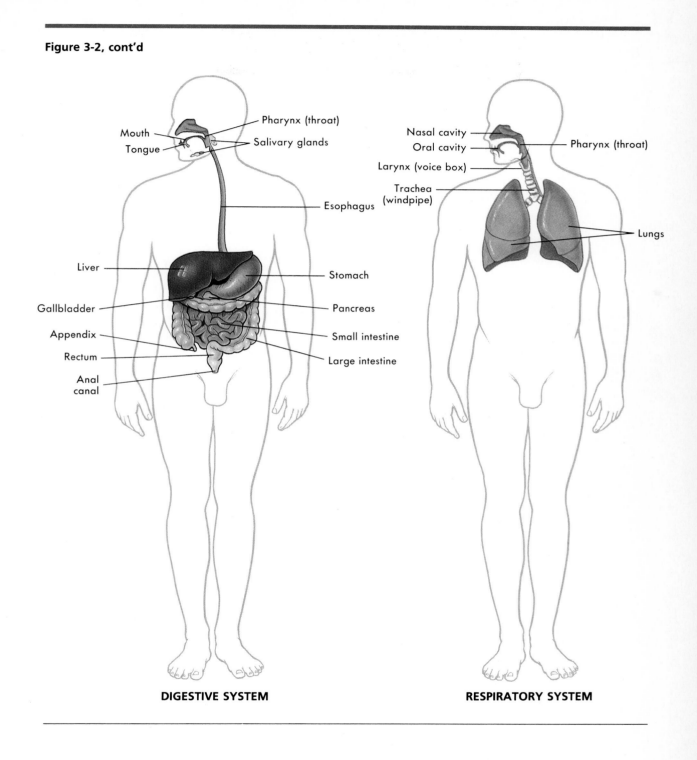

DIGESTIVE SYSTEM   **RESPIRATORY SYSTEM**

Figure 3-2, cont'd

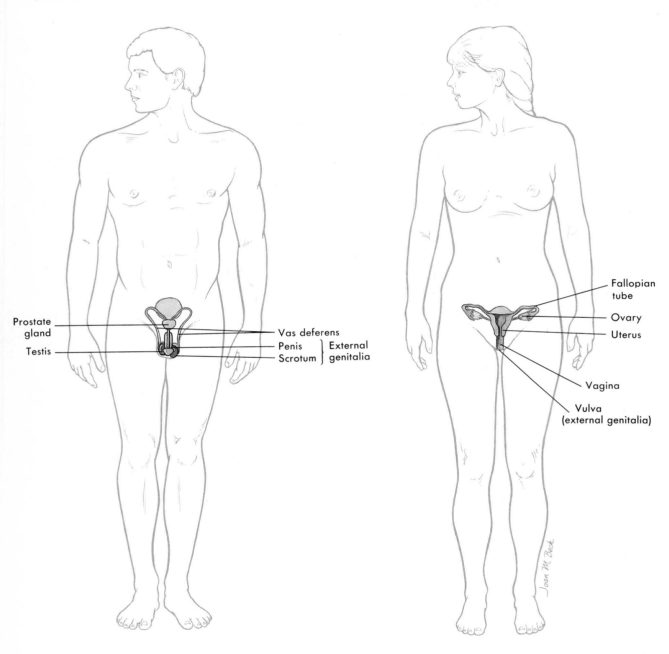

Prostate gland

Testis

Vas deferens

Penis
Scrotum } External genitalia

Fallopian tube

Ovary

Uterus

Vagina

Vulva (external genitalia)

MALE REPRODUCTIVE SYSTEM

FEMALE REPRODUCTIVE SYSTEM

SKELETAL SYSTEM

The sternum or breastbone and the humerus and femur shown in Figure 3-2 are examples of the 206 individual organs (bones) found in the **skeletal system.** The system as a whole includes not only bones but also related tissues such as cartilage and ligaments that together provide the body with a rigid framework for support and protection. In addition, the skeletal system, through the existence of **joints** between bones, makes the movements of body parts possible. Without joints between our bones, we could make no movements; our bodies would be rigid, immobile hulks. Bones also serve as storage areas for such important minerals as calcium and phosphorus. The formation of blood cells in the red marrow of certain bones is another crucial function of the skeletal system.

MUSCULAR SYSTEM

Individual muscles are the organs of the **muscular system.** In addition to **voluntary** or **skeletal** muscles, which have the ability to contract when stimulated and are under conscious control, the muscular system also contains **smooth** or **involuntary** muscles found in such organs as the stomach and small intestine. The third type of muscle tissue is the **cardiac muscle** of the heart. Muscles not only produce movement and maintain body posture; they are also responsible for generating the heat required for maintenance of a constant core body temperature.

The tendon labeled in Figure 3-2 is shown as an example of how muscles attach to bones. When stimulated by a nervous impulse, muscle tissue will shorten or contract. Meaningful movement of the body occurs when skeletal muscles contract because of the way muscles are attached to bones and the way bones articulate or join together with one another in joints.

NERVOUS SYSTEM

The brain, spinal cord, and nerves are the organs of the **nervous system.** As you can see in Figure 3-2, nerves extend from the brain and spinal cord to every area of the body. The extensive networking of the components of the nervous system makes it possible for this complex system to perform its primary functions. These include:

1 Communication between various body functions
2 Integration of body functions
3 Control of body functions

These functions are accomplished by specialized signals called **nerve impulses.** In general, the functions of the nervous system result in very rapid activity that lasts usually for a short duration. For example, we can chew our food normally, walk, and perform coordinated muscular movements only if our nervous system is functioning properly. It is the nerve impulse that permits the rapid and precise control of diverse body functions. In addition, elements of the nervous system can recognize certain **stimuli** (STIM-u-li)—such as heat, light, pressure, or temperature—that affect the body. Nervous impulses may then be generated to convey this information to the brain, where it can be analyzed and where appropriate action can be initiated. Other types of nerve impulses cause glands to secrete.

ENDOCRINE SYSTEM

The **endocrine system** is composed of specialized glands that secrete chemicals known as **hormones** directly into the blood. Sometimes called *ductless glands,* the organs of the endocrine system perform the same general functions as the nervous system: communication, integration, and control. The nervous system provides rapid, brief control by fast-traveling nerve impulses. The endocrine system provides slower but longer-lasting control by the secretion of hormones; for example, secretion of the growth hormone controls the rate of development over long periods of gradual growth.

In addition to controlling growth, hormones are the main regulators of metabolism, reproduction, and many other body activities. They play roles of the utmost importance in such areas as fluid and electrolyte balance, acid-base balance, and energy metabolism.

As you can see in Figure 3-2, the endocrine glands are widely distributed in the body. The **pituitary** (pi-TU-i-TAR-e) **gland, pineal** (PIN-e-al) **gland,** and **hypothalamus** (hi-po-THAL-ah-mus) are located in the skull. The **thyroid** (THI-roid) and **parathyroid** (PAR-ah-THI-roid) **glands** are in the neck, and the **thymus** (THI-mus) **gland** is in the thoracic cavity. Recall from Chapter 1 that the space between the lungs is actually a specialized subdivision of the thoracic cavity called the *mediastinum.* The thymus is located in the mediastinum. The **adrenal** (ah-DRE-nal) **glands** and **pancreas** (PAN-kre-as) are found in the abdominal cavity. Note in Figure 3-2 that the ovaries in the female and the testes in the male also function as endocrine glands.

CIRCULATORY SYSTEM

Note in Figure 3-2 that the **circulatory system** is subdivided into the **cardiovascular** (KAR-de-o-VAS-ku-lar) and **lymphatic** (lim-FAT-ik) subdivisions.

Cardiovascular Subdivision

The **cardiovascular** subdivision consists of the heart and a closed system of vessels called arteries, veins, and capillaries. As the name implies, blood contained in the circulatory system is pumped by the heart around a closed circle, or circuit, of vessels as it passes through the body.

The primary function of the cardiovascular subdivision of the circulatory system is *transportation.* The need for an efficient transportation system in the body is obvious. Critical transportation needs include movement of oxygen and carbon dioxide, nutrients, hormones, and other important substances on a continuing basis.

Lymphatic Subdivision

The **lymphatic** subdivision of the circulatory system is composed of **lymph nodes, lymphatic vessels,** and specialized lymphatic organs such as the **thymus** and **spleen.** Note that the thymus functions both as an endocrine and as a lymphatic gland. Instead of containing blood, the lymphatic vessels are filled with a whitish, watery fluid that contains lymphocytes, proteins, and some fatty molecules. No red blood cells are present. The lymph is formed from the fluid surrounding the body cells and diffuses into the lymph vessels. However, unlike blood, lymph does not circulate repeatedly through a closed circuit or "loop" of vessels. Instead, lymph flowing through lymphatic vessels eventually enters the cardiovascular subdivision of the circulatory system. It does so by passing through large ducts, including the **thoracic duct** shown in Figure 3-2, which in turn connect with veins in the upper area of the thoracic cavity. Collections of lymph nodes can be seen in the axillary (arm pit) and inguinal (groin) areas of the body in Figure 3-2. The formation and movement of lymph will be discussed in Chapter 11.

The functions of the lymphatic system include movement of fluids and certain large molecules from the tissue spaces surrounding the cells and movement of fat-related nutrients from the digestive tract back to the blood. The lymphatic system is also involved in the overall functioning of the immune system, which plays a critical role in the defense mechanism of the body against disease.

URINARY SYSTEM

The organs of the **urinary system** include the two **kidneys,** the **ureters** (u-RE-ters), the **bladder,** and the **urethra** (u-RE-thrah).

The kidneys function to "clear" or clean the blood of the many waste products that are continually produced by the metabolism of foodstuff in the body cells. The kidneys also play an important role in maintaining the electrolyte, water, and acid-base balance in the body.

The waste product produced by the kidneys is called **urine** (U-rin). Once produced by the kidneys, it flows out of the kidneys through the two ureters into the urinary bladder where it is stored. Urine passes from the bladder to the outside of the body through the urethra. In the male the urethra passes through the penis, which has a double function: it serves to transport both urine and semen or seminal fluid. Therefore it has both a urinary and a reproductive purpose. In the female the urinary and reproductive passages are completely separate and distinct and so the urethra performs only a urinary function.

DIGESTIVE SYSTEM

The organs of the **digestive system** are often separated into two groups: the *primary organs* and the *secondary* or *accessory organs* of the digestive system (see Figure 3-1). They work together to ensure proper digestion and absorption of nutrients. The primary organs include the mouth, pharynx, esophagus, stomach, small intestine, large intestine, rectum, and anal canal. The accessory organs of digestion include the teeth, salivary glands, tongue, liver, gallbladder, pancreas, and appendix.

The primary organs of the digestive system form a tube, open at both ends, called the **gastrointestinal** (GAS-tro-in-TES-ti-nal) or **GI tract.** Food that enters the tract is digested, its nutrients are absorbed, and the undigested residue is eliminated from the body as waste material called **feces** (FE-seez). The accessory organs assist in the mechanical or chemical breakdown of ingested food. The appendix, although classified as an accessory organ of digestion and physically attached to the digestive tube, is not functionally important in the digestive process. However, inflammation of the appendix, called **appendicitis** (ah-PEN-di-SI-tis) is a very serious clinical condition, one that frequently requires surgery.

RESPIRATORY SYSTEM

The organs of the **respiratory system** include the nose, **pharynx** (FAR-inks), **larynx** (LAR-inks), **trachea** (TRA-ke-ah), **bronchi** (BRON-ki), and lungs. Together these organs permit the movement of air into the tiny, thin-walled sacs of the lungs called **alveoli** (al-VE-o-li). It is in the alveoli that oxygen from the air is exchanged for the waste product carbon dioxide, which is carried to the lungs by the blood so that it can be eliminated from the body.

REPRODUCTIVE SYSTEM

The importance of normal **reproductive system** function is notably different from the end result of "normal function" as measured in any other organ system of the body. The proper functioning of the reproductive system ensures survival, not of the individual but of the species—the human race. In addition, production of the hormones that permit the development of sexual characteristics occurs as a result of normal reproductive system activity.

Male Subdivision

The male reproductive structures shown in Figure 3-2 include: the **gonads** (GO-nads) called **testes** (TES-teez), which produce the sex cells or **sperm;** one of the important **genital ducts** called the **vas deferens** (vas DEF-er-enz); and the **prostate** (PROS-tate), which is classified as an **accessory gland** in the male. The **penis** (PE-nis) and **scrotum** (SKRO-tum) are called **supporting structures** and together are known as the **genitalia** (JEN-e-tal-yah). The urethra, which is identified in Figure 3-2 as part of the urinary system, passes through the penis. It serves as both a genital duct that carries sperm to the exterior and as a passageway for the elimination of urine. Functioning together, these structures produce, transfer, and ultimately introduce sperm into the female reproductive tract where fertilization can occur. Sperm produced by the testes travels through a number of genital ducts, including the vas deferens, in order to exit from the body. The prostate and other accessory glands, which add fluid and nutrients to the sex cells as they pass through the ducts and the supporting structures (especially the penis), permit transfer of sex cells into the female reproductive tract.

Female Subdivision

The female **gonads** are the **ovaries.** The **accessory organs** shown in Figure 3-2 include the **uterus** (U-ter-us), **uterine** (U-ter-in) or **fallopian tubes,** and the **vagina** (vuh-JI-nah). In the female the term **vulva** (VUL-vah) is used to describe the external genitalia. The breasts or **mammary glands** are also classified as external accessory sex organs in the female.

The reproductive organs in the female are intended to produce the sex cells or **ova,** receive the male sex cells (sperm), permit fertilization and transfer of the sex cells to the uterus, and allow for the development, birth, and nourishment of offspring.

• • •

As you study the more detailed structure and function of the organ systems in the chapters that follow, always relate the system and its component organs to the body as a whole. No one body system functions entirely independently of other systems. Instead, you will find that they are structurally and functionally interrelated and interdependent.

OUTLINE SUMMARY

Definitions and Concepts

A Organ—a structure made up of two or more kinds of tissues organized in such a way that together they can perform a more complex function than can any one tissue alone

B Organ system—a group of organs arranged in such a way that together they can perform a more complex function than can any one organ alone

C A knowledge of individual organs and how they are organized into groups makes more meaningful the understanding of how a particular organ system functions as a whole

Organ Systems

A Integumentary system
  1 Structure—organs
    a Skin
    b Hair
    c Nails
    d Sense receptors
    e Sweat glands
    f Oil glands
  2 Functions
    a Protection
    b Regulation of body temperature
    c Synthesis of chemicals and hormones
    d Serves as a sense organ
B Skeletal system
  1 Structure
    a Bones
    b Joints

  2 Functions
    a Movement (with joints and muscles)
    b Storage of minerals
    c Blood formation
C Muscular system
  1 Structure
    a Muscles
      (1) Voluntary or striated
      (2) Involuntary or smooth
      (3) Cardiac
  2 Functions
    a Movement
    b Maintains body posture
    c Produces heat
D Nervous system
  1 Structure
    a Brain
    b Spinal cord
    c Nerves
  2 Functions
    a Communication
    b Integration
    c Control
  3 System functions by production of nerve impulses caused by stimuli such as heat and pressure
  4 Control is fast-acting and of short duration
E Endocrine system
  1 Structure
    a Pituitary gland
    b Pineal gland
    c Hypothalamus
    d Thyroid gland

    e Parathyroid glands
    f Thymus gland
    g Adrenal glands
    h Pancreas (islet tissue)
    i Ovaries (female)
    j Testes (male)
  2 Functions
    a Secrete special substances called hormones directly into the blood
    b Same overall functions as nervous system—communication, integration, control
    c Control is slow and of long duration
    d Examples of hormone regulation:
      (1) Growth
      (2) Metabolism
      (3) Reproduction
      (4) Fluid and electrolyte balance

F Circulatory system
  1 Structure
    a Cardiovascular subdivision
      (1) Heart
      (2) Blood vessels
    b Lymphatic subdivision
      (1) Lymph nodes
      (2) Lymph vessels
      (3) Thymus
      (4) Spleen
  2 Functions
    a Transportation
    b Immune system (body defense)

G Urinary system
  1 Structure
    a Kidneys
    b Ureters
    c Urinary bladder
    d Urethra
  2 Functions
    a "Clear" or clean blood of waste products—waste product excreted from body is called *urine*
    b Electrolyte balance
    c Water balance
    d Acid-base balance
    e In male, urethra has both urinary and reproductive functions

H Digestive system
  1 Structure
    a Primary organs
      (1) Mouth
      (2) Pharynx
      (3) Esophagus
      (4) Stomach
      (5) Small intestine
      (6) Large intestine
      (7) Rectum
      (8) Anal canal
    b Accessory organs
      (1) Teeth
      (2) Salivary glands
      (3) Tongue
      (4) Liver
      (5) Gallbladder
      (6) Pancreas
      (7) Appendix
  2 Functions
    a Mechanical and chemical breakdown (digestion) of food
    b Absorption of nutrients
    c Undigested waste product that is excreted is called *feces*
    d Appendix is a structural but not a functional part of digestive system
    e Inflammation of appendix called *appendicitis (malfunction)*

I Respiratory system
  1 Structure
    a Nose
    b Pharynx
    c Larynx
    d Trachea
    e Bronchi
    f Lungs
  2 Functions
    a Exchange of waste gas (carbon dioxide) for oxygen in the lungs
    b Area of gas exchange in the lungs called *alveoli*

J Reproductive system
  1 Structure
    a Male
      (1) Gonads—testes
      (2) Genital ducts—epididymis; vas deferens; ejaculatory ducts; urethra

(3) Accessory glands—seminal vesicles; prostate; bulbourethral (Cowper's) glands

(4) Supporting structures—genitalia (penis and scrotum) and spermatic cords

b Female

(1) Gonads—ovaries

(2) Accessory organs—uterus; uterine (fallopian) tubes; vagina

(3) Supporting structures—genitalia (vulva); mammary glands (breasts)

2 Functions

a Survival of species

b Production of sex cells (male: sperm; female: ova)

c Allows transfer of sex cells and fertilization to occur

d Permits development and birth of offspring

e Nourishment of offspring

f Production of sex hormones

NEW WORDS

Review the names of organ systems and individual organs in Figures 3-1 and 3-2.

appendicitis	hormone
cardiovascular	integumentary
endocrine	lymphatic
feces	nerve impulse
gastrointestinal (GI) tract	stimuli
genitalia	urine
gonads	

CHAPTER TEST

1. The circulatory system is subdivided into the _____ and _____ subdivisions.

2. The largest and most important organ of the integumentary system is the _____ .

3. The 206 individual organs of the skeletal system are called _____ .

4. The three primary types of muscle tissue are called: _____ , _____ , and _____ .

5. The three primary functions of the nervous system are _____ , _____ , and _____ of body functions.

6. The organs of the endocrine system secrete substances called _____ into the blood.

7. The thymus and spleen are classified as _____ organs.

8. The waste product produced by the kidneys is called _____ .

9. The tongue, liver, and gallbladder are all classified as
_____ organs of digestion.
10. Oxygen from the air is exchanged for the waste product carbon dioxide in thin-walled sacs in the lungs called _____ .

Select the most appropriate answer in column B for each item in column A. (Only one answer is correct.)

Column A	Column B
11. ____ Appendage of skin	a. Hormones
12. ____ Storage site for calcium	b. Lymphatic system
13. ____ Function to generate heat	c. Muscles
14. ____ Permit rapid communication	d. Feces
15. ____ Secreted by ductless glands	e. Hair
16. ____ Primary function of cardiovascular subdivision of circulatory system	f. Vulva
17. ____ Important in body defense mechanism	g. Testes
18. ____ Undigested residue in GI tract	h. Bones
19. ____ Female external genitalia	i. Transportation
20. ____ Male gonads	j. Nerve impulses

REVIEW QUESTIONS

1 Give brief definitions of the terms *organ* and *organ system*.
2 List the 10 major organ systems of the body.
3 Discuss the structure and generalized functions of the integumentary system.
4 Explain the functional interaction that occurs between the skeletal, muscular, and nervous systems.
5 Compare the generalized functions of the nervous and endocrine systems. How are they similar? How do they differ?

6 What is the relationship between a stimulus and a nerve impulse?
7 Define the term *hormone*.
8 Compare the cardiovascular and lymphatic divisions of the circulatory system.
9 Identify the waste products associated with the urinary and digestive systems.
10 List the primary and accessory organs of the digestive system.
11 Compare the structure and function of the male and female reproductive systems.

CHAPTER
4

The Integumentary System and Body Membranes

OBJECTIVES

After you have completed this chapter, you should be able to:

1 Describe the structure and function of the epidermis and dermis of the skin.

2 List and briefly describe each of the accessory organs of the skin.

3 List and discuss the three primary functions of the integumentary system.

4 Classify burns and describe how to estimate the extent of a burn injury.

5 Classify, compare the structure of, and give examples of each type of body membrane.

In Chapter 1 the concept of progressive organization of body structures from simple to complex was established. Complexity in body structure and function progresses from cells to tissues and then to organs and organ systems. This chapter discusses the skin and its **appendages**—the hair, nails, and skin glands—as an organ system. That system is called the **integumentary system. Integument** (in TEG-u-ment) is another name for the skin, and the skin itself is the principal organ of the integumentary system. The skin is one of a group of anatomically simple but functionally important sheetlike organs called **membranes.** After the integumentary system is discussed, the body membranes as a group will be classified and discussed. Ideally, you should study the skin and its appendages before proceeding to the more traditional organ systems in the chapters that follow in order to improve your understanding of how structure is related to function.

THE SKIN

The brief description of the skin in Chapter 3 identified it not only as the primary organ of the integumentary system but as the largest and one of the most important organs in the body. Architecturally the skin is a marvel. Consider the incredible number of structures fitted into 1 square inch of skin: 500 sweat glands; over 1,000 nerve endings; yards of tiny blood vessels; nearly 100 oil or sebaceous (see-BAY-shus) glands; 150 sensors for pressure, 75 for heat, 10 for cold; and literally millions of cells.

STRUCTURE OF THE SKIN

The skin itself is sometimes called the **cutaneous** (ku-TA-ne-us) **membrane.** It is a sheetlike organ composed of two layers of distinct tissue (Figure 4-1).
1 The **epidermis** is the outermost layer of the skin. It is a relatively thin sheet of stratified squamous epithelium.
2 The **dermis** is the deeper of the two layers. It is thicker than the epidermis and is made up largely of connective tissue.

Subcutaneous injection

Although the subcutaneous layer is not part of the skin itself, it carries the major blood vessels and nerves to the skin above. The rich blood supply and loose spongy texture of the subcutaneous layer make it an ideal site for the rapid and relatively pain-free absorption of injected material. Liquid medicines such as insulin, and pelleted implant materials are often administered by **subcutaneous injection** into this spongy and porous layer beneath the skin.

As you can see in Figure 4-1, the layers of the skin are supported by a thick layer of loose connective tissue and fat called **subcutaneous** (sub ku-TA-ne-us) **tissue.** Fat in the subcutaneous layer serves to insulate the body from extremes of heat and cold. It also serves as a stored source of energy for the body and can be used as a food source if required. In addition, the subcutaneous tissue acts as a shock-absorbing pad and helps protect underlying tissues from potential injury caused by bumps and blows to the body surface.

Epidermis

The tightly packed epithelial cells of the epidermis are arranged in many distinct layers. The cells of the innermost layer of the epidermis, called the **stratum germinativum,** undergo mitosis and reproduce themselves (Figure 4-1). This ability of epidermal cells to undergo mitosis and reproduce is of critical clinical significance. It enables the skin to repair itself if injured. The self-repairing characteristic of normal skin makes it possible for the body to maintain an effective barrier against infection even when subjected to injury and normal wear and tear. As new cells are produced in the deep layer of the epidermis, they move up toward the surface. As they approach the surface, the cytoplasm is replaced by one of nature's most unique proteins—a substance called **keratin** (KER-ah-tin). Keratin is a

Figure 4-1 Microscopic view of the skin in longitudinal section. The epidermis is shown raised at one corner to reveal the ridges in the dermis.

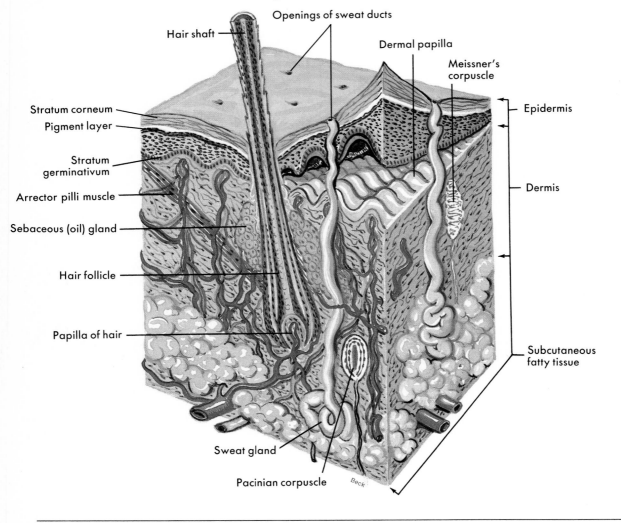

tough waterproof material that provides cells in the outer layer of the skin with a horny, abrasion-resistant, and protective quality. The tough outer layer of the epidermis is called the **stratum corneum** (KOR-ne-um). Cells filled with keratin are continually pushed to the surface of the epidermis. Note in the photomicrograph of the skin shown in Figure 4-2 that many of the outermost cells of the stratum corneum have been dis-lodged. These dry dead cells filled with keratin "flake off" by the thousands onto our clothes, our bathwater, and things we handle. Millions of epithelial cells reproduce daily to replace the millions shed—just one example of the work our bodies do without our knowing it, even when they seem to be resting.

Another cell layer of the epidermis identified in Figure 4-1 is the **pigment layer.** The term *pig-*

Figure 4-2 Photomicrograph of skin. Many dead cells of the stratum corneum have flaked off from the surface of the epidermis. Note that the epidermis is very cellular. The dermis has fewer cells and more connective tissue.

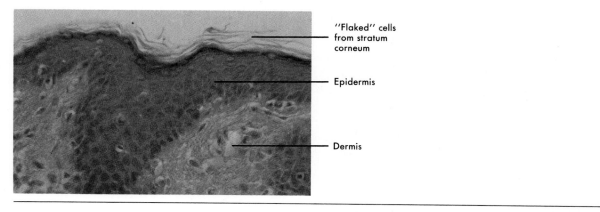

"Flaked" cells from stratum corneum

Epidermis

Dermis

ment comes from a Latin word meaning "paint." It is an appropriate name for the epidermal layer that gives color to the skin. The brown pigment **melanin** (MEL-ah-nin) is produced by specialized cells in the pigment layer. These cells are called **melanocytes** (MEL-ah-no-cytes). The higher the concentration of melanin, the deeper is the color of skin. The amount of melanin in your skin depends first on certain skin color genes you have inherited. In a word, heredity determines how dark or light your basic skin color is. However, other factors such as sunlight, can modify this hereditary effect. Prolonged exposure to sunlight in light-skinned people darkens the exposed area because it leads to increased melanin deposits in the epidermis. If the skin contains little melanin, a change in color can occur if the volume of blood in the skin changes significantly or if the amount of oxygen in the blood is increased or decreased. In these individuals increased blood flow to the skin or increased blood oxygen levels can cause a pink "flush" to appear. However, if blood oxygen levels decrease or if actual blood flow is reduced dramatically, the skin turns a bluish gray color— a condition called **cyanosis** (SI-ah-NO-sis). In general, the less abundant the melanin deposits in the skin, the more visible are the changes in

color that are caused by the change in skin blood volume or oxygen level. Conversely, the richer the skin's pigmentation, the less noticeable such changes will be.

The cells of the epidermis are packed tightly together. They are held firmly to one another and to the dermis below by specialized junctions that exist between the membranes of adjacent cells. If these specialized linkages, sometimes described as "spot welds," are weakened or destroyed, the skin literally falls apart. When this occurs because of burns, friction injuries, or exposure to irritants, **blisters** may result. The blisters shown in Figure 4-3 were caused by the irritant chemicals in poison ivy that produced cell injury and death.

The junction that exists between the thin epidermal layer of the skin above and the dermal layer below is called the **dermal-epidermal junction.** The area of contact between dermis and epidermis functions to "glue" them together and provide support for the epidermis, which is attached to its upper surface. Blister formation will also occur if this junction is damaged or destroyed. The junction is visible in Figure 4-1, which shows the epidermis raised on one corner to reveal the underlying dermis more clearly.

Figure 4-3   Blisters resulting from contact with poison ivy.

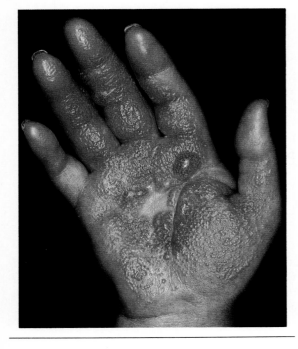

Dermis

The dermis is the deeper of the two primary skin layers and is much thicker than the epidermis. It is composed largely of connective tissue. Instead of cells being crowded close together like the epithelial cells of the epidermis, they are scattered far apart, with many fibers in between. Some of the fibers are tough and strong (collagenous or white fibers), and others are stretchable and elastic (elastic fibers).

The upper region of the dermis is characterized by parallel rows of peglike projections called **dermal papillae** (pah-PIL-e), which are visible in Figure 4-1. These upward projections are interesting and useful structural features. They form an important part of the dermal-epidermal junction that helps to bind the two skin layers together. In addition, they form the ridges and grooves that make possible fingerprinting as a means of identification.

You can observe these ridges on the tips of the fingers and on the skin covering the palms of your hands. Observe in Figure 4-1 how the epidermis follows the contours of the dermal papillae. These ridges develop sometime before birth. Not only is their pattern unique in each individual, but also it never changes except to grow larger—two facts that explain why our fingerprints or footprints positively identify us. Many hospitals identify newborn babies by footprinting them soon after birth.

The deeper area of the dermis is filled with a dense network of interlacing fibers. Most of the fibers in this area are of the collagenous type that gives toughness to the skin. However, elastic fibers are also present. These make the skin stretchable and elastic (able to rebound). As we age, the number of elastic fibers in the dermis decreases and the amount of fat stored in the subcutaneous tissue is reduced. Wrinkles develop as the skin loses elasticity, sags, and becomes less soft and pliant.

In addition to connective tissue elements, the dermis contains a specialized network of nerves and nerve endings to process sensory information such as pain, pressure, touch, and temperature. At various levels of the dermis, there are muscle fibers, hair follicles, sweat and sebaceous glands, and many blood vessels.

APPENDAGES OF THE SKIN
Hair

The human body is literally covered with millions of hairs. Indeed, at the time of birth most of the specialized structures called **follicles** (FOL-li-kls) that are required for hair growth are already present. They develop early in fetal life and by the time of birth are present in most parts of the skin. The hair of a newborn infant is extremely fine and soft; it is called **lanugo** (lah-NU-go) from the Latin word meaning "down." In premature infants lanugo may be noticeable over most of the body, but soon after birth the lanugo that remains is lost and replaced by new hair that is stronger and more pigmented. Although only a few areas of the skin are hairless—notably the lips, the palms of the hands, and the soles of the feet—most body hair remains almost invisible. Hair is most visible on the

Figure 4-4 Hair follicle. Relationship of a hair follicle and related structures to the epidermal and dermal layers of the skin.

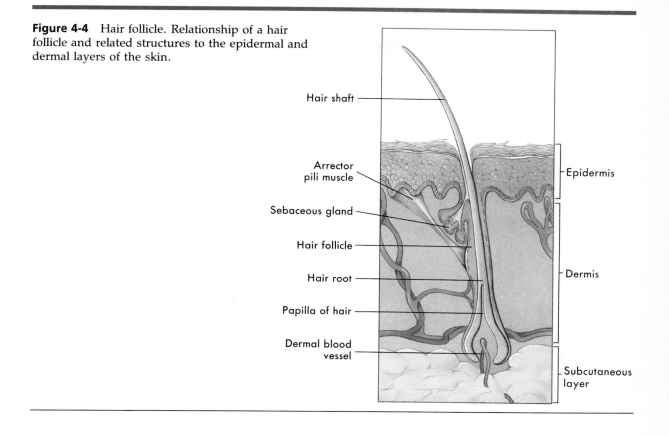

scalp, eyelids, and eyebrows. The coarse hair that first appears in the pubic and axillary regions at the time of puberty develops in response to the secretion of hormones.

Hair growth begins when cells of the epidermal layer of the skin grow down into the dermis, forming a small tube called the **hair follicle.** The relationship of a hair follicle and its related structures to the epidermal and dermal layers of the skin is shown in Figure 4-4. Hair growth begins from a small cap-shaped cluster of cells called the **hair papilla** (pah-PIL-ah), which is located at the base of the follicle. The papilla is nourished by a dermal blood vessel. Note in Figure 4-4 that part of the hair, namely the **root,** lies hidden in the follicle. The visible part of a hair is called the **shaft.**

As long as cells in the papilla of the hair follicle remain alive, new hair will replace any that is cut or plucked. Contrary to popular belief, frequent cutting or shaving does not make hair grow faster or become coarser. Why? Because neither process affects the epithelial cells that form the hairs, since they are embedded in the dermis.

Depilatories

   Depilatories (de-PIL-ah-toe-res), such as Neet, are used to remove unwanted hair. They act by dissolving the protein in hair shafts that extend above the skin surface. Since the follicle is not affected, regrowth of hair continues at a normal rate.

Figure 4-5 Skin receptors. Receptors are specialized nerve endings that make it possible for the skin to act as a sense organ. **A,** Meissner's corpuscle. **B,** Pacinian corpuscle. (See also Figure 4-1.)

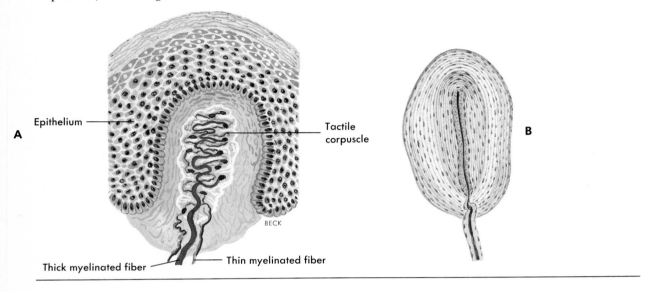

A tiny smooth (involuntary) muscle can be seen in Figure 4-4. It is called an **arrector pili** (ah-REK-tor PIE-li) muscle. Note that it is attached to the base of a dermal papilla above and to the side of a hair follicle below. Generally these muscles contract only when we are frightened or cold. When contraction occurs, each muscle simultaneously pulls on its two points of attachment—that is, up on a hair follicle but down on a part of the skin. This produces little raised places, called "goose pimples," between the depressed points of the skin and at the same time pulls the hairs up until they are more or less straight. The name **arrector pili** describes the function of these muscles; it is Latin for "erectors of the hair." We unconsciously recognize these facts in expressions such as "I was so frightened my hair stood on end."

Receptors

Receptors in the skin make it possible for the body surface to act as a sense organ, relaying messages to the brain concerning such sensa-tions as touch, pain, temperature, and pressure. Receptors differ in structure from the highly complex to the very simple. Figure 4-5 shows enlarged views of a **Meissner's** (MIS-nerz) corpuscle and a **pacinian** (pah-SIN-e-an) **corpuscle.** Look again at Figure 4-1 and find these receptors. The pacinian corpuscle is shown deep in the dermis. It is capable of detecting *pressure* on the skin surface. The Meissner's corpuscle is generally located rather close to the skin surface. It is capable of detecting sensations of *light touch*. Both of these specialized receptors are widely distributed in skin.

Nails

Nails are classified as accessory organs of the skin and are produced by cells in the epidermis. They form when epidermal cells over the terminal ends of the fingers and toes fill with keratin and become hard and platelike. The component parts of a typical fingernail and its associated structures are shown in Figure 4-6. In this illustration the fingernail of the index finger

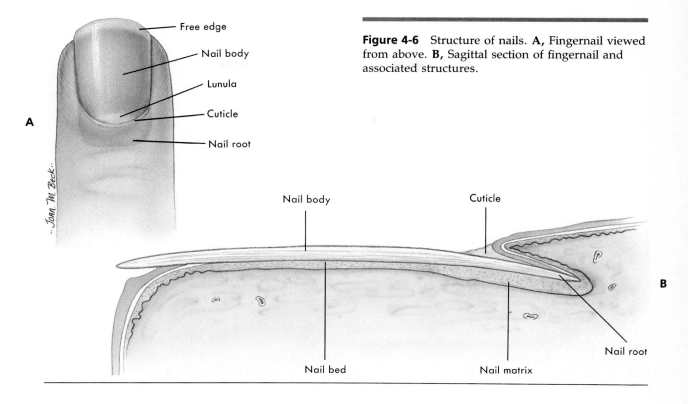

..Joan. W. Beck..

Figure 4-6   Structure of nails. **A,** Fingernail viewed from above. **B,** Sagittal section of fingernail and associated structures.

is viewed from above and in sagittal section. (Recall that a sagittal section divides a body part into right and left portions.) Look first at the nail as seen from above. The visible part of the nail is called the **nail body.** The rest of the nail, namely, the **root,** lies in a groove and is hidden by a fold of skin called the **cuticle** (KU-te-kl). In the sagittal section you can see the nail root from the side and note its relationship to the cuticle, which is folded back over its upper surface. The nail body nearest the root has a crescent-shaped white area known as the **lunula** (LU-nu-lah) or "little moon." You should be able to identify this area quite easily on your own nails; it is most noticeable on the thumbnail. Under the nail lies a layer of epithelium called the **nail bed,** which is labeled on the sagittal section in Figure 4-6. Because it contains abundant blood vessels, it appears pink in color through the translucent nail bodies. If blood oxygen levels drop and cyanosis develops, the nail bed will turn blue.

Skin Glands

The skin glands include the two varieties of **sweat** or **sudoriferous** (su-do-RIF-er-us) **glands** and the microscopic **sebaceous** (see-BA-shus) **glands.**

Sweat (sudoriferous) glands   Sweat or **sudoriferous glands** are the most numerous of the skin glands. They can be classified into two groups—**eccrine** (EK-rin) and **apocrine** (AP-o-krin)—based on type of secretion and location. **Eccrine sweat glands** are by far the more numerous, important, and widespread sweat glands in the body. They are quite small and with few exceptions are distributed over the total body surface. They function throughout life to produce a transparent watery liquid called **perspiration** or **sweat.** Sweat serves to assist in the elimination of certain waste products such as ammonia and uric acid. In addition to elimination of waste, sweat plays a critical role in helping the body maintain a constant temperature.

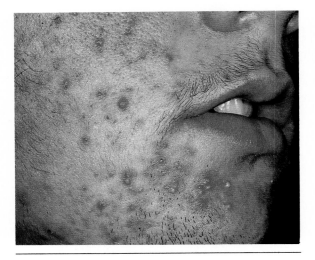

Figure 4-7 Acne.

Anatomists estimate that a single square inch of skin on the palms of the hands contains about 3,000 eccrine sweat glands. With a magnifying glass you can locate the pinpoint-sized openings on the skin that you probably call **pores.** The pores are outlets of small ducts from the eccrine sweat glands.

Apocrine sweat glands are found primarily in the skin in the armpit (axilla) and in the pigmented skin areas around the genitals. They are larger than the eccrine glands and, instead of watery sweat, they secrete a thicker milky secretion. The odor that is associated with apocrine gland secretion is not caused by the secretion itself. Instead, it is caused by the contamination and decomposition of the secretion by skin bacteria. Apocrine glands enlarge and begin to function at puberty.

Sebaceous glands Sebaceous glands secrete oil for the hair and skin. Wherever hairs grow, oil or sebaceous glands also grow. Their tiny ducts open into hair follicles (Figure 4-4) so that their secretion, called **sebum** (SEE-bum), lubricates the hair as well as the skin. Someone aptly described sebum as "nature's skin cream." Sebum secretion increases during adolescence, stimulated by the increased blood levels of the sex hormones. Frequently sebum accumulates in and enlarges some of the ducts of the sebaceous glands, forming white pimples. This sebum often darkens, forming a **blackhead.**

Common **acne** (AK-ne) (Figure 4-7) occurs most frequently in adolescence as a result of overactive secretion by the sebaceous glands with blockage and inflammation of their ducts. There is more than a fivefold increase in the rate of sebum secretion in individuals between the ages of 10 and 19. As a result, sebaceous gland ducts may become plugged with skin cells and sebum that is contaminated with bacteria. The inflamed plug is called a **comedo** (KOM-e-do) and is the most characteristic sign of acne. Pus-filled **pimples** result from secondary infections within or beneath the epidermis, often in a hair follicle or sweat pore.

FUNCTIONS OF THE SKIN

The skin or cutaneous membrane serves three critically important functions that contribute to survival itself. The most important functions are:

1 Protection
2 Temperature regulation
3 Sense organ activity

Protection

The skin as a whole is often described as our "first line of defense" against a multitude of hazards. It protects us against the daily invasion of deadly microbes. The tough keratin-filled cells of the stratum corneum also resist the entry of harmful chemicals and protect against physical tears and cuts. Because it is waterproof, **keratin** also protects the body from excessive fluid loss. Melanin in the pigment layer of the skin prevents the sun's harmful ultraviolet rays from penetrating the interior of the body.

Temperature Regulation

The skin plays a key role in regulating the body's temperature. Incredible as it seems, on a hot and humid day the skin can serve as a means for releasing almost 3,000 calories of body heat—enough heat energy to boil over 5 gallons of water! It accomplishes this feat in two ways: by regulating sweat secretion and by regulating the flow of blood close to the body surface. When sweat evaporates from the body surface, heat is also lost. The principle of heat loss through evaporation is basic to many cooling systems. When increased quantities of blood are allowed to fill the vessels close to the skin, heat will also be lost by radiation. Blood supply to the skin far exceeds the amount needed by the skin. Such an abundant blood supply primarily enables the regulation of body temperature.

Sense Organ Activity

The skin functions as an enormous sense organ. Its millions of nerve endings serve as antennas or receivers for the body, keeping it informed of changes in its environment. The specialized receptors shown in Figures 4-1 and 4-5 make it possible for the body to detect sensations of light touch (Meissner's corpuscles) and pressures (pacinian corpuscles). Other receptors make it possible for us to respond to the sensations of pain, heat, and cold.

BURNS

Burns constitute one of the most serious and frequent problems that affect the skin. Typically, we think of a burn as an injury caused by fire or by contact of the skin with a hot surface. However, overexposure to ultraviolet light (sunburn) or contact of the skin with an electric current or a harmful chemical such as an acid can also cause burns.

Estimating Body Surface Area

When burns involve large areas of the skin, treatment and the possibility for recovery depend in large part on the **total area involved** and the **severity of the burn.** The severity of a burn is determined by the depth of the injury, as well as by how much of the body surface area is affected.

The **"rule of nines"** is one of the most frequently used methods of determining the extent of a burn injury. With this technique (Figure 4-8) the body is divided into 11 areas of 9% each, with the area around the genitals representing the additional 1% of body surface area. As you can see in Figure 4-8, in the adult 9% of the skin covers the head and each upper extremity, including front and back surfaces. Twice as much, or 18%, of the total skin area covers the front and back of the trunk and each lower extremity, including front and back surfaces.

Classification of Burns

The classification system used to describe the severity of burns is based on how many tissue layers of the skin are involved. The most severe burns destroy not only layers of the skin and subcutaneous tissue but underlying tissues as well.

First-degree burns   A **first-degree burn**, a good example of which is a typical sunburn, will cause minor discomfort and some reddening of the skin. Although the surface layers of the epidermis may peel in 1 to 3 days, no blistering occurs and actual tissue destruction is minimal.

Second-degree burns   A **second-degree burn** involves the deep epidermal layers and always causes injury to the upper layers of the dermis. Although deep second-degree burns damage sweat glands, hair follicles, and sebaceous glands, complete destruction of the dermis does not occur. Blisters, severe pain, generalized swelling, and fluid loss characterize this type of burn. Scarring is common.

Figure 4-8 The "rule of nines" used to estimate the amount of skin surface burned in an adult.

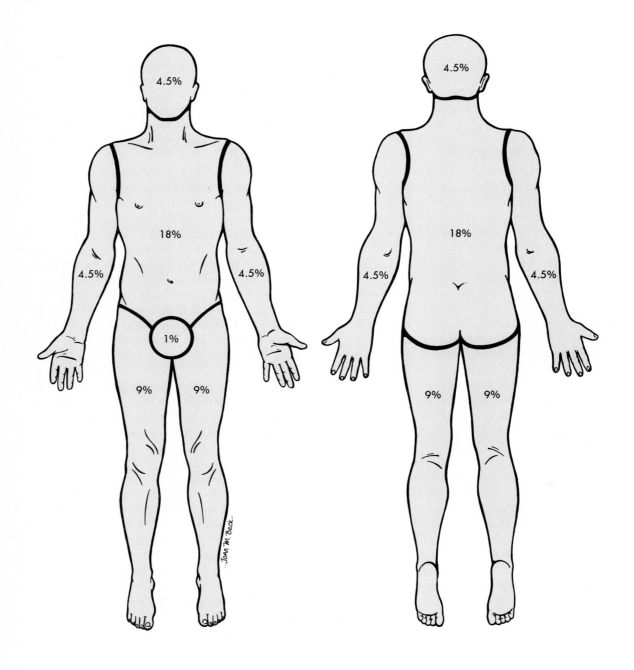

Third-degree burns A **third-degree burn** is characterized by complete destruction of both the epidermis and dermis. In addition, tissue death extends below the primary skin layers into the subcutaneous tissue. Third-degree burns often involve underlying muscles and even bone. One distinction between second- and third-degree burns is that third-degree lesions are insensitive to pain immediately after injury because of the destruction of nerve endings. The scarring that results is a very serious problem.

TYPES OF BODY MEMBRANES

The term **membrane** refers to a thin sheetlike structure that may serve a number of important functions in the body. Membranes cover and protect the body surface, line body cavities, and cover the inner surfaces of the hollow organs such as the digestive, reproductive, and respiratory passageways. Some membranes anchor organs to each other or to bones and others cover the internal organs. In certain areas of the body, membranes secrete lubricating fluids that reduce friction during such organ movements as the beating of the heart or lung expansion and contraction. Membrane lubricants also decrease the friction between bones in joints.

There are two major categories or types of body membranes:
1 **Epithelial membranes,** which are composed of epithelial tissue and an underlying layer of specialized connective tissue
2 **Connective tissue membranes** that are composed exclusively of various types of connective tissue. No epithelial cells are present in this type of membrane

EPITHELIAL MEMBRANES

There are three types of epithelial membranes in the body:
1 Cutaneous membrane
2 Serous membranes
3 Mucous membranes

Cutaneous Membrane

We have completed our study of the **cutaneous membrane.** As you know, the cutaneous membrane is the primary organ of the integu-mentary system and is known more commonly as the **skin.** It fulfills the requirements necessary for an epithelial membrane. The outer epidermis is composed of epithelial cells, and the underlying dermis is made up of connective tissue.

Serous Membranes

Like all epithelial membranes, a **serous** (SE-rus) **membrane** is composed of two distinct layers of tissue. The epithelial sheet is a thin layer of simple squamous epithelium. The connective tissue layer forms a very thin and delicate **basement membrane** that supports the epithelial cells.

There are two types of serous membranes: (1) the first type can be found lining body cavities, and (2) the second type can be found covering the organs in those cavities. The serous membrane, which lines the walls of a body cavity much like wallpaper covers the walls of a room, is called the **parietal** (pah-RI-it-tal) **portion.** The other type of serous membrane, which covers the surface of organs found in body cavities, is called the **visceral** (VIS-er-al) **portion.**

The serous membranes of the thoracic and abdominal cavities are identified in Figure 4-9. In the thoracic cavity the serous membranes are called **pleura** (PLOOR-ah), and in the abdominal cavity they are called **peritoneum** (per-i-to-NE-um). Look again at Figure 4-9 to note the placement of the **parietal** and **visceral pleura** and the **parietal** and **visceral peritoneum.** In both cases the parietal layer forms the lining of the body cavity, and the visceral layer covers the organs found in that cavity.

Serous membranes secrete a thin, watery fluid that helps to reduce friction and serves as a lubricant when organs rub against one another and against the walls of the cavities that contain them. **Pleurisy** (PLOOR-i-se) is a pathological and very painful condition characterized by inflammation of the serous membranes (pleura) that line the chest cavity and cover the lungs. Pain is caused by irritation and friction as the lungs rub against the walls of the chest cavity. In severe cases the inflamed surfaces of the pleura fuse, and permanent damage may develop. The term **peritonitis** (per-i-to-NI-tis) is

Figure 4-9   Types of body membranes. **A,** Epithelial membranes: *(1),* Cutaneous membrane (skin); *(2),* Serous membranes (parietal and visceral pleura and peritoneum); *(3),* mucous membranes. **B,** Connective tissue membranes: *(1),* synovial membranes. See text for explanation.

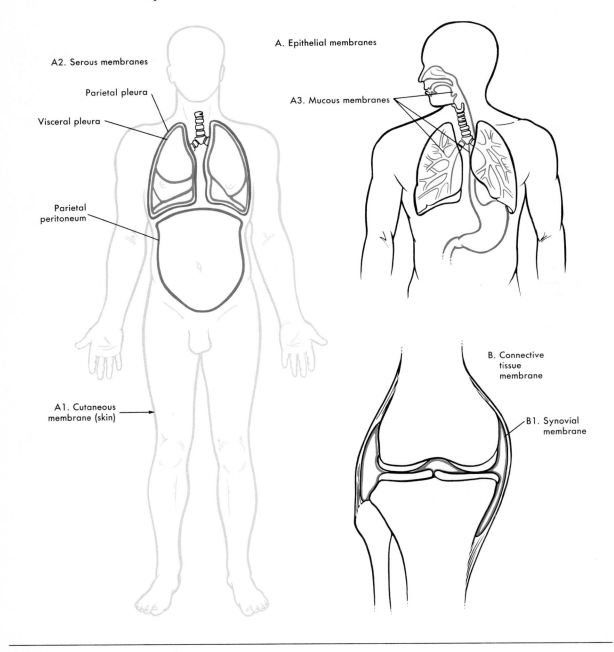

used to describe inflammation of the serous membranes in the abdominal cavity. Peritonitis is sometimes a serious complication of an infected appendix.

Mucous Membranes

Mucous (MU-kus) **membranes** are epithelial membranes that are found lining the body surfaces that open directly to the exterior. Examples of mucous membranes include those that line the respiratory, digestive, urinary, and reproductive tracts. The epithelial component of a mucous membrane will vary depending on its location and function. In the esophagus, for example, a tough abrasion-resistant stratified squamous epithelium is found. A thin layer of simple columnar epithelium covers the walls of the lower segments of the digestive tract.

The epithelial cells of most mucous membranes secrete a thick slimy material called **mucus** (MU-kus), which keeps the membranes moist and soft.

CONNECTIVE TISSUE MEMBRANES

Unlike cutaneous, serous, and mucous membranes, connective tissue membranes do not contain epithelial components. The **synovial** (si-NO-ve-al) **membranes** that line the joint spaces between bones at joints that move are classified as connective tissue membranes. These membranes are smooth and slick and secrete a thick and colorless lubricating fluid called **synovial fluid.** The membrane itself, together with its specialized fluid, helps reduce friction between the opposing surfaces of bones in movable joints. Synovial membranes also line the small cushionlike sacs called **bursae** (BER-sae), that are found between moving body parts.

OUTLINE SUMMARY

The Skin and Integumentary System

The skin (Figures 4-1 and 4-2)

A Structure—two primary layers called epidermis and dermis
  1 Epidermis
    a Outermost and thinnest primary layer of skin
    b Composed of several layers of stratified squamous epithelium
    c Innermost layer of cells continually reproduces, and new cells move up toward the surface
    d As cells approach surface, they are filled with a tough, waterproof protein called keratin, and they eventually flake off
    e Stratum corneum—outermost layer of keratin-filled cells
    f Pigment layer—epidermal layer that contains pigment cells called melanocytes, which produce the brown pigment melanin

  g Blisters (Figure 4-3) caused by breakdown of union between cells or primary layers of skin
  h Dermal-epidermal junction—specialized area between two primary skin layers
  2 Dermis
    a Deeper and thicker of the two primary skin layers composed of connective tissue
    b Upper area of dermis characterized by parallel rows of peglike dermal papillae
    c Ridges and grooves in dermis form pattern unique to each individual (basis of fingerprinting)
    d Deeper areas of dermis filled with network of tough collagenous and stretchable elastic fibers
    e Number of elastic fibers decreases with age and contributes to wrinkle formation

    f Dermis also contains nerve endings, muscle fibers, hair follicles, sweat and sebaceous glands, and many blood vessels

B Appendages of the skin

  1 Hair (Figure 4-4)

    a Soft hair of fetus and newborn called lanugo

    b Hair growth requires epidermal tube-like structure called hair folicle

    c Hair growth begins from hair papilla

    d Hair root lies hidden in follicle and visible part of hair called shaft

    e Arrector pili—specialized smooth muscle that produces "goose pimples" and causes hair to stand up straight

  2 Receptors (Figure 4-5)

    a Specialized nerve endings—make it possible for skin to act as a sense organ

    b Meissner's corpuscle—capable of detecting light touch

    c Pacinian corpuscle—capable of detecting pressure

  3 Nails (Figure 4-6)

    a Produced by epidermal cells over terminal ends of fingers and toes

    b Visible part called nail body

    c Root lies in a groove and is hidden by cuticle

    d Crescent-shaped area nearest root called lunula

    e Nail bed may change color with change in blood flow

  4 Skin glands

    a Types

      (1) Sweat or sudoriferous

      (2) Sebaceous

    b Sweat or sudoriferous glands

      (1) Types

        (a) Eccrine sweat glands

          • Most numerous, important, and widespread of the sweat glands

          • Produce perspiration or sweat, which flows out through pores on skin surface

          • Function throughout life and assist in body heat regulation

        (b) Apocrine sweat glands

          • Found primarily in axilla and around genitalia

          • Secrete a thicker milky secretion quite different from perspiration

          • Breakdown of secretion by skin bacteria produces odor

    c Sebaceous glands

      (1) Secrete oil or sebum for hair and skin

      (2) Level of secretion increases during adolescence

      (3) Amount of secretion regulated by sex hormones

      (4) Sebum in sebaceous gland ducts may darken to form a blackhead

      (5) Acne is inflammation of sebaceous gland and related structures (Figure 4-7)

C Functions of the skin

  1 Protection

    a First line of defense against infection by microbes

    b Against ultraviolet rays from sun

    c Against harmful chemicals

    d Against cuts and tears

  2 Temperature regulation

    a Skin can release almost 3,000 calories of body heat per day

      (1) Mechanisms of temperature regulation

        (a) Regulation of sweat secretion

        (b) Regulation of flow of blood close to the body surface

  3 Sense organ activity

    a Skin functions as an enormous sense organ

    b Receptors serve as receivers for the body, keeping it informed of changes in its environment

D Burns

  1 Treatment and recovery or survival depend on total area involved and severity or depth of the burn

2 Estimating body surface area using the "rule of nines" (Figure 4-8) in adults
  a Body divided into 11 areas of 9% each
  b Additional 1% of body surface area around genitals
3 Classification of burns
  a First-degree burns—only surface layers of epidermis involved
  b Second-degree burns—involve the deep epidermal layers and always cause injury to the upper layers of the dermis
  c Third-degree burns—characterized by complete destruction of both the epidermis and dermis
    (1) May involve underlying muscle and bone
    (2) Lesion is insensitive to pain because of destruction of nerve endings immediately after injury—intense pain is soon experienced

Types of Body Membranes

A Classification of body membranes (Figure 4-9)
  1 Epithelial membranes—composed of epithelial tissue and an underlying layer of connective tissue
  2 Connective tissue membranes—composed exclusively of various types of connective tissue
B Epithelial membranes
  1 Cutaneous membrane—the skin

2 Serous membranes—simple squamous epithelium on a connective tissue basement membrane
  a Types
    (1) Parietal—line walls of body cavities
    (2) Visceral—cover organs found in body cavities
  b Examples
    (1) Pleura—parietal and visceral layers line walls of thoracic cavity and cover the lungs
    (2) Peritoneum—parietal and visceral layers line walls of abdominal cavity and cover the organs in that cavity
  c Diseases
    (1) Pleurisy—inflammation of the serous membranes that line the chest cavity and cover the lungs
    (2) Peritonitis—inflammation of the serous membranes in the abdominal cavity that line the walls and cover the abdominal organs
3 Mucous membranes
  a Line body surfaces that open directly to the exterior
  b Produce mucus—a thick secretion that keeps the membranes soft and moist
C Connective tissue membranes
  1 Do not contain epithelial components
  2 Produce a lubricant called synovial fluid
  3 Examples are the synovial membranes that are present in the spaces between joints and in the lining of bursal sacs

NEW WORDS

acne	dehydration	Meissner's corpuscle	peritonitis
apocrine sweat gland	depilatories	melanin	pleura
arrector pili	dermis	melanocyte	pleurisy
blister	eccrine sweat gland	mucous membrane	sebaceous gland
bursa	epidermis	mucus	serous
comedo	follicle	pacinian corpuscle	stratum corneum
cutaneous	keratin	papilla	subcutaneous
cuticle	lanugo	parietal	sudoriferous gland
cyanosis	lunula	peritoneum	synovial
			visceral

CHAPTER TEST

1. The principal organ of the integumentary system is the

   _____ .
2. The skin is classified as a _____ membrane.
3. The two principal layers of the skin are called the _____ and the _____ .
4. The tough waterproof material that provides cells in the outer layer of the skin with a protective quality is called _____ .
5. The brown pigment that gives color to the skin is known as

   _____ .
6. The upper region of the dermis is characterized by parallel rows of peglike projections called _____ _____ .
7. The part of a hair that lies hidden in the follicle is called the

   _____ .
8. The pacinian corpuscle is capable of detecting _____ on the skin surface.
9. Sweat or sudoriferous glands are classified into two groups, _____ and _____ , based on type of secretion and location.
10. Specialized glands that secrete oil for the hair and skin are known as _____ glands.
11. The most common condition that results from overactive secretion of the sebaceous glands during adolescence is called _____ .
12. The "Rule of Nines" is used to aid in the treatment and prognosis of

    _____ .
13. There are two major categories of body membranes called _____ membranes and _____ _____ membranes.
14. The membrane that lines the walls of a body cavity is called the _____ portion of the membrane, and the portion that covers the surface of organs found in body cavities is called the _____ portion of that membrane.
15. The connective tissue membranes that line joint spaces are called _____ membranes.

Select the most correct answer from Column B for each statement in Column A. (Only one answer is correct.)

Column A	Column B
16. ____ Another name for skin	a. Keratin
17. ____ Outermost layer of skin	b. Lanugo
18. ____ Tough, protective protein	c. Arrector pili
19. ____ Contains melanocytes	d. Shaft
20. ____ Fine, soft hair of newborn	e. Eccrine
21. ____ Visible part of a hair	f. Integument
22. ____ Produce "goose pimples"	g. Sebaceous glands
23. ____ Type of sweat gland	h. Pigment layer
24. ____ Produce oil for skin	i. Bursae
25. ____ Lined with synovial membrane	j. Epidermis

REVIEW QUESTIONS

1 How do the terms *integument* and *integumentary system* differ in meaning?
2 Identify and compare the two main layers of the skin. How are these layers related to the subcutaneous layer?
3 List the appendages of the skin.
4 Discuss the three primary functions of the skin.
5 Classify the skin glands. Locate each type of gland and compare the secretions of each.
6 What are the two major factors that determine the treatment of and possibility for recovery from burn injuries?
7 Classify and compare the three major types of burn injuries.
8 Define the term *membrane* and discuss a variety of functions that membranes serve in the body.
9 Classify body membranes.
10 Discuss the two types of serous membranes and give examples of each.
11 Compare mucous and synovial membranes.

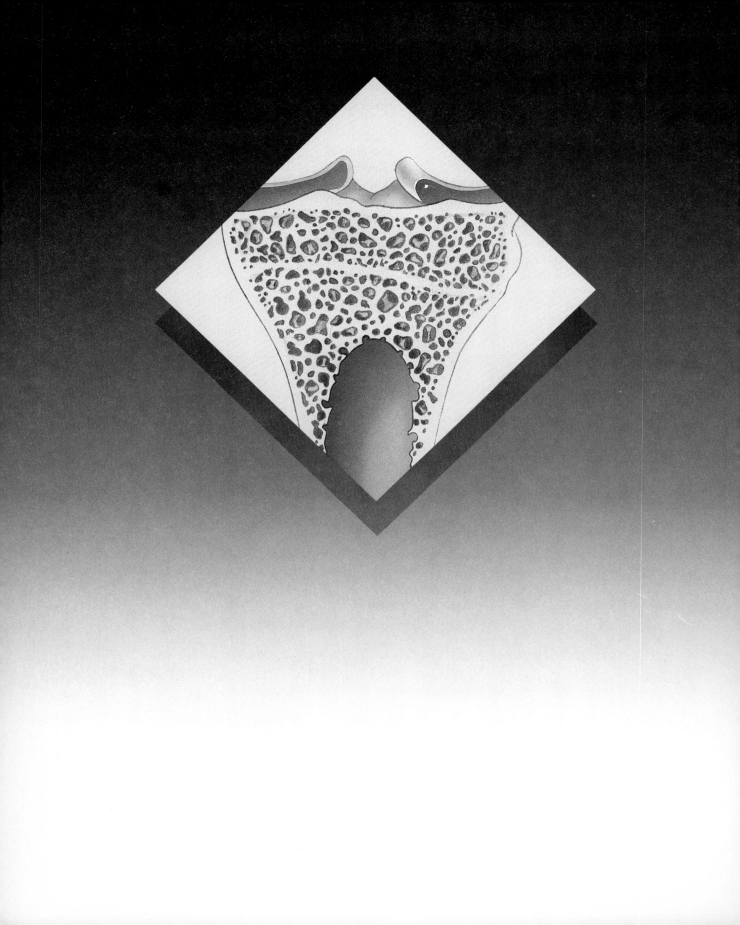

UNIT TWO

Systems That Form the Framework of the Body and Move It

CHAPTER
5   The Skeletal System

CHAPTER OUTLINE

OBJECTIVES

After you have completed this chapter, you should be able to:

1 List and discuss the generalized functions of the skeletal system.

2 Discuss the microscopic structure of bone and cartilage, including the identification of specific cell types and structural features.

3 Identify the major anatomical structures found in a typical long bone and discuss bone formation and growth.

4 Identify the two major subdivisions of the skeleton and list the bones found in each area.

5 List and compare the major types of joints found in the body and give an example of each type.

Unit two consists of two chapters. This chapter discusses the skeletal system, the system that provides the body with a rigid framework. It lies buried within the muscles and other soft tissues, thus providing a support structure for the whole body. In this respect the skeletal system functions like steel girders in a building, but in other ways it functions quite differently. Unlike steel girders, bones can move toward or away from each other and in some cases can even move around in circles.

Bones are also living organs. Because they are alive, bones can change and help the body respond to a changing environment. For example, specialized cells make it possible for bones to grow in size and change in shape as we grow taller or as our body weight fluctuates. This ability of bones to change allows our bodies to grow and change as well. We will conclude Chapter 5 with a discussion of **joints** or **articulations** (ar-tic-u-LA-shuns). Chapter 6 deals with the muscular system—the system that moves bones.

FUNCTIONS

The following functions are performed by the skeletal system.

SUPPORT

Bones form the body's supporting framework much as steel girders form the supporting framework of modern buildings.

PROTECTION

Hard, bony "boxes" protect delicate structures enclosed within them. For example, the skull protects the brain. The breastbone and ribs protect vital organs (heart and lungs) and also a vital tissue (red bone marrow, the blood cell-forming tissue).

MOVEMENT

Muscles are anchored firmly to bones. As muscles contract and shorten, they pull on bones and thereby move them.

STORAGE

Bones play an important part in maintaining homeostasis of blood calcium, a vital substance. They serve as a safety-deposit box for calcium. When the amount of calcium in blood increases above normal, calcium moves out of the blood and into the bones for storage. Conversely, when blood calcium decreases below normal, calcium moves in the opposite direction. It comes out of storage in bones and enters the blood.

HEMOPOIESIS

The term **hemopoiesis** (he-mo-poi-E-sis) is used to describe the process of blood cell formation. It is a combination of two Greek words: *hemo* (HE-mo) meaning "blood" and *poiesis* (poi-E-sis) meaning "to make." Blood cell formation is a vital process carried on in **red bone marrow.** As you can see in Figures 5-1 and 5-6, red marrow is found in the spongy type of bone located

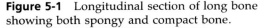

Figure 5-1 Longitudinal section of long bone showing both spongy and compact bone.

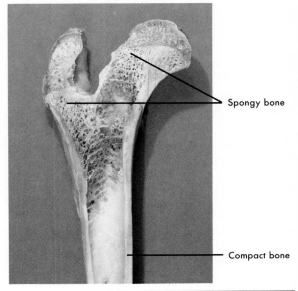

Spongy bone

Compact bone

Figure 5-2 Microscopic structure of bone Haversian systems, several of which are shown here, compose compact bone. Note the structures that make up one Haversian system: concentric lamellae, lacunae, canaliculi, and a Haversian canal. Shown bordering the compact bone on the left is spongy bone, a name descriptive of the many open spaces that characterize it.

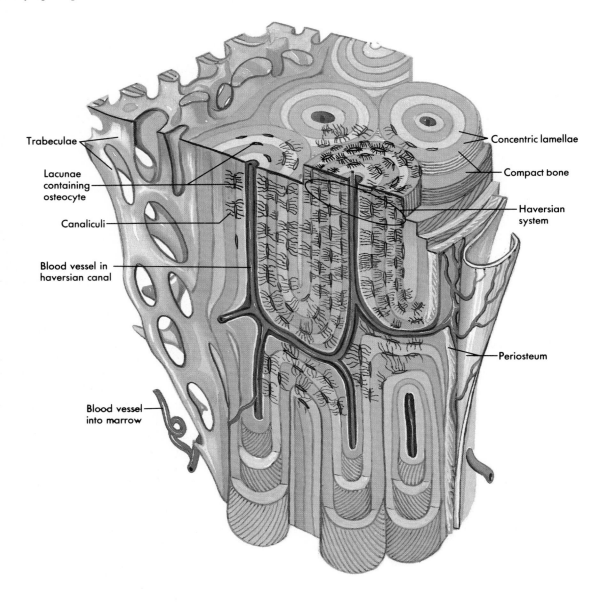

Trabeculae

Lacunae containing osteocyte

Canaliculi

Blood vessel in haversian canal

Blood vessel into marrow

Concentric lamellae

Compact bone

Haversian system

Periosteum

Figure 5-3   Photomicrograph of compact bone showing Haversian system of organization.

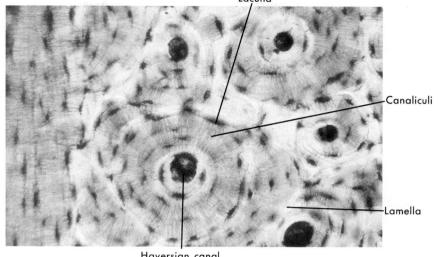

Lacuna

Canaliculi

Lamella

Haversian canal

in the ends of some long bones. In an infant's or child's body there is more red marrow than in an adult's. As an individual ages, much of the red marrow becomes transformed into **yellow bone marrow,** which is an inactive fatty tissue. Yellow bone marrow is shown in the **medullary cavity** of the long bone in Figure 5-6.

MICROSCOPIC STRUCTURE OF BONE AND CARTILAGE

The skeletal system contains two major types of connective tissue: **bone** and **cartilage.** Bone has a different appearance and texture, depending on where it is located. Note in Figure 5-1 that the outer layer of bone is hard and dense. Bone of this type is called **dense** or **compact bone.** The porous bone seen in the end of the long bone in Figure 5-1 is called **spongy bone.** As the name implies, spongy bone contains many spaces that may be filled with marrow. Compact or dense bone appears solid to the naked eye. Figure 5-2 shows the microscopic appearance of both spongy and compact bone. The needlelike threads of spongy bone that surround

a network of spaces are called **trabeculae** (trah-BEK-u-le). Spongy bone is located on the left side of the illustration.

As you can see in Figures 5-2 and 5-3, compact or dense bone does not contain a network of open spaces. Instead, the matrix is organized into numerous structural units called **Haversian systems.** Each circular and tubelike haversian system is composed of multiple layers of calcified matrix arranged in layers resembling the rings of an onion. Each ring of bone is called a **concentric lamella** (lah-MEL-ah). The circular rings or lamellae surround the **Haversian canal,** which contains a blood vessel.

Bones are not lifeless structures. Within their hard, seemingly lifeless matrix are many living bone cells called **osteocytes** (OS-te-o-cytes). Osteocytes lie imprisoned between the hard layers of the lamellae in little spaces called **lacunae** (lah-KU-ne). In Figures 5-2 and 5-3 note that tiny passageways or canals called **canaliculi** (kan-ah-LIK-u-li) connect the lacunae with each other and with the central canal in each Haversian system. Nutrients pass from the blood vessel in the Haversian canal through the canaliculi to the

Figure 5-4 Photomicrograph of cartilage tissue.

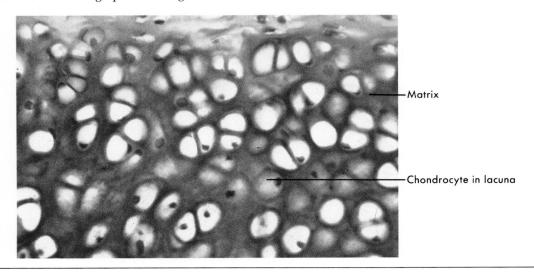

Matrix

Chondrocyte in lacuna

osteocytes. Note also in Figure 5-2 that numerous blood vessels from the outer **periosteum** (per-e-OS-te-um) enter the bone and eventually pass through the Haversian canals.

In addition to the bone-forming osteocytes, specialized bone-absorbing cells called **osteoclasts** (OS-te-o-klasts) are also present in bone. These large cells make it possible for bone tissue already present to be absorbed and removed. Both types of cells are required for bones to grow and repair themselves if injured.

Cartilage both resembles and differs from bone. Like bone, it consists more of intercellular substance than of cells. Innumerable collagenous fibers reinforce the matrix of both tissues. But in cartilage the fibers are embedded in a firm gel instead of in a calcified cement substance as they are in bone; hence cartilage has the flexibility of a firm plastic material rather than the rigidity of bone. Note in Figure 5-4 that the cartilage cells called **chondrocytes** (kon-dro-cytes), like the osteocytes of bone, are located in lacunae. In cartilage the lacunae are suspended in the cartilage matrix much like air bubbles in a block of firm gelatin. Since there are no blood vessels in cartilage, nutrients must diffuse through the matrix to reach the cells.

STRUCTURE OF LONG BONES

Figures 5-5 and 5-6 will help you learn the names of the main parts of a long bone. Identify each of the following:

1 **Diaphysis** or shaft—a hollow tube made of hard compact bone, hence a rigid and strong structure that is light enough in weight to permit easy movement

Epiphyseal fracture

The point of articulation between the epiphysis and diaphysis of a growing long bone is susceptible to injury if overstressed—especially in the young child or preadolescent athlete. In these individuals the epiphyseal plate can be separated from the diaphysis or epiphysis, causing an **epiphyseal fracture** like the one seen in Figure 5-7. The epiphyseal line can be seen in both external and cutaway views of a juvenile long bone in Figure 5-6.

Figure 5-5 Structure of a long bone as seen in longitudinal section.

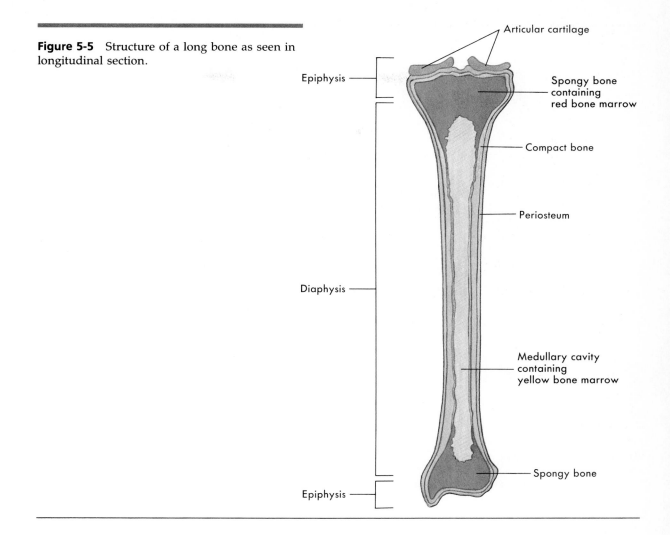

Articular cartilage

Epiphysis

Spongy bone containing red bone marrow

Compact bone

Periosteum

Diaphysis

Medullary cavity containing yellow bone marrow

Spongy bone

Epiphysis

2 Medullary cavity—the hollow area inside the diaphysis of a bone; contains soft yellow bone marrow

3 Epiphyses or the ends of the bone—red bone marrow fills in small spaces in the spongy bone composing the epiphyses

4 Articular cartilage—a thin layer of cartilage covering each epiphysis; functions like a small rubber cushion would if it were placed over the ends of bones where they form a joint

5 Periosteum—a strong fibrous membrane covering a long bone except at joint surfaces, where it is covered by articular cartilage

BONE FORMATION AND GROWTH

When the skeleton first forms in a baby before its birth, it consists not of bones but of cartilage and fibrous structures shaped like bones. Gradually these cartilage "models" become transformed into real bones when the cartilage is replaced with calcified bone matrix deposited by specialized bone-forming cells called **osteoblasts** (OS-te-o-blasts). This process of constantly "remodeling" a growing bone as it changes from a small cartilage precursor to the characteristic shape and proportion of the adult bone requires continuous activity by the bone-forming osteoblasts and bone-resorbing osteoclasts. The lay-

Figure 5-6 **A,** External view of epiphyseal lines on a juvenile long bone. **B,** Longitudinal section of long bone showing structural details.

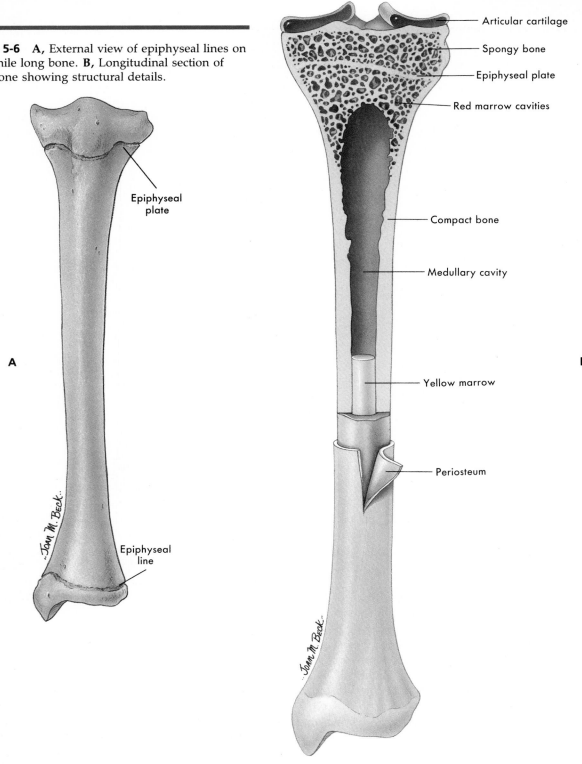

Articular cartilage

Spongy bone

Epiphyseal plate

Red marrow cavities

Compact bone

Medullary cavity

Yellow marrow

Periosteum

Epiphyseal plate

Epiphyseal line

A

B

Figure 5-7 Epiphyseal fracture in a long bone of a young boy. Note the separation of the diaphysis and epiphysis at the level of the epiphyseal plate.

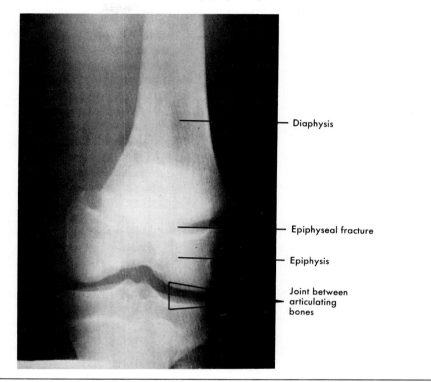

— Diaphysis

— Epiphyseal fracture

— Epiphysis

Joint between articulating bones

ing down of calcium salts in the gel-like matrix of the newly forming bones is an ongoing process. This calcification process is what makes bones as "hard as bone." It is the combined action of the osteoblasts and osteoclasts that sculpts bones into their adult shapes.

A long bone grows from a center in its diaphysis and from other centers in the layers of cartilage that separate each of its epiphyses from the diaphysis. The diaphysis grows in both directions toward both epiphyses, and both epiphyses grow toward the diaphysis. As long as any epiphyseal cartilage remains, this two-way growth continues. It ceases when all the epiphyseal cartilage has become transformed into bone. Physicians sometimes use this knowledge to determine whether a child is going to grow any more. They have the child's wrist x-rayed,

and if it shows a layer of epiphyseal cartilage, they know that additional growth will occur. If, however, it shows no epiphyseal cartilage, they know that growth has stopped and that the individual has attained adult height. Epiphyseal plates are visible in the long bone shown in Figure 5-6.

DIVISIONS OF SKELETON

The human skeleton has two divisions: the **axial skeleton** and the **appendicular skeleton.** Bones of the skull, spine, and chest and the hyoid bone in the neck make up the axial skeleton. The appendicular skeleton consists of the bones of the upper extremities (shoulder, pectoral girdles, arms, wrists, and hands) and the lower extremities (hip, pelvic girdles, legs, an-

Figure 5-8 **A,** Skeleton, anterior view. Axial skeleton is shown in blue. Appendicular system is bone colored. **B,** Photograph showing anterior aspect of the right half of the thoracic, upper limb, abdominal, and pelvic skeleton.

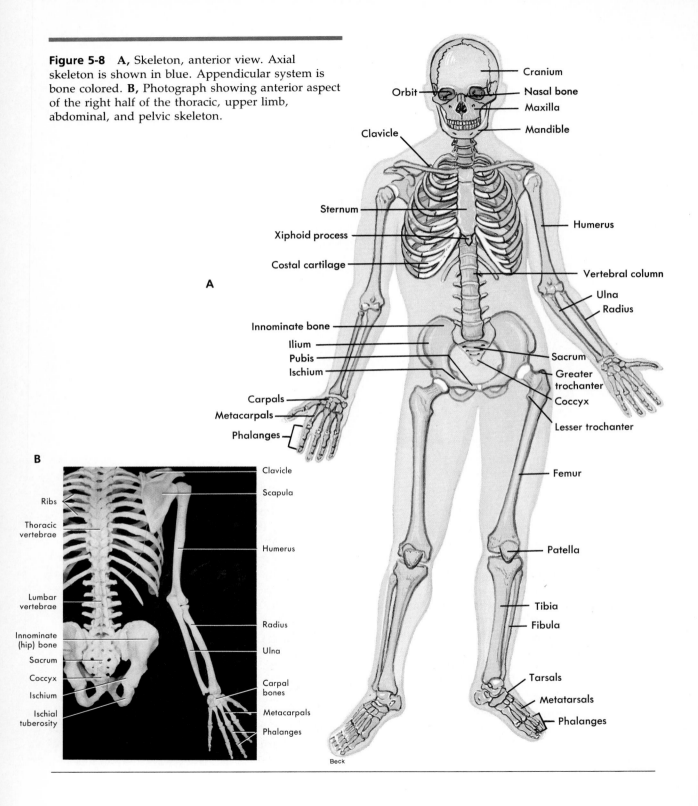

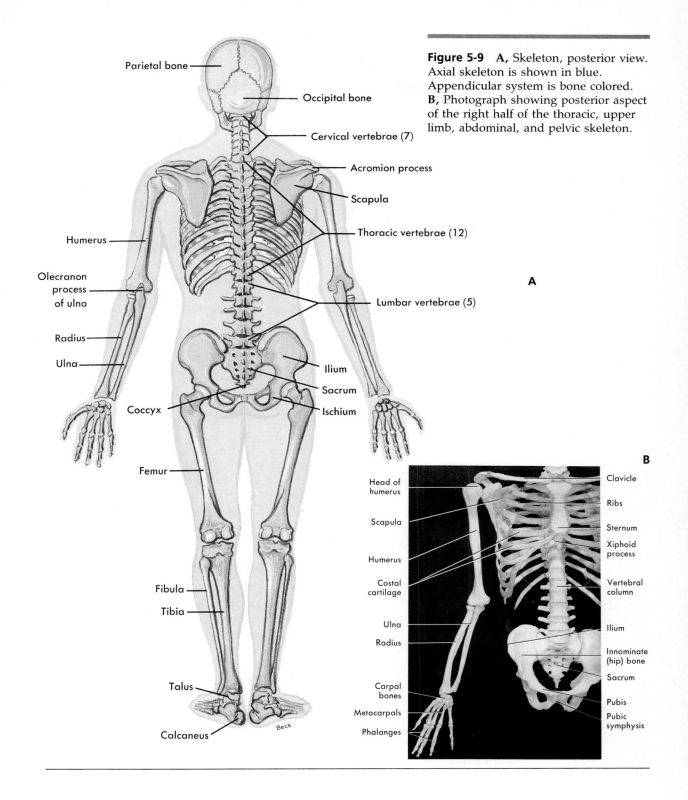

Figure 5-9 A, Skeleton, posterior view. Axial skeleton is shown in blue. Appendicular system is bone colored. **B,** Photograph showing posterior aspect of the right half of the thoracic, upper limb, abdominal, and pelvic skeleton.

Parietal bone
Occipital bone
Cervical vertebrae (7)
Acromion process
Scapula
Thoracic vertebrae (12)
Humerus
Olecranon process of ulna
Lumbar vertebrae (5)
Radius
Ulna
Ilium
Sacrum
Coccyx
Ischium
Femur
Fibula
Tibia
Talus
Calcaneus
Beck

A

B

Head of humerus
Clavicle
Ribs
Scapula
Sternum
Humerus
Xiphoid process
Costal cartilage
Vertebral column
Ulna
Radius
Ilium
Innominate (hip) bone
Sacrum
Carpal bones
Pubis
Metacarpals
Pubic symphysis
Phalanges

Table 5-1 Main parts of the skeleton	
Axial skeleton	**Appendicular skeleton**
Skull	Upper extremities
Cranium	Shoulder (pectoral) girdle
Ear bones	Arms
Face	Wrists
	Hands
Spine	Lower extremities
Vertebrae	Hip (pelvic) girdle
	Legs
	Ankles
	Feet
Thorax	
Ribs	
Sternum	
Hyoid bone	

Mastoiditis

   Mastoiditis (mas-toi-DI-tis), or inflammation of the air spaces within the mastoid portion of the temporal bone, can produce very serious medical problems unless treated promptly. Infectious material frequently finds its way into the mastoid air cells from middle ear infections. The mastoid air cells do not drain into the nose as do the paranasal sinuses. As a result, infectious material that accumulates may erode the thin bony partition that separates the air cells from the cranial cavity that surrounds the brain. Should this occur, the inflammation may spread to the brain itself or to its covering membranes. Locate the mastoid process of the temporal bone just behind the external ear opening or meatus in Figure 5-10. You should be able to feel or **palpate** (PAL-pate) this process through the skin on your own skull just behind and below the ear.

kles, and feet). Read Table 5-1. Then locate the various parts of the axial skeleton and the appendicular skeleton in Figures 5-8 and 5-9.

AXIAL SKELETON
Skull

   The skull consists of eight bones that form the **cranium,** fourteen bones that form the **face,** and six tiny bones in the **middle ear.** You will probably want to learn the names and locations of these bones. These are given in Table 5-2. Find as many of them as you can on Figure 5-10. Feel their outlines in your own body where possible. Examine them on a skeleton if you have access to one.

   "My sinuses give me so much trouble." Have you ever heard this complaint or perhaps uttered it yourself? **Sinuses** are spaces, or cavities, inside some of the cranial bones. Four pairs of them (those in the frontal, maxillary, sphenoid, and ethmoid bones) have openings into the nose

and so are referred to as **paranasal sinuses.** Sinuses give trouble when the mucous membrane that lines them becomes inflamed, swollen, and painful. For example, inflammation in the frontal sinus (*frontal sinusitis*) often starts from a common cold. The letters "-itis" added to a word mean "inflammation of."

Spine (Vertebral Column)

   The term *vertebral column* may conjure up a mental picture of the spine as a single long bone shaped like a column in a building, but this is far from true. The vertebral column consists of a series of separate bones or **vertebrae** connected in such a way that they form a flexible curved rod. Different sections of the spine have different names: cervical region, thoracic region, lumbar region, sacrum, and coccyx. They are illustrated in Figure 5-11 and described in Table 5-2.

   Although individual vertebrae are small bones, irregular in shape, they have several well-defined parts. Note, for example, in Figure 5-13, the body of the lumbar vertebra shown there, its spinous process (or spine), its two transverse processes, and the hole in its center, called the vertebral foramen. To feel the tip of the spinous process of one of your own vertebrae, simply bend your head forward and run your fingers down the back of your neck until you feel a projection of bone at shoulder level. This is the tip of the seventh cervical vertebra's long spinous process. The seven cervical vertebrae form the supporting framework of the neck.

Have you ever noticed the four curves in your spine? Your neck and the small of your back curve slightly inward or forward, whereas the chest region of the spine and the lowermost portion curve in the opposite direction. The cervical and lumbar curves of the spine are called concave curves, and the thoracic and sacral curves are called convex curves. This is not true, however, of a newborn baby's spine. It forms a continuous convex curve from top to bottom. Gradually, as the baby learns to hold up his head, a reverse or concave curve develops in his neck, (cervical region). Later, as the baby learns to stand, the lumbar region of his spine also becomes concave.

The normal curves of the spine serve important functions. They give it enough strength to support the weight of the rest of the body. They also provide the balance necessary for us to stand and walk on two feet instead of having to crawl on all fours. A curved structure has more strength than a straight one of the same size and materials. (The next time you pass a bridge, look to see whether or not its supports form a curve.) Clearly the spine needs to be a strong structure. It supports the head balanced on top of it, the ribs and internal organs suspended from it in front, and the hips and legs attached to it below. Poor posture or disease often causes the lumbar curve to become abnormally exaggerated, a condition commonly called "swayback" or, technically, **lordosis** (lor-DO-sis). Another abnormal curvature is **kyphosis** (ki-FO-sis), known to most of us as "hunchback." Abnormal side-to-side curvature is called **scoliosis** (sko-le-O-sis).

Figure 5-12 illustrates the abnormal spinal curvatures.

Thorax

Twelve pairs of ribs, the sternum (breastbone), and the thoracic vertebrae form the bony cage known as the **thorax** or **chest.** Each of the twelve pairs of ribs attaches posteriorly to a vertebra. Also, all the ribs except the lower two pairs attach to the sternum and so have an anterior as well as a posterior anchorage. Look closely at Figure 5-8 and you can see that the first seven pairs of ribs (sometimes referred to as the *true ribs*) attach to the sternum by means of costal cartilage. The eighth, ninth, and tenth pairs of ribs attach to the cartilage of the seventh ribs and are sometimes called *false ribs*. The last two pairs of ribs, in contrast, do not attach to any costal cartilage but seem to float free in front—hence their descriptive name, *floating ribs.*

APPENDICULAR SKELETON

Of the 206 bones that constitute the skeleton as a whole, 126 are contained in the appendicular subdivision. Look again at Table 5-1 and Figures 5-8 and 5-9 to identify the appendicular components of the skeleton. Note that the bones in the shoulder or pectoral girdle serve to attach the bones of the arm, forearm, wrist, and hands to the axial skeleton of the thorax, and the hip or pelvic girdle attaches the bones of the thigh, leg, ankle, and foot to the axial skeleton of the pelvis.

Upper Extremity

The **scapula** (SKAP-u-la), or shoulder blade, and the **clavicle** (KLAV-ik-l), or collar bone, compose the *shoulder* or *pectoral girdle*. This device functions to attach the upper extremity to the axial skeleton. The only direct point of attachment between bones occurs at the **sternoclavicular** (ster-no-klah-VIK-u-lar) **joint** between the clavicle and the sternum or breastbone. As you can see in Figure 5-8, this joint is very small. Since the upper extremity is capable of a wide range of motion, great pressures can occur at or near the joint. As a result, fractures of the clavicle are very common. *Text continued on p. 105.*

Table 5-2 Bones of the skeleton

Name	Number	Description
Cranial bones		
Frontal	1	Forehead bone; also forms front part of floor of cranium and most of upper part of eye sockets; cavity inside bone above upper margins of eye sockets (orbits) called *frontal sinus*; lined with mucous membrane
Parietal	2	Form bulging topsides of cranium
Temporal	2	Form lower sides of cranium; contain *middle* and *inner ear structures*; *mastoid sinuses* are mucosa-lined spaces in *mastoid process,* the protuberance behind ear; *external auditory canal* is tube leading into temporal bone
Occipital	1	Forms back of skull; spinal cord enters cranium through large hole *(foramen magnum)* in occipital bone
Sphenoid	1	Forms central part of floor of cranium; pituitary gland located in small depression in sphenoid called sella turcica *(Turkish saddle)*
Ethmoid	1	Complicated bone that helps form floor of cranium, side walls and roof of nose and part of its middle partition (nasal septum), and part of orbit; contains honeycomb-like spaces, the *ethmoid sinuses; superior* and *middle turbinate bones* (conchae) are projections of ethmoid bone; forms "ledges" along side wall of each nasal cavity
Face bones		
Nasal	2	Small bones that form upper part of bridge of nose
Maxillary	2	Upper jawbones; also help form roof of mouth, floor, and side walls of nose and floor of orbit; large cavity in maxillary bone is *maxillary sinus*
Zygoma (malar)	2	Cheek bones; also help form orbit
Mandible	1	Lower jawbone
Lacrimal	2	Small bone; helps form medial wall of eye socket and side wall of nasal cavity
Palatine	2	Form back part of roof of mouth and floor and side walls of nose and part of floor of orbit
Inferior turbinate	2	Form curved "ledge" along inside of side wall of nose, below middle turbinate
Vomer	1	Forms lower, back part of nasal septum
Ear bones		
Malleus	2	Malleus, incus, and stapes are tiny bones in middle ear cavity in temporal bone; malleus means "hammer"—shape of bone
Incus	2	Incus means "anvil"—shape of bone
Stapes	2	Stapes means "stirrup"—shape of bone
Hyoid bone	1	U-shaped bone in neck at base of tongue
Vertebral column		
Cervical vertebrae	7	Upper seven vertebrae, in neck region; first cervical vertebra called *atlas;* second called *axis*
Thoracic vertebrae	12	Next twelve vertebrae; ribs attach to these
Lumbar vertebrae	5	Next five vertebrae; those in small of back

Figure 5-12 Abnormal spinal curvatures. **A,** Kyphosis. **B,** Lordosis. **C,** Scoliosis.

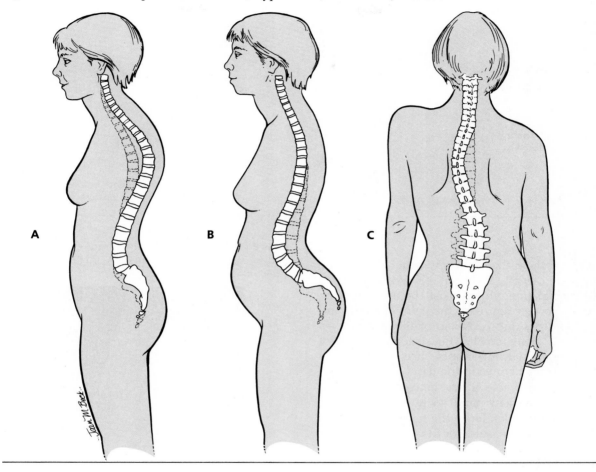

The **humerus** (HU-mer-us) is the long bone of the arm and the second longest bone in the body. It attaches to the scapula at its proximal end and articulates with the two bones of the forearm at the elbow joint. The bones of the forearm are the **radius** and the **ulna.** The anatomy of the elbow is a good example of how structure determines function. Note in Figure 5-14 that the large bony process of the ulna called the **olecranon** (o-LEK-rah-non) **process** fits nicely into a large depression on the posterior surface of the humerus called the **olecranon**

fossa. This structural relationship makes movement at the joint possible.

The radius and the ulna of the forearm articulate with each other and with the distal end of the humerus at the elbow joint. In addition, they also touch each another distally where they articulate with the bones of the wrist. In the anatomical position with the arm at the side and the palm facing forward, the radius runs along the lateral side of the forearm, and the ulna is located along the medial border.

The wrist and the hand have more bones in

Figure 5-13 **A,** Third lumbar vertebra viewed from above. **B,** Third lumbar vertebra viewed from the side.

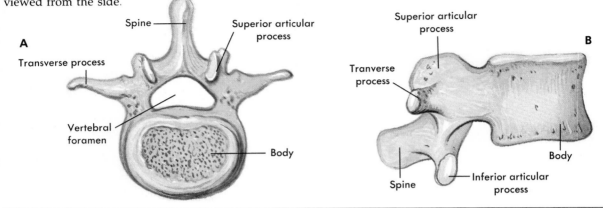

them for their size than any other part of the body—8 **carpal** (KAR-pal) or wrist bones, 5 **metacarpal** (met-ah-KAR-pal) bones that form the support structure for the palm of hand, and 14 **phalanges** (fah-LAN-jez) or finger bones—a total of 27 bones in all. This structural fact is of great functional importance. It is the presence of many small bones in the hand and wrist and the many movable joints between them that makes the human hand so highly dexterous. Some anatomists refer to the hand and wrist as the functional "reason" for the upper extremity. Refer to Figure 5-15 to see the relationships between the bones of the wrist and hand.

Lower Extremity

It is the *hip* or *pelvic girdle* that attaches the legs to the trunk. The hip girdle as a whole consists of two large **innominate** (in-NOM-i-nate) bones, one located on each side of the pelvis. These two bones, together with the sacrum and coccyx behind, provide a strong base of support for the torso and serve to attach the lower extremities to the axial skeleton. In an infant's body each innominate bone consists of three separate bones—**ilium** (ILL-e-um), **ischium** (IS-ke-um), and **pubis** (PU-bis). Later these bones grow together to become one bone in an adult (Figures 5-8 and 5-19). *Text continued on p. 112.*

Figure 5-14 Bones of the right arm, elbow joint, and forearm—posterior aspect. **A,** Right elbow skeleton.

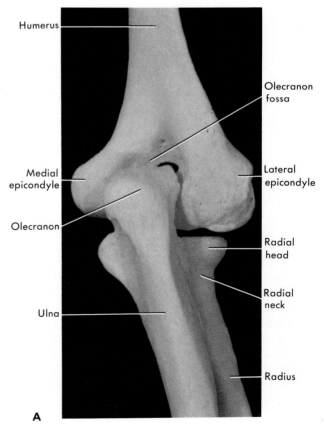

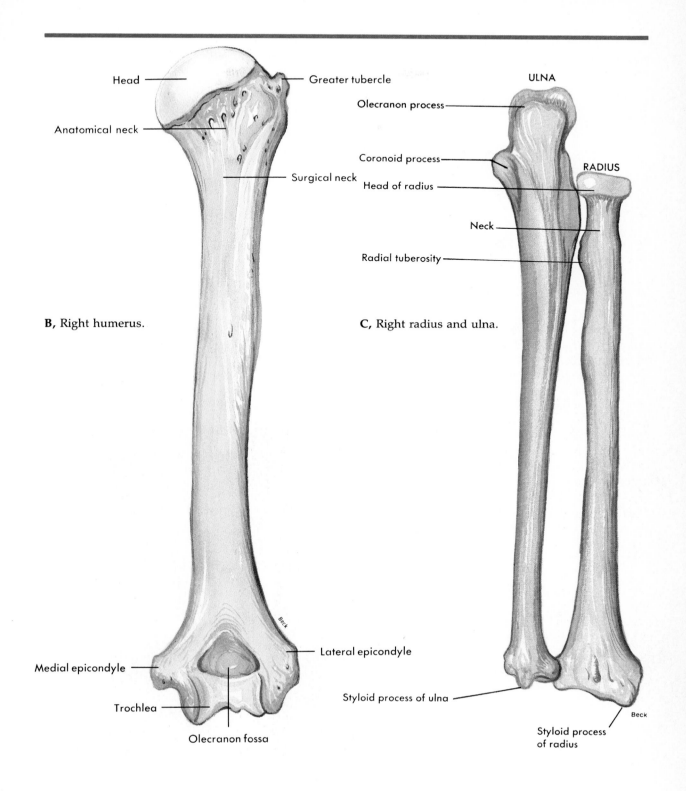

B, Right humerus.

C, Right radius and ulna.

Figure 5-15 Illustrations and photographs showing the bones of the right hand and wrist.

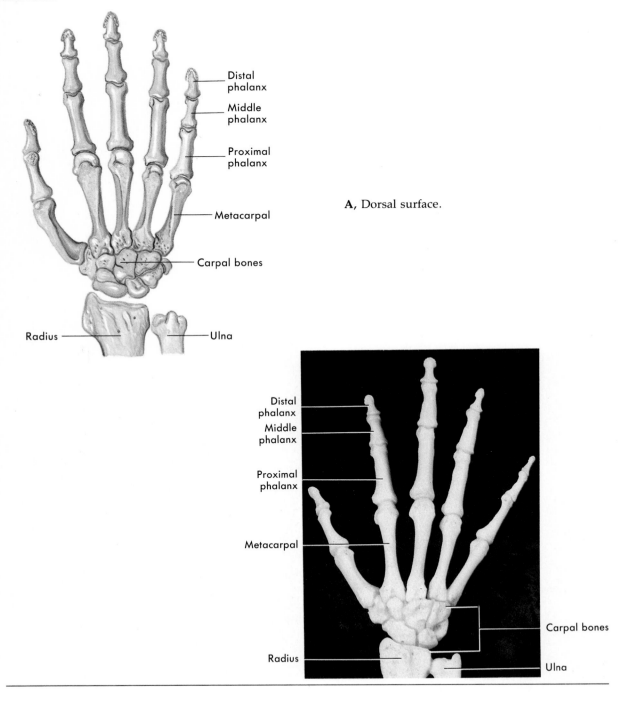

Distal phalanx

Middle phalanx

Proximal phalanx

Metacarpal

Carpal bones

Radius

Ulna

A, Dorsal surface.

Distal phalanx

Middle phalanx

Proximal phalanx

Metacarpal

Carpal bones

Radius

Ulna

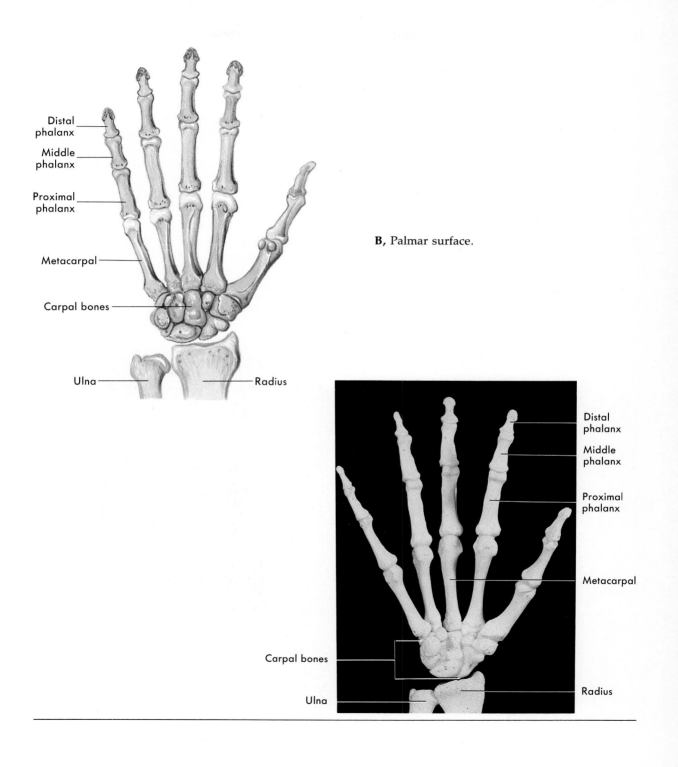

Distal phalanx

Middle phalanx

Proximal phalanx

Metacarpal

Carpal bones

Ulna

Radius

B, Palmar surface.

Distal phalanx

Middle phalanx

Proximal phalanx

Metacarpal

Carpal bones

Ulna

Radius

Anterior aspect of the right forearm and hand.

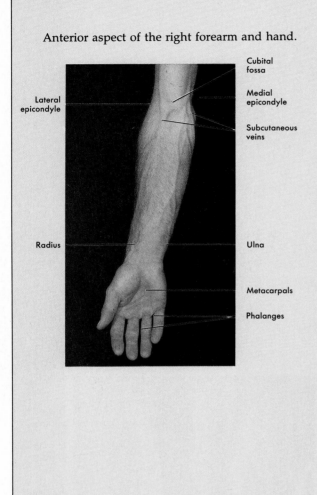

Lateral epicondyle

Radius

Cubital fossa

Medial epicondyle

Subcutaneous veins

Ulna

Metacarpals

Phalanges

Palpable bony landmarks

Health professionals often identify **externally palpable bony landmarks** in dealing with the sick and injured. **Palpable** bony landmarks are simply bones that can be touched and identified through the skin. They serve as reference points useful in identifying other body structures. There are several externally palpable bony landmarks on the upper and lower extremities. Examples from the upper extremity include the medial and lateral epicondyles of the humerus, the olecranon process of the ulna, and the distal ends of both the ulna and the radius at the wrist. The joints between the metacarpal bones and phalanges (knuckles) are easily palpated on the dorsum or posterior aspect of the hand. In the lower ex-

Posterior aspect of the right forearm and hand.

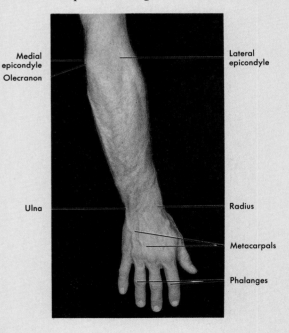

Medial epicondyle
Olecranon

Lateral epicondyle

Ulna

Radius

Metacarpals

Phalanges

tremity the medial malleolus of the tibia and the lateral malleolus of the fibula are prominent at the ankle. The calcaneus, or heel bone, is easily palpated on the posterior aspect of the foot. On the anterior aspect of the lower extremity, examples of palpable bony landmarks include the patella, or knee cap, the medial and lateral condyles of the femur and anterior border of the tibia or shin bone, and the metatarsals and phalanges of the toes. Try to identify as many of the externally palpable bones of the skeleton as possible on your own body. Using these as points of reference will make it easier for you to "visualize" the placement of other bones that cannot be touched or palpated through the skin.

Posterior aspect of the right leg and foot.

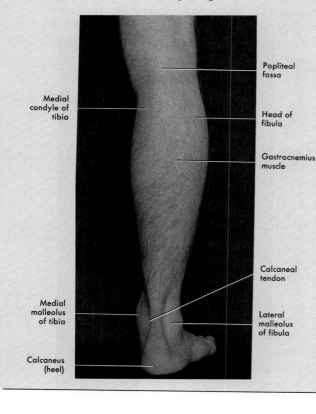

Anterior aspect of the right leg and foot.

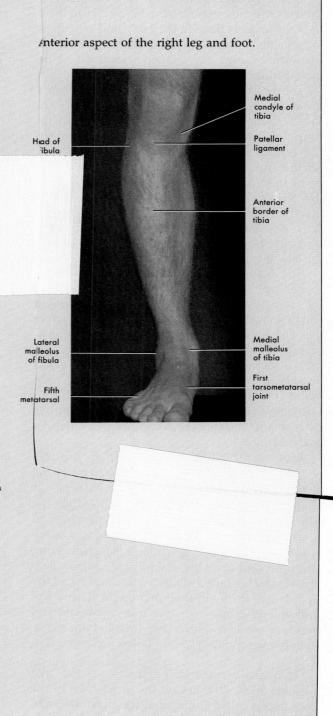

Like the humerus in the arm, the **femur** (FE-mur) is the only bone in the thigh (Figure 5-16). It is the longest bone in the body and articulates proximally (toward the hip) with the innominate bone in a deep, cup-shaped socket called the **acetabulum** (as-e-TAB-u-lum). The articulation of the head of the femur in the acetabulum is much more stable than the articulation of the head of the humerus with the scapula in the upper extremity. As a result, dislocation of the hip occurs far less often than does disarticulation of the shoulder. Distally, the femur articulates with the knee cap or **patella** (pah-TEL-ah) and the **tibia** or "shinbone." It is the tibia that forms a rather sharp edge along the front of your lower leg. A slender nonweight-bearing and rather fragile bone named the **fibula** lies along the outer or lateral border of the lower leg.

Toe bones have the same name as finger bones—**phalanges.** There are the same number of toe bones as finger bones, a fact that might surprise you, since toes are so much shorter than fingers. Foot bones comparable to the metacarpals and carpals of the hand have slightly different names. They are called **metatarsals** and **tarsals** in the foot (Figure 5-17). Just as each hand contains five metacarpal bones, each foot contains five metatarsal bones. But the foot has only seven tarsal bones, in contrast to the hand's eight carpals.

Feet are for standing on, so certain features of their structure make them able to support the body's weight. The great toe, for example, is considerably more solid and less mobile than the thumb. The foot bones are held together in such a way as to form springy lengthwise and cross-wise arches. These provide great supporting strength and a highly stable base. Strong ligaments and leg muscle tendons normally hold the foot bones firmly in their arched positions. Not infrequently, however, the foot ligaments and tendons weaken. The arches then flatten, a condition appropriately called fallen arches or flat-feet.

Two arches extend lengthwise in the foot (Figure 5-18). One lies on the inside part of the foot and is called the **medial longitudinal arch.** The other lies along the outer edge of the foot and is named the **lateral longitudinal arch.** An-other arch extends across the ball of the foot; it is called the **transverse** or **metatarsal arch.**

DIFFERENCES BETWEEN A MAN'S AND A WOMAN'S SKELETON

A man's skeleton and a woman's skeleton differ in several ways. Were you to examine a male skeleton and a female skeleton placed side by side, you would probably notice first the difference in their sizes. Most male skeletons are larger than most female skeletons—a structural fact that seems to have no great functional importance. Structural differences between the male and female hipbones, however, do have functional importance. The female pelvis is made so that the body of a baby can be cradled in it before birth and can pass through it during the birth journey. Although the individual male hipbones (innominate bones) are generally larger than the individual female hipbones, to-

Figure 5-16 Bones of the right thigh, knee joint, and leg—anterior aspect. **A,** Right knee skeleton.

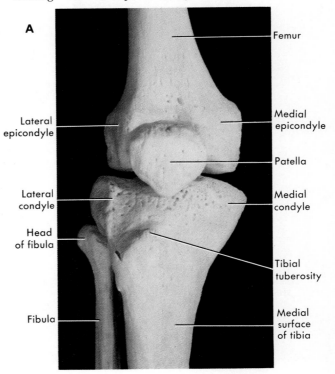

A

Femur

Lateral epicondyle

Medial epicondyle

Patella

Lateral condyle

Medial condyle

Head of fibula

Tibial tuberosity

Fibula

Medial surface of tibia

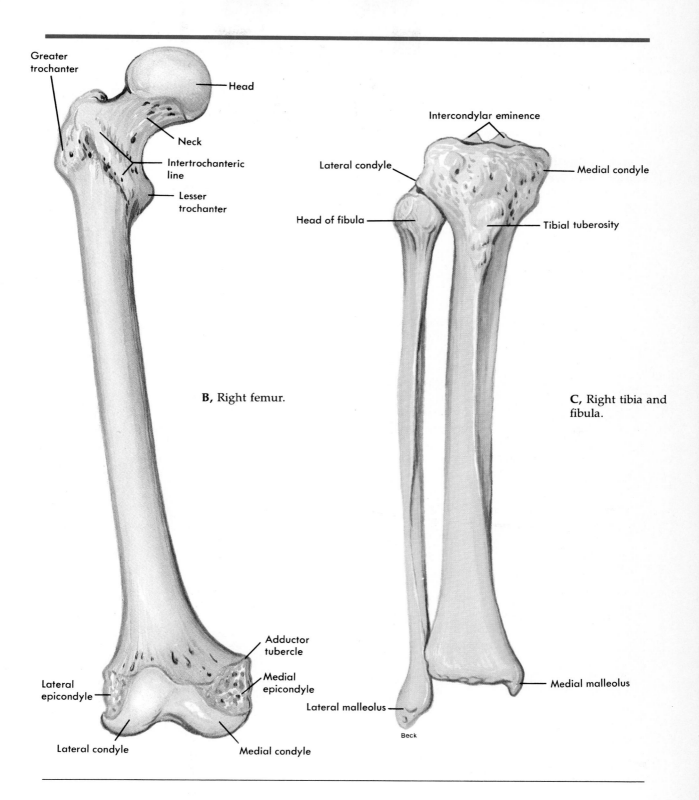

Greater
trochanter

Head

Neck

Intertrochanteric
line

Lesser
trochanter

B, Right femur.

Adductor
tubercle

Medial
epicondyle

Lateral
epicondyle

Lateral condyle

Medial condyle

Intercondylar eminence

Lateral condyle

Medial condyle

Head of fibula

Tibial tuberosity

C, Right tibia and
fibula.

Medial malleolus

Lateral malleolus

Beck

gether the male hipbones form a narrower structure than do the female hipbones. A man's pelvis is shaped something like a funnel, but a woman's pelvis has a broader, shallower shape, more like a basin. (Incidentally, the word *pelvis* means "basin.") Another difference is that the pelvic inlet, or brim, is normally much wider in the female than in the male. Figures 5-19 and 5-20 show this difference clearly. In these figures you can also see how much wider the angle is at the front of the female pelvis where the two pubic bones join than it is in the male.

Osteoporosis

Osteoporosis (os-te-o-po-RO-sis) is one of the most common and most serious of all bone diseases. It is characterized by excessive loss of both calcified matrix and collagenous fibers from bone. Osteoporosis occurs most frequently in white, elderly females. Although both white and black males are also susceptible, black women are seldom affected by it.

Since sex hormones play important roles in stimulating osteoblast activity after puberty, decreasing levels of these hormones in the blood of elderly persons reduces new bone growth and the maintenance of existing bone mass. Therefore some resorption of bone and subsequent loss of bone mass is an accepted consequence of advancing years. However, bone loss in osteoporosis goes far beyond the modest decrease normally seen in old age. The result is a dangerous pathological condition resulting in bone degeneration, increased susceptibility to "spontaneous fractures," and pathological curvature of the spine. Treatment may include sex hormone therapy and dietary supplements of calcium and vitamin D to replace deficiencies or to offset intestinal malabsorption.

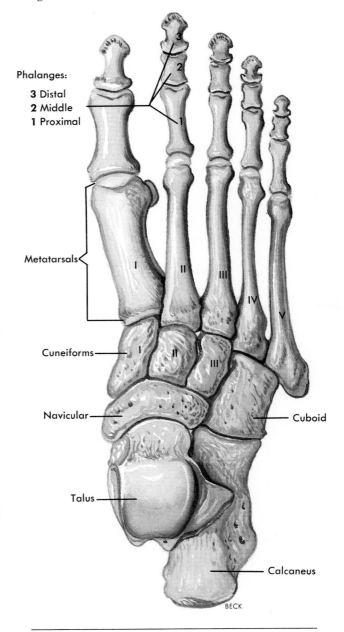

Figure 5-17 Bones of the right foot viewed from above. Tarsal bones consist of cuneiform bones, navicular bone, talus, cuboid bone, and calcaneus. Compare the names and numbers of foot bones shown here with those of the hand bones shown in Figure 5-15.

Phalanges:

3 Distal
2 Middle
1 Proximal

Metatarsals

Cuneiforms

Navicular

Talus

Cuboid

Calcaneus

BECK

Table 5-3 Identification of bone markings

Bone	Marking	Description
Frontal	Supraorbital margin	Arched ridge just below eyebrows
	Frontal sinuses	Cavities inside bone just above supraorbital margin; lined with mucosa; contain air
Temporal	Mastoid process	Protuberance just behind ear
	Mastoid air cells	Air-filled, mucosa-lined spaces within mastoid process
	External auditory meatus (or canal)	Opening into ear and tube extending into temporal bone
	Zygomatic process	Projection that articulates with malar (or zygomatic) bone
	Mandibular fossa	Oval depression anterior to external auditory meatus; forms socket for condyle of mandible
Occipital	Foramen magnum	Hole through which spinal cord enters cranial cavity
	Condyles	Convex, oval processes on either side of foramen magnum; articulate with depressions on first cervical vertebra
Sphenoid	Body	Hollow, cubelike central portion
	Sella turcica (or Turkish saddle)	Saddle-shaped depression on upper surface of sphenoid body; contains pituitary gland
	Sphenoid sinuses	Irregular air-filled, mucosa-lined spaces within central part of sphenoid
Ethmoid	Horizontal (cribriform) plate	Olfactory nerves pass through numerous holes in this plate
	Crista galli	Meninges attach to this process
	Perpendicular plate	Forms upper part of nasal septum
	Ethmoid sinuses	Honeycombed, mucosa-lined air spaces within lateral masses of bone
	Superior and middle turbinates (conchae)	Help to form lateral walls of nose
Mandible	Body	Main part of body; forms chin
	Condyle (or head)	Part of each ramus that articulates with mandibular fossa of temporal bone
	Alveolar process	Teeth set into this arch
Maxilla	Alveolar process	Arch containing teeth
	Maxillary sinus or antrum of Highmore	Large air-filled mucosa-lined cavity within body of each maxilla; largest of sinuses
Special features of skull	Sutures	Immovable joints between skull bones
	1 Sagittal	**1** Joint between two parietal bones
	2 Coronal	**2** Joint between parietal bones and frontal bone
	3 Lambdoidal	**3** Joint between parietal bones and occipital bone

Table 5-3 Identification of bone markings—cont'd

Bone	Marking	Description
Special features of skull— cont'd	Fontanels	"Soft spots" where ossification is incomplete at birth; allow some compression of skull during birth; also important in determining position of head before delivery; six such areas located at angles of parietal bones
	1 Anterior (or frontal)	**1** At intersection of sagittal and coronal sutures (juncture of parietal bones and frontal bone); diamond shaped; largest of fontanels; usually closed by 1½ years of age
	2 Posterior (or occipital)	**2** At intersection of sagittal and lambdoidal sutures (juncture of parietal bones and occipital bone); triangular in shape; usually closed by second month
	Wormian bones	Small islands of bones within suture
Sternum	Body	Main central part of bone
	Manubrium	Flaring, upper part
	Xiphoid process	Projection of cartilage at lower border of bone
Scapula (Figure 5-9)	Spine	Sharp ridge running diagonally across posterior surface of shoulder-blade
	Acromion process	Slightly flaring projection at lateral end of scapular spine; may be felt as tip of shoulder; articulates with clavicle
	Coracoid process	Projection on anterior from upper border of bone; may be felt in groove between deltoid and pectoralis major muscles about 1 inch below clavicle
	Glenoid cavity	Arm socket
Humerus	Head	Smooth, hemispherical enlargement at proximal end of humerus
Ulna (Figure 5-14)	Olecranon process	Elbow
	Styloid process	Sharp protuberance at distal end; can be seen from outside on posterior surface
Radius (Figure 5-14)	Head	Disk-shaped process forming proximal end of radius; articulates with capitulum of humerus and with radial notch of ulna
	Styloid process	Protuberance at distal end on lateral surface (with forearm supinated as in anatomical position)
Innominate (hip)	Ilium	Upper, flaring portion
	Ischium	Lower, posterior portion
	Pubic bone or pubis	Medial, anterior section
	Acetabulum	Hip socket; formed by union of ilium, ischium, and pubis
	Iliac crests	Upper, curving boundary of ilium
	Anterosuperior spine	Prominent projection at anterior end of iliac crest; can be felt externally as "point" of hip

Table 5-3 Identification of bone markings—cont'd

Bone	Marking	Description
Innominate — cont'd	Ischial tuberosity	Large, rough, quadrilateral process forming posterior part of ischium; in erect sitting position body rests on these tuberosities
	Symphysis pubis	Cartilaginous, amphiarthrotic joint between pubic bones
	Obturator foramen	Large hole in anterior surface of os coxae; formed by pubis and ischium; largest foramen in body
	Pelvic brim (or inlet)	Boundary of opening leading into true pelvis; size and shape of this inlet has great obstetrical importance, since if any of its diameters is too small, infant skull cannot enter true pelvis for natural birth
	True (or lesser) pelvis	Space below pelvic brim; true "basin" with bone and muscle walls and muscle floor; pelvic organs located in this space
	False (or greater) pelvis	Broad, shallow space above pelvic brim inlet; name "false pelvis" is misleading since this space is actually part of abdominal cavity, not pelvic cavity
Femur (Figure 5-16)	Head	Rounded, upper end of bone; fits into acetabulum
	Greater trochanter	Protuberance located interiorly and laterally to head
	Lesser trochanter	Small protuberance located inferiorly and medially to greater trochanter
	Condyles	Large, rounded bulges at distal end of femur; one on medial and one on lateral surface
Tibia (Figure 5-16)	Crest	Sharp ridge on anterior surface
	Medial malleolus	Rounded downward projection at distal end of tibia; forms prominence on outer surface of ankle
Fibula (Figure 5-16)	Lateral malleolus	Rounded prominence at distal end of fibula; forms prominence on outer surface of ankle
Tarsals (Figure 5-17)	Calcaneus	Heel bone
	Talus	Uppermost of tarsals; articulates with tibia and fibula; boxed in by medial and lateral malleoli
	Longitudinal arches	Tarsals and metatarsals so arranged as to form arch from front to back of foot
	1 Inner	**1** Formed by calcaneus, navicular, cuneiforms, and three medial metatarsals
	2 Outer	**2** Formed by calcaneus, cuboid, and two lateral metatarsals
	Transverse (or metatarsal) arch	Metatarsals and distal row of tarsals (cuneiforms and cuboid) so articulated as to form arch across foot; bones kept in two arched positions by means of powerful ligaments in sole of foot and by muscles and tendons

Figure 5-18 Medial and lateral longitudinal arches of the foot.

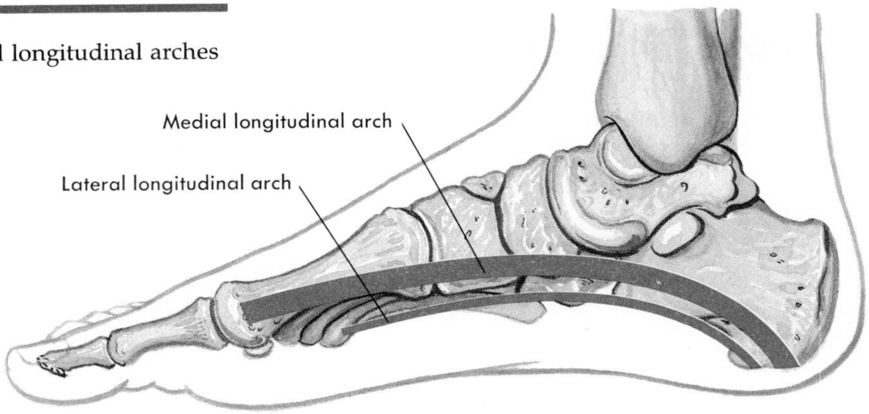

Medial longitudinal arch

Lateral longitudinal arch

JOINTS (ARTICULATIONS)

Every bone in the body, except one, connects to at least one other bone. In other words, every bone but one forms a joint with some other bone. (The exception is the hyoid bone in the neck, to which the tongue anchors.) Most of us probably never think much about our joints unless something goes wrong with them and they do not function properly. Then their tremendous importance becomes painfully clear. Joints perform two functions: they hold our bones together securely, and at the same time they make it possible for movement to occur between the bones—between most of them, that is. Without joints we could not move our arms, legs, or any other of our body parts. Our bodies would, in short, be rigid, immobile hulks. Try, for example, to move your arm at your shoulder joint in as many directions as you can. Try to do the same thing at your elbow joint. Now examine the shape of the bones at each of these joints on a skeleton or in Figures 5-8, 5-9 and 5-21, C. Do you see why you cannot move your arm at your elbow in nearly as many directions as you can at your shoulder?

KINDS OF JOINTS

One method classifies joints according to the degree of movement they allow into three types:
1 Synarthroses (no movement)
2 Amphiarthroses (slight movement)
3 Diarthroses (free movement)

Differences in the structure of joints account for differences in the degree of movement they make possible.

Synarthroses

A synarthrosis is a joint in which fibrous connective tissue grows between the articulating (joining) bones holding them close together. The joints between cranial bones are synarthroses, commonly called sutures.

Amphiarthroses

An amphiarthrosis is a joint in which cartilage connects the articulating bones. The symphysis pubis, the joint between the two pubic bones, is an amphiarthrosis, as are the joints between the bodies of the vertebrae.

Diarthroses

Fortunately most of our joints by far are diarthroses. All such joints allow considerable movement—sometimes in many directions and sometimes in only one or two directions.

Structure of diarthroses (freely movable joints) Freely movable joints are all made alike in certain ways. All of them have a joint capsule, a joint cavity, and a layer of cartilage over the ends of two joining bones (Figure 5-22). The **joint capsule** is made of the body's strongest and toughest material—fibrous connective tissue—and is lined with smooth, slippery synovial membrane. The capsule fits over the ends of the two bones somewhat like a sleeve. Be-

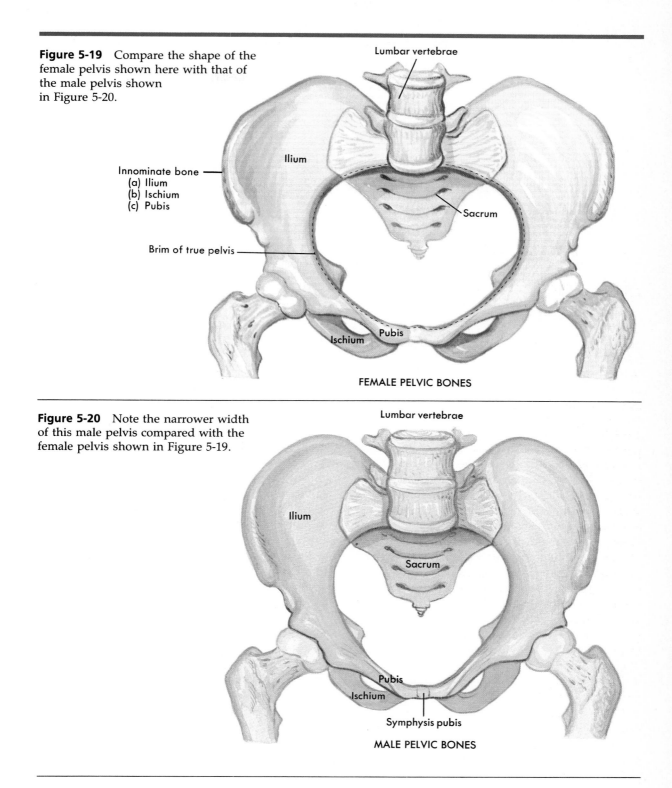

Figure 5-19 Compare the shape of the female pelvis shown here with that of the male pelvis shown in Figure 5-20.

Lumbar vertebrae

Ilium

Innominate bone
(a) Ilium
(b) Ischium
(c) Pubis

Sacrum

Brim of true pelvis

Pubis

Ischium

FEMALE PELVIC BONES

Figure 5-20 Note the narrower width of this male pelvis compared with the female pelvis shown in Figure 5-19.

Lumbar vertebrae

Ilium

Sacrum

Pubis

Ischium

Symphysis pubis

MALE PELVIC BONES

Figure 5-21 Typical joints. **A,** Synarthrotic (fibrous) joint; **B,** amphiarthrotic (cartilagenous) joint.

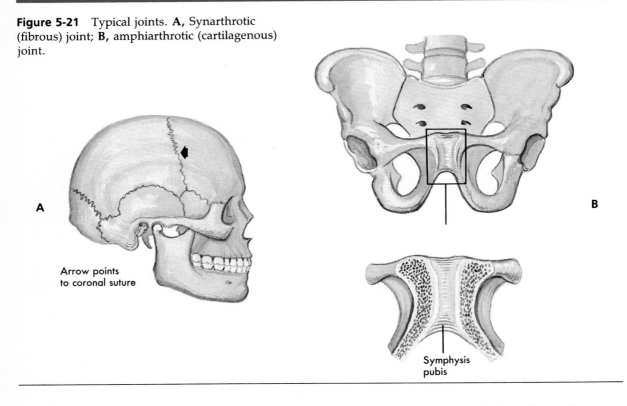

A

B

Arrow points
to coronal suture

Symphysis
pubis

cause it attaches firmly to the shaft of each bone to form its covering (called the periosteum; peri means "around" and osteon means "bone"), the joint capsule holds the bones together securely but at the same time permits movement at the joint. The structure of the joint capsule, in other words, helps make possible the joint's function.

Ligaments (cords or bands made of the same strong fibrous connective tissue as the joint capsule) also grow out of the periosteum and lash the two bones together even more firmly.

The layer of **articular cartilage** over the joint ends of bones acts like a rubber heel on a shoe—it absorbs jolts. The **synovial membrane** secretes a lubricating fluid (synovial fluid), which allows easier movement with less friction.

There are several types of diarthroses, namely, ball-and-socket, hinge, pivot, saddle, and certain others. Because they differ somewhat in structure, they differ also in their possible range of movement. In a ball-and-socket

joint, a ball-shaped head of one bone fits into a concave socket of another bone. Shoulder and hip joints, for example, are ball-and-socket joints. Of all the joints in our bodies, these permit the widest range of movements. Think for a moment how many ways you can move your upper arms. You can flex them (move them forward), you can extend them (move them backward), you can abduct them (move them away from the sides of your body), and you can adduct them (move them back down to your sides). You can also circumduct them (move them around so as to describe a circle with your hands).

Hinge joints, like the hinges on a door, allow movements in only two directions, namely, flexion and extension. Flexion is bending a part, extension is straightening it out. Elbow and knee joints and the joints in the fingers are hinge joints.

Pivot joints are those in which a small projec-

Figure 5-21, cont'd   Typical joints. **C,** Diarthrotic joints.

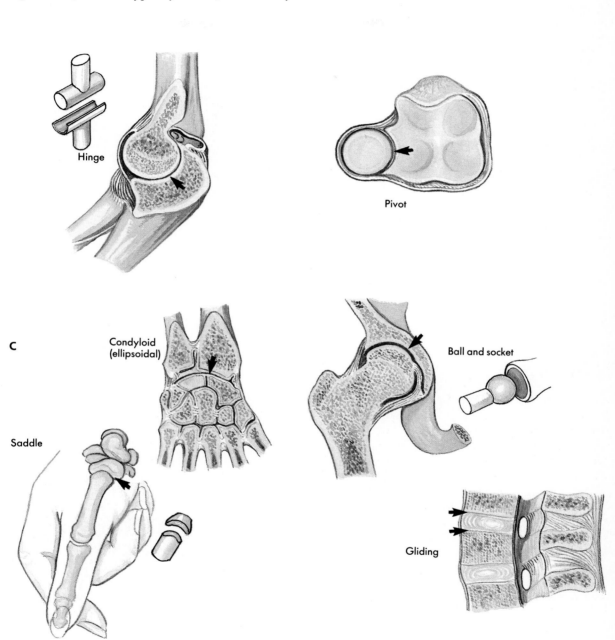

Figure 5-22 Structure of a freely movable (diarthrotic) joint. Note these typical features: joint capsule, joint cavity lined with synovial membrane, and articular cartilage covering the end surfaces of the bones within the joint capsule.

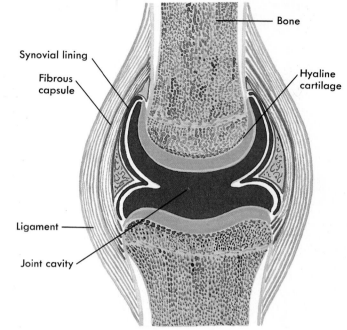

Figure 5-23 Sagittal section of vertebrae showing both normal and herniated disks.

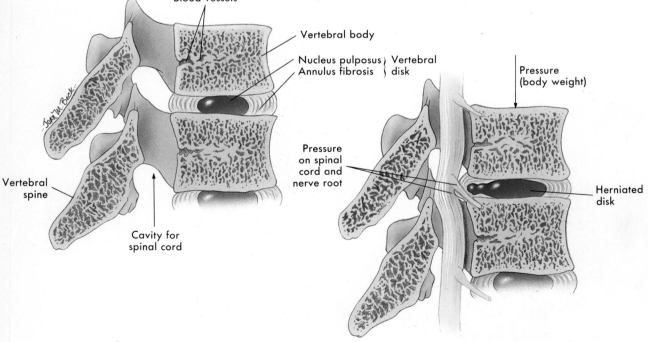

tion of one bone pivots in an arch of another bone. For example, a projection of the axis, the second vertebra in the neck, pivots in an arch of the atlas, the first vertebra in the neck. This rotates the head, which rests on the atlas.

Only one pair of saddle joints exists in the body—between the metacarpal bone of each thumb and a carpal bone of the wrist (the name of this carpal bone is the *trapezium*). Because the articulating surfaces of these bones are saddle-shaped, they make possible the human thumb's great mobility, a mobility no animal's thumb possesses. We can flex, extend, abduct, adduct, and circumduct our thumbs, and most important of all, we can move our thumbs to touch the tip of any one of our fingers. (This movement is called opposing the thumb to the fingers.) Without the saddle joints at the base of each of our thumbs, we could not do such a simple act as picking up a pin or grasping a pencil between thumb and forefinger.

Joints between the bodies of the vertebrae are amphiarthroses. These joints make it possible to flex the trunk forward or sideways and even to circumduct and rotate it. Strong ligaments connect the bodies of the vertebrae, and fibrous disks lie between them. The central core of these intervertebral disks consists of a pulpy, elastic substance that loses some of its resiliency with age. It may then be compressed by sudden exertion or injury, with fragments protruding into the spinal canal and pressing on spinal nerve routes of the spinal cord. Severe pain results. Medical terminology calls this a herniated disk; popular language calls it a "slipped disk" (Figure 5-23.)

OUTLINE SUMMARY

Functions
A Supports and gives shape to body
B Protects internal organs
C Helps make movements possible
D Stores calcium
E Hemopoiesis or blood cell formation

Microscopic Structure of Bone and Cartilage
A Bone types (Figure 5-1 and 5-2)
  1 Spongy
    a Texture results from needlelike threads on bone called *trabeculae* surrounded by a network of open spaces
    b Found in epiphyses of bones
    c Spaces contain red bone marrow
  2 Compact
    a Structural unit is haversian system—composed of concentric lamella, lacunae containing osteocytes, and canaliculi, all covered by periosteum
B Cartilage (Figure 5-4)
  1 Cell type called *chondrocyte*
  2 Matrix is gel-like and lacks blood vessels

Structure of Long Bones
A Structural components (Figures 5-5 and 5-6)
  1 Diaphysis or shaft
  2 Medullary cavity containing yellow marrow
  3 Epiphyses or ends of the bone; spongy bone contains red bone marrow
  4 Articular cartilage—covers epiphyses as a cushion
  5 Periosteum—strong membrane covering bone except at joint surfaces

Bone Formation and Growth
A Sequence of development—early cartilage models replaced by calcified bone matrix
B Osteoblasts are required to form new bone, and osteoclasts serve to resorb bone

Divisions of Skeleton
Skeleton composed of the following two main divisions and their subdivisions:
A Axial skeleton
  1 Skull

2 Spine

3 Thorax

4 Hyoid bone

B Appendicular skeleton

    1 Upper extremities, including shoulder girdle

    2 Lower extremities, including hip girdle

C Location and description of bones—see Figures 5-8 to 5-20 and Table 5-2

Differences Between a Man's and a Woman's Skeleton

A In size—male skeleton generally larger

B In shape of pelvis—male pelvis deep and narrow, female pelvis broad and shallow

C In size of pelvic inlet—female pelvic inlet generally wider, normally large enough for baby's head to pass through

D In pubic angle—angle between pubic bones of female generally wider

Joints (Articulations)

A Kinds of joints

    1 Synarthroses (no movement)—fibrous connective tissue grows between articulating bones; for example, sutures of skull

    2 Amphiarthroses (slight movement)—cartilage connects articulating bones; for example, symphysis pubis

    3 Diarthroses (free movement)—most joints of the body belong to this class

        a Structures of freely movable joints—joint capsule and ligaments hold adjoining bones together but permit movement at joint

        b Articular cartilage—covers joint ends of bones and absorbs joints

        c Synovial membrane—lines joint capsule and secretes lubricating fluid

        d Joint cavity—space between joint ends of bones

B Types of freely movable joints—ball-and-socket, hinge, pivot, saddle, and certain others

NEW WORDS

amphiarthroses	Haversian system	periosteum
appendicular skeleton	hemopoiesis	red bone marrow
articular cartilage	kyphosis	scoliosis
articulation	lacunae	sinus
axial skeleton	lamella	skull
canaliculi	lordosis	spine
chondrocytes	mastoiditis	synarthroses
compact bone	medullary cavity	synovial membrane
diaphysis	osteoclasts	trabeculae
diarthroses	osteocytes	thorax
epiphyseal fracture	pectoral girdle	yellow bone marrow
epiphyses	pelvic girdle	

CHAPTER TEST

1. The process of blood cell formation in bone marrow is called _____ .

2. Spongy bone contains needlelike threads of bone known as _____ .

3. The structural units of compact bone are called _____ _____ .

4. Bone-forming cells are called _____ , whereas bone-absorbing cells are called _____ .

5. The hollow shaft of a long bone is also known as the _____ .

6. The human skeleton has two main divisions called the _____ skeleton and the _____ skeleton.

7. Spaces or cavities located inside some of the cranial bones are called _____ .

8. Abnormal side-to-side curvature of the spine is called _____ .

9. The supporting framework of the neck is formed by the seven _____ vertebrae.

10. The shoulder blade is also known as the _____ , and the collar bone is called the _____ .

11. The long bones of the forearm are the _____ and the _____ .

12. The large bony process of the ulna that forms the "elbow" is called the _____ process.

13. There are 8 _____ bones in the wrist and 14 _____ or finger bones in each hand.

14. Each innominate or hip bone consists of three separate bones called the _____ , _____ , and _____ , which fuse in the adult to become one bone.

15. The femur of the leg articulates distally with the "knee cap" or _____ and the "shinbone" or _____ .

16. The metatarsal and tarsal bones are located in the _____ .

17. The bone disease characterized by excessive loss of both calcified matrix and collagenous fibers is called _____ .

18. The foramen magnum is a large opening in the occipital bone of the _____ .

19. The body, manubrium, and xiphoid process are all components of the _____ .

20. Joints that permit free movement are called _____ joints.

Circle the "T" before each true statement and the "F" before each false statement.

T F 21. Adult bones do not contain living cells.

T F 22. Red bone marrow functions in hemopoiesis or blood cell formation.

T F 23. The hollow cylindrical portion of a long bone is called the epiphysis.

T F 24. Haversian systems are components of spongy bone.

T F 25. Lacunae are found in both compact bone and cartilage.

T F 26. Epiphyseal fractures occur only in mature adults.

T F 27. There are more bones in the axial than in the appendicular skeleton.

T F 28. Lordosis, kyphosis, and scoliosis are all abnormal or pathological conditions that affect the spine.

T F 29. The term *phalanges* is used to describe the bones of both the fingers and the toes.

T F 30. Osteoporosis is a bone disease that occurs only in males.

T F 31. "Soft spots" in the skull at birth are called fontanels.

T F 32. Synarthroses are freely movable joints.

T F 33. Synovial membranes are found in diarthrotic joints.

T F 34. Normally, teeth can be found in both the mandible and maxilla.

T F 35. The glenoid cavity is located on the innominate bone.

REVIEW QUESTIONS

1 List and discuss the generalized functions of the skeletal system.

2 Compare the structures of compact and spongy bone.

3 What is a Haversian system, and what are its component parts?

4 Discuss the following cell types: osteocyte, osteoblast, osteoclast, chondrocyte.

5 List the major structural components of a typical long bone, and briefly describe the function of each.

6 Discuss the mechanism of bone formation and growth.

7 Is it possible to tell whether a child is going to grow any taller? If so, how can a doctor tell this?

8 What is an epiphyseal fracture? At what age is this type of fracture most common? Why?

9 What are the two major subdivisions of the skeleton? What are the major body areas in each subdivision?

10 Give the correct anatomical name for each of the following:
collarbone
breastbone
wrist bones
finger bones
forearm bones
thigh bone
hip bone
knee cap
ankle bones
neck vertebrae

11 Compare the structures and functions of the:
a Arms and legs
b Pectoral and pelvic girdles
c Shoulder and hip joints
d Hands and feet

12 Describe one functionally important difference between a male and a female adult skeleton.

13 Classify the major types of joints.

14 List and compare the types of freely movable joints.

CHAPTER
6   The Muscular System

BOXED ESSAY

Intramuscular injections

OBJECTIVES

After you have completed this chapter, you should be able to:

1 List, locate in the body, and compare the structure and function of the three major types of muscle tissue.

2 Discuss the microscopic structure of a skeletal muscle sarcomere and motor unit.

3 Discuss how a muscle is stimulated and compare the major types of skeletal muscle contractions.

4 Define the terms *all or none, graded response, origin, insertion, prime mover, posture,* and *oxygen debt.*

5 List and explain the most common types of movement produced by skeletal muscles.

The muscular system consists of more than 500 muscles that move us about in many ways, varying in complexity from blinking an eye or smiling to climbing a mountain or ski jumping. Not many of our body structures have as great an importance for happy, useful living as do our voluntary muscles, and only a few have greater importance for life itself. A great deal is known about muscles—enough, in fact, to fill several books. The plan for this chapter is to investigate first the different types of muscle tissue, then to note some general facts about the structure and function of skeletal muscles and the types of muscle contractions, next to present some specific facts about certain key muscles, and finally to consider some muscle disorders.

MUSCLE TISSUE

If you weigh 120 pounds, about 50 pounds of your weight comes from your muscles, the "red meat" attached to your bones. Under the microscope these muscles appear as bundles of fine threads with many crosswise stripes. Each fine thread is a muscle cell or, as it is usually called, a muscle fiber. This type of muscle tissue has three names: *striated muscle*—because of its cross stripes or striae; *skeletal muscle*—because it attaches to bone; and *voluntary muscle*—because its contractions can be controlled voluntarily.

Besides **skeletal muscle,** the body also contains two other kinds of muscle tissue: cardiac muscle and nonstriated, smooth, or involuntary muscle. **Cardiac muscle,** as its name suggests, composes the bulk of the heart. As you can see in Figure 6-1, cardiac muscle cells branch frequently. Nonstriated or **smooth muscle** lacks the cross stripes or striae seen in skeletal muscle. It has a "smooth," even appearance when viewed through a microscope. It is called involuntary because we normally do not have control over its contractions. Smooth or involuntary muscle forms an important part of blood vessel walls and of many hollow internal organs (viscera) such as the gut (Figure 6-1).

Muscle cells specialize in the function of contraction, or shortening. Every movement we make is produced by contractions of skeletal muscle cells. Contractions of cardiac muscle cells keep the blood circulating through its vessels, and smooth muscle contractions do many things—for instance, move food into and through the stomach and intestines and make a major contribution to the maintenance of normal blood pressure.

SKELETAL MUSCLES (ORGANS)
STRUCTURE

A skeletal muscle is an organ composed mainly of striated muscle cells and connective tissue. Most skeletal muscles attach to two bones that have a movable joint between them. In other words, most muscles extend from one bone across a joint to another bone. Also, one of the two bones moves less easily than the other. The muscle's attachment to this more stationary bone is called its **origin.** Its attachment to the more movable bone is called the muscle's **insertion.** The rest of the muscle (all of it except its two ends) is called the *body* of the muscle. *Tendons* anchor muscles firmly to bones. Made of dense fibrous connective tissue in the shape of heavy cords, tendons have great strength. They do not tear or pull away from bone easily. Yet any emergency room nurse or physician sees many tendon injuries—severed tendons and tendons torn loose from bones.

Small fluid-filled sacs called **bursae** lie between some tendons and the bones beneath them (Figure 6-2). These small sacs are made of connective tissue and lined with **synovial membrane.** The synovial membrane secretes a slippery, lubricating fluid (synovial fluid) that fills the bursa. Like a small, flexible cushion, a bursa

Both bursae and tendon sheaths can become inflamed. Inflammation of a bursa—a relatively common ailment, particularly in older people—is called **bursitis** (ber-SI-tis). Inflammation of a tendon sheath is **tenosynovitis** (ten-o-sin-o-VI-tis).

Figure 6-1 Muscle tissues of the human body. **A,** Striated (voluntary), or skeletal, muscle tissue; left, a microscopic view showing cross striations and multinuclei in each cell; right, a macroscopic view of skeletal muscle organs. **B,** Nonstriated (involuntary), smooth, muscle tissue; left, a microscopic view; right, a loop of intestine, one of many internal organs whose walls contain smooth muscle. **C,** Branching, or cardiac, muscle tissue; left, a microscopic view showing cross-striations and branching cells; right, the heart, the only organ made of cardiac muscle tissue.

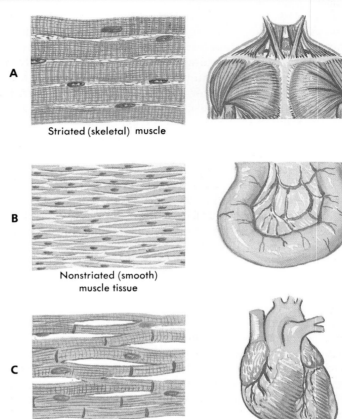

A

Striated (skeletal) muscle

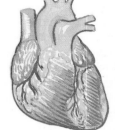

B

Nonstriated (smooth)
muscle tissue

C

Branching (cardiac)
muscle tissue

Figure 6-2 Bursa of the elbow joint. The bursa acts as a cushion to relieve pressure between the olecranon process of the ulna and the humerus during movement at the elbow joint.

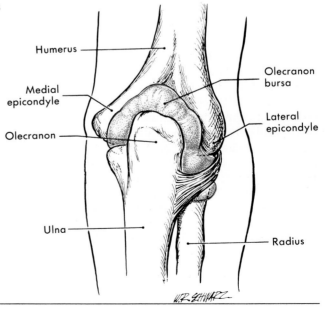

Humerus

Medial
epicondyle

Olecranon

Ulna

Olecranon
bursa

Lateral
epicondyle

Radius

makes it easier for a tendon to slide over a bone when the tendon's muscle shortens. **Tendon sheaths** enclose certain tendons. Because these tube-shaped structures are also lined with synovial membrane and moistened with synovial fluid, they, like the bursas, facilitate movement.

MICROSCOPIC STRUCTURE

Muscle tissue consists of specialized contractile cells or **muscle fibers** that are grouped together and arranged in a highly organized way. Each skeletal muscle fiber is itself filled with two kinds of very fine and threadlike structures called **thick** and **thin myofilaments** (mi-o-FIL-a-ments). The thick myofilaments are formed from a protein called **myosin,** and the thin myofilaments are composed of the protein **actin.** Find the label **sarcomere** (SAR-ko-meer) in Figure 6-3. Think of the sarcomere as the basic functional or *contractile unit* of skeletal muscle. Recall that the haversian system served as the basic building block of compact bone; the sarcomere serves that function in skeletal muscle. The submicroscopic structure of a sarcomere consists of numerous actin and myosin myofilaments arranged so that when viewed under a microscope, dark and light stripes or cross-striae are seen. The repetitive units or sarcomeres are separated from each other by dark bands called **Z lines.**

Look at the structure of a sarcomere in both Figures 6-3 and 6-4. Note in Figure 6-3 that the thick and thin myofilaments overlap each other to form a dark area called the **A band.** A light area between the Z line and the A band is called the **I band.** Each A band is composed of both thick and thin myofilaments, and each I band is composed only of thin myofilaments. The sarcomeres that you see in Figures 6-3 and 6-4 are from muscle fibers that are relaxed. During the contraction process energy obtained from ATP molecules enables the two types of myofilaments to slide toward each other and shorten the sarcomere and, eventually, the entire muscle. This explanation of muscle contraction resulting from the movement of thick and thin myofibrils toward one another is called the **sliding-filament theory.** This theory is now accepted by most scientists as the best explanation of how muscles shorten.

Figure 6-4 is an electron photomicrograph showing a portion of skeletal muscle enlarged 25,000 times. Compare the bands labeled in the photomicrograph with those in Figure 6-3. Electron microscopy of skeletal muscle has revolutionized our concept of both its structure and function.

FUNCTIONS

The three primary functions of the muscular system are:
1 Movement
2 Posture or muscle tone
3 Heat production

MOVEMENT

Muscles move bones by pulling on them. Because the length of a skeletal muscle becomes shorter as its fibers contract, the bones to which the muscle attaches move closer together. As a rule, only the insertion bone moves. The shortening of the muscle pulls the insertion bone toward the origin bone. The origin bone stays put, holding firm, while the insertion bone moves toward it (Figure 6-5). One tremendously important function of skeletal muscle contractions, therefore, is to produce movements. Remember this rule: a muscle's insertion bone moves toward its origin bone. It can help you understand muscle actions.

Voluntary muscular movement is normally smooth and free of spastic jerks and tremors because skeletal muscles generally work in coordinated teams, not singly. Several muscles contract while others relax at the same time to produce almost any movement you can think of. Of all the muscles contracting simultaneously, the one mainly responsible for producing a particular movement is called the **prime mover** for that movement. The other muscles that help in producing the movement are called **synergists** (SIN-er-jists). As prime movers and synergist muscles at a joint contract, other muscles called **antagonists** (an-TAG-o-nists) relax. When antagonist muscles contract, they pro-

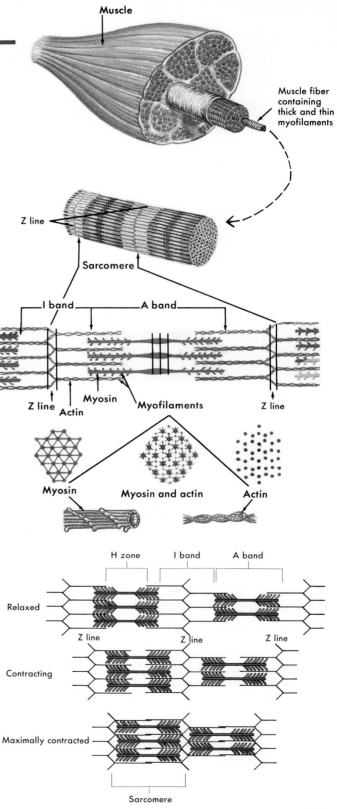

Figure 6-3 Structure of skeletal muscle. See text for discussion.

Figure 6-4   Electron photomicrograph of skeletal muscle. (×25,000)

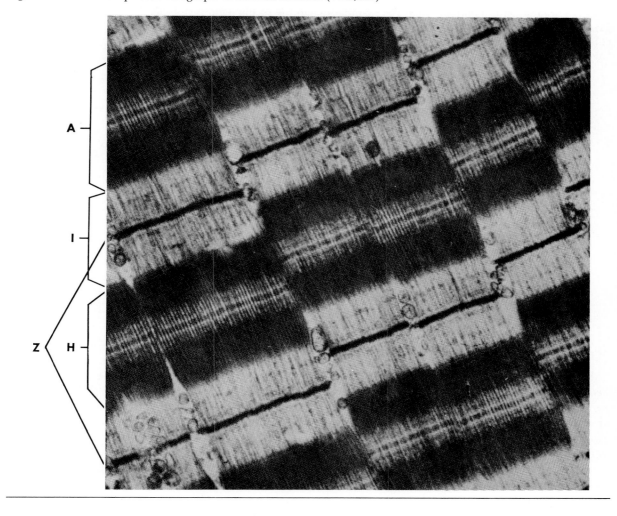

duce a movement opposite to that of prime movers and their synergist muscles.

Locate the biceps brachii, brachialis, and triceps brachii muscles in Figures 6-6 and 6-7. All the muscles in these figures are involved in flexion and extension of the forearm at the elbow joint. The biceps brachii is the prime mover during flexion, and the brachialis is its helper or synergist muscle. When the biceps brachii and brachialis muscles flex the forearm, the triceps brachii relaxes. Therefore during flexion of the forearm the triceps brachii is the antagonistic muscle. During extension of the forearm these three muscles continue to work as a team. However, during extension, the triceps brachii becomes the prime mover and the biceps and brachialis become the antagonistic muscles. This combined and coordinated activity is what makes our muscular movements smooth and graceful.

Figure 6-5 Diagram illustrates the principle that when a muscle contracts, it pulls its insertion bone toward its origin bone. For example, when the rectus femoris muscle (part of the quadriceps femoris muscle group) contracts, it pulls its insertion bone, the tibia, toward its origin, the hipbone. This straightens the knee joint and extends the lower leg as shown.

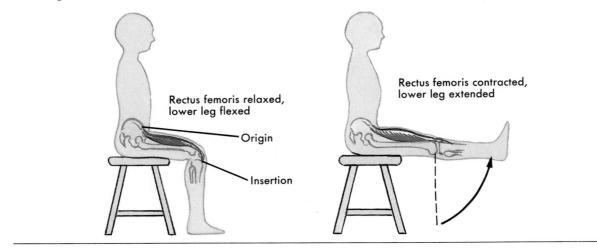

Rectus femoris relaxed, lower leg flexed

Origin

Insertion

Rectus femoris contracted, lower leg extended

POSTURE

We are able to maintain our body position because of a specialized type of skeletal muscle contraction called **tonic contraction.** Because relatively few of a muscle's fibers shorten at one time in a tonic contraction, the muscle as a whole does not shorten, and no movement occurs. Consequently, tonic contractions do not move any body parts. What they do is hold them in position. In other words, muscle tone maintains **posture.** Good posture means that body parts are held in the positions that favor best function. These are positions that balance the distribution of weight and that therefore put the least strain on muscles, tendons, ligaments, and bones. To have good posture in a standing position, for example, you must stand with your head and chest held high, your chin, abdomen, and buttocks pulled in, and your knees bent slightly.

To judge for yourself how important good posture is, consider some of the effects of poor posture. Besides detracting from appearance, poor posture makes a person tire more quickly. It puts an abnormal pull on ligaments, joints, and bones and therefore often leads to deformities. Poor posture crowds the heart, making it harder for it to contract. Poor posture crowds the lungs, decreasing their breathing capacity.

Skeletal muscle tone maintains posture by counteracting the pull of gravity. Gravity tends to pull the head and trunk down and forward, but the tone in certain back and neck muscles pulls just hard enough in the opposite direction to overcome the force of gravity and hold the head and trunk erect. The tone in thigh and leg muscles puts just enough pull on thigh and leg bones to counteract the pull of gravity on them that would otherwise collapse the hip and knee joints and cause us to fall in a heap.

HEAT PRODUCTION

Healthy survival depends on our ability to maintain a constant body temperature. A fever or elevation in body temperature of only a degree or two above 98.6° F will result in illness. Just as serious is any fall in body temperature. Any decrease below normal, a condition called **hypothermia** (hi-po-THER-me-ah), will drasti-

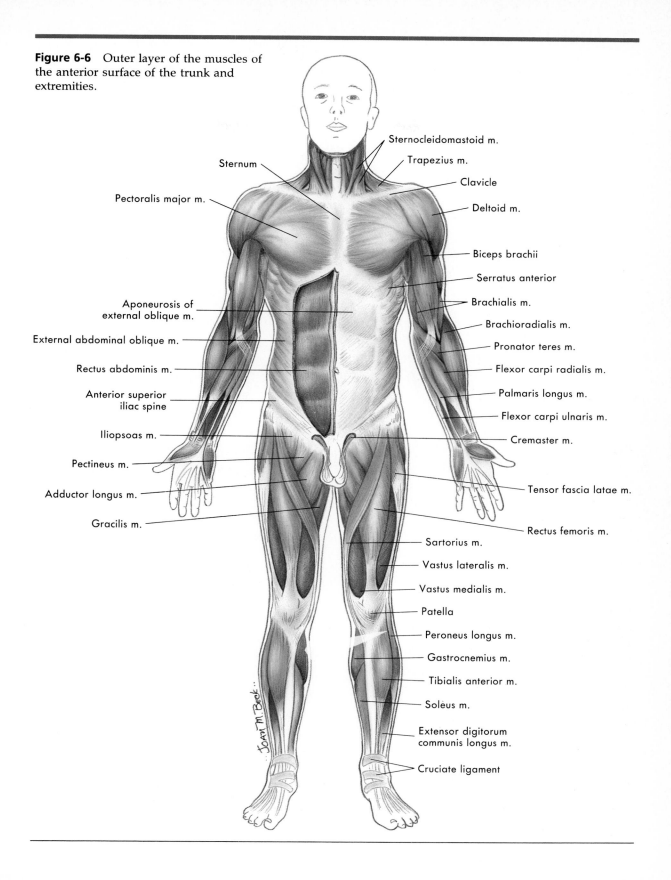

Figure 6-6 Outer layer of the muscles of the anterior surface of the trunk and extremities.

Sternocleidomastoid m.

Trapezius m.

Sternum

Clavicle

Pectoralis major m.

Deltoid m.

Biceps brachii

Serratus anterior

Brachialis m.

Aponeurosis of external oblique m.

Brachioradialis m.

External abdominal oblique m.

Pronator teres m.

Flexor carpi radialis m.

Rectus abdominis m.

Palmaris longus m.

Anterior superior iliac spine

Flexor carpi ulnaris m.

Iliopsoas m.

Cremaster m.

Pectineus m.

Adductor longus m.

Tensor fascia latae m.

Gracilis m.

Rectus femoris m.

Sartorius m.

Vastus lateralis m.

Vastus medialis m.

Patella

Peroneus longus m.

Gastrocnemius m.

Tibialis anterior m.

Soleus m.

Extensor digitorum communis longus m.

Cruciate ligament

JoAnn M. Beck

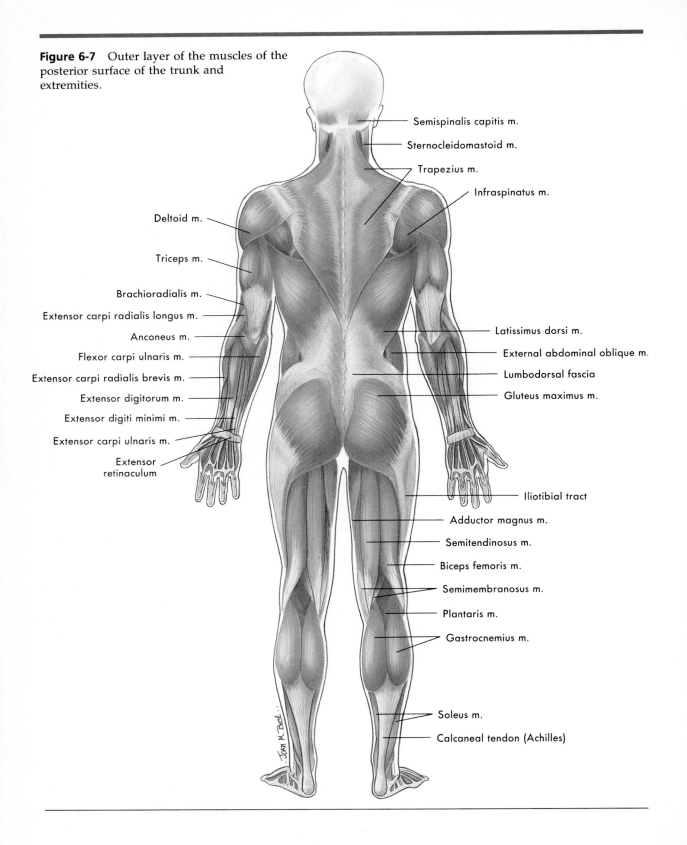

Figure 6-7 Outer layer of the muscles of the posterior surface of the trunk and extremities.

Semispinalis capitis m.

Sternocleidomastoid m.

Trapezius m.

Infraspinatus m.

Deltoid m.

Triceps m.

Brachioradialis m.

Extensor carpi radialis longus m.

Anconeus m.

Flexor carpi ulnaris m.

Extensor carpi radialis brevis m.

Extensor digitorum m.

Extensor digiti minimi m.

Extensor carpi ulnaris m.

Extensor retinaculum

Latissimus dorsi m.

External abdominal oblique m.

Lumbodorsal fascia

Gluteus maximus m.

Iliotibial tract

Adductor magnus m.

Semitendinosus m.

Biceps femoris m.

Semimembranosus m.

Plantaris m.

Gastrocnemius m.

Soleus m.

Calcaneal tendon (Achilles)

Figure 6-8 Motor unit. A motor unit consists of one motor neuron and the muscle cells supplied by its branches.

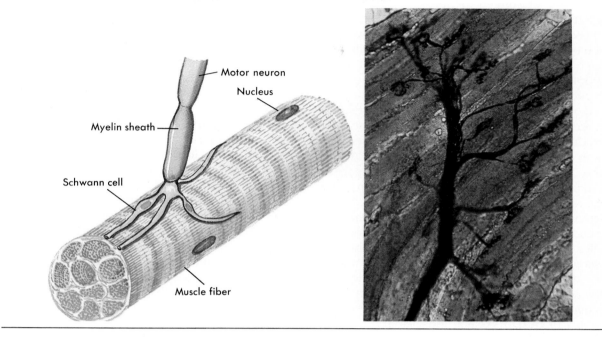

cally affect cellular activity and normal body function. The contraction of muscle fibers produces most of the heat required to maintain our body temperature. Energy required to produce a muscle contraction is obtained from ATP. Most of the energy released during the breakdown of ATP during a muscular contraction is used to shorten the muscle fibers; however, some of the energy is lost as heat during the reaction. It is this heat that helps us to maintain our body temperature at a constant level.

It is important to remember that muscles do not function alone. Other structures such as bones and joints must function with them. Skeletal muscles (except for a few in the face) bring about movements by pulling on bones across movable joints. But before a skeletal muscle can contract and pull on a bone to move it, the muscle must first be stimulated by nerve impulses.

MOTOR UNIT

Muscle cells are stimulated or "shocked" by a nervous impulse that enters the muscle fiber through a specialized nerve called a **motor neuron** (Figure 6-8). The point of contact between the nerve ending and the muscle fiber is called a **neuromuscular junction.** When a nervous impulse passes through this junction, specialized chemicals, which cause the muscle to contract or shorten, are released. A motor neuron together with the muscle cells it innervates is called a **motor unit** (Figure 6-8).

FATIGUE

If muscle cells are stimulated repeatedly without adequate periods of rest, the strength of muscle contraction will decrease resulting in **fatigue.** If repeated stimulation occurs, the strength of contraction will continue to decrease, and eventually the muscle will lose its ability to contract.

During prolonged periods of exercise the stored ATP required for muscle contraction becomes depleted. In addition, the rapid utilization of oxygen and nutrients during exercise will often outstrip the ability of the muscle's blood supply to replenish them. When oxygen supplies run low, muscle cells produce lactic acid and other waste products during the contraction process. It is this build-up of lactic acid that often produces muscle soreness after exercise. The term **oxygen debt** is used to describe the depletion of oxygen in muscle cells during vigorous and prolonged exercise. This "debt" is repaid by the rapid and deep breathing that follows such activity. The respiratory, circulatory, nervous, muscular, and skeletal systems all play essential roles in producing normal movements. This fact has great practical importance. For example, a person might have perfectly normal muscles and still not be able to move normally. He might have a nervous system disorder that shuts off impulses to certain skeletal muscles and thereby paralyzes them. Multiple sclerosis is one great enemy that acts in this way, but so do some other conditions—a brain hemorrhage, a brain tumor, or a spinal cord injury, to mention a few. Skeletal system disorders, especially arthritis, can have disabling effects on movement. Muscle functioning, then, depends on the functioning of many other parts of the body. This fact illustrates a principle repeated often in this book. It can be simply stated: no part of the body lives by or for itself alone. Each part depends on all other parts for its healthy survival. Each part contributes something to the healthy survival of all other parts.

MUSCLE STIMULUS

If a muscle cell is "shocked" by an adequate stimulus, it will contract completely. It is because of this fact that muscle cells are said to respond "**all or none.**" However, it is important to remember that a muscle is composed of many individual muscle cells. As a result, all the cells in a muscle do not contract simultaneously. This is the "**principle of graded response.**" Physiologists use this principle to explain that although each individual muscle cell will contract "all or none" if subjected to a minimal or threshold stimulus, the strength of contraction of an entire muscle will show a "graded response" as more and more of its cells are stimulated and contract. These important facts about muscle physiology are illustrated in Figure 6-9. Note in this illustration that very low–level stimuli, called **subminimal stimuli,** do not cause any contraction of muscle cells. As you can see in the upper portion of the illustration, the strength of contraction remains at zero if the level of stimulation is subminimal, even if the stimulus is repeated several times. However, if the strength of the stimulus is gradually increased, it will reach a level that will cause some muscle cells to contract for the first time. This is called a **minimal** or **threshold stimulus.** As the strength of the stimulus increases, more and more muscle cells contract, and the strength of contraction of the muscle as a whole also increases according to the principle of graded response. However, the strength of contraction will increase only up to a certain point. The **maximal stimulus** shown in Figure 6-9 will produce the maximum contraction. In other words, a maximal stimulus will cause every cell in the entire muscle to respond "all or none," and the muscle as a whole will contract to the limits of its capacity. Increasing the stimulus strength even more—a **supramaximal stimulus**—will have no additional effect on the strength of contraction. Once all the fibers in the muscle are contracting, increasing the stimulus strength is futile, and the principle of graded response is no longer applicable.

TYPES OF SKELETAL MUSCLE CONTRACTION

In addition to the specialized tonic contraction of muscle that maintains muscle tone and posture, other types of contraction also occur. Additional types of muscle contraction include:

1 Isometric contraction
2 Isotonic contraction

Figure 6-9 Variation of strength of contraction of muscle with strength of stimulus.

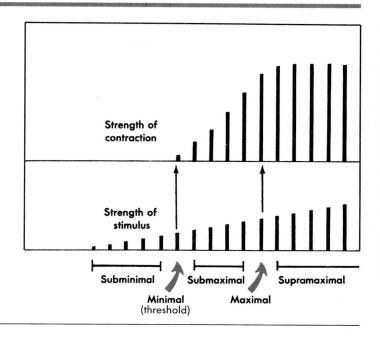

3 Twitch contraction
4 Tetanic (tetanus) contraction

ISOMETRIC CONTRACTION

Contraction of a skeletal muscle does not produce movement. Sometimes it increases the tension within a muscle but does not decrease the length of the muscle. When the muscle does not shorten and no movement results, it is called an isometric contraction. The word *isometric* comes from Greek words that mean "equal measure." In other words, a muscle's length during an isometric contraction and during relaxation is about equal. Although muscles do not shorten (and therefore do not produce movement) during isometric contractions, the tension within them increases. Because of this, repeated isometric contractions tend to make muscles grow larger and stronger—hence the popularizing in recent years of isometric exercises as great muscle builders. Pushing against a wall or other immovable object is a good example of isometric exercise. Although no movement occurs and the muscle does not shorten, its internal tension increases dramatically.

ISOTONIC CONTRACTION

In most cases isotonic contraction of muscle produces movement at a joint. In this type of contraction the muscle shortens, and the insertion end of the muscle moves toward the point of origin. Exceptions to this rule are the isotonic contraction of the facial muscles (Figure 6-10). Walking, running, breathing, and lifting and twisting movements are all examples of isotonic contraction.

TWITCH CONTRACTION

A twitch contraction is a quick, jerky contraction in response to a single stimulus. Figure 6-11 shows a record of such a contraction. It reveals that the muscle does not shorten at the instant of stimulation, but rather a fraction of a second later, and that it reaches a peak of shortening and then gradually resumes its former length. These three phases of contraction are spoken of, respectively, as the *latent period*, the *contraction phase*, and the *relaxation phase*. The entire twitch usually lasts less than $\frac{1}{10}$ of a second. Normally, twitch contractions rarely occur

Figure 6-10 Muscles of facial expression. Most muscles of facial expression surround the eyes, nose, and mouth. Contraction of these muscles can produce a wide variety of facial expressions and convey numerous emotions.

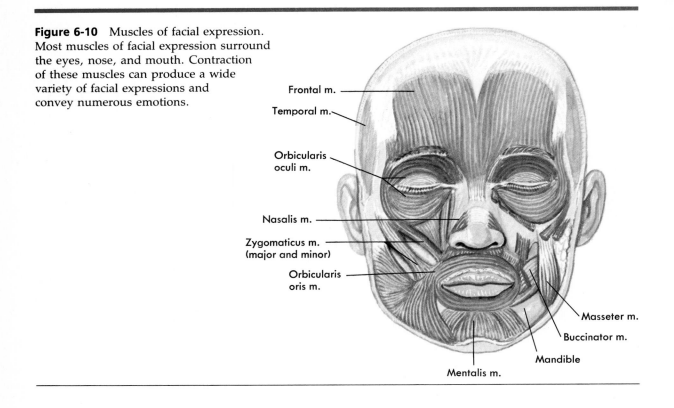

Frontal m.

Temporal m.

Orbicularis oculi m.

Nasalis m.

Zygomaticus m. (major and minor)

Orbicularis oris m.

Masseter m.

Buccinator m.

Mandible

Mentalis m.

in the body. Laboratory studies of twitch contractions, however, have provided much information about basic muscle physiology.

TETANIC CONTRACTION (TETANUS)

The tetanic type of contraction is more sustained than a twitch. It is produced by a series of stimuli bombarding the muscle in rapid succession. About 30 stimuli per second, for example, will evoke a tetanic contraction in certain types of skeletal muscle. The rate of stimuli required to produce tetanus will vary for different muscles and different conditions. Figure 6-12 shows records of incomplete and complete tetanus. Normal movements are said to be produced by incomplete tetanic contractions. A disease caused by bacteria that sometimes enter the body in puncture wounds is also called tetanus or "lockjaw." The name is appropriate because it results in severe cramping with almost continuous tetanic type muscle contractions.

Perhaps the most important function of skeletal muscle is movement itself. This basic function is even more important for health than is good posture. Most of us believe that "exercise is good for us," even if we have no idea what or how many specific benefits can come from it. The benefits of exercise are almost legion. Professional journals carry articles about exercising and so do many popular magazines and books. Some of the good consequences of regular, properly practiced exercise are greatly improved muscle tone, better posture, more efficient heart and lung functioning, less fatigue, and, last but not least, looking and feeling better.

Sick people naturally move about less than well people. Medically speaking, they become less mobile or even immobile. Too little mobility inevitably has bad effects on virtually all body structures and functions. These effects often threaten survival and sometimes lead to death. Good health care, however, can do much to de-

Figure 6-11   Simple muscle twitch and its time components.

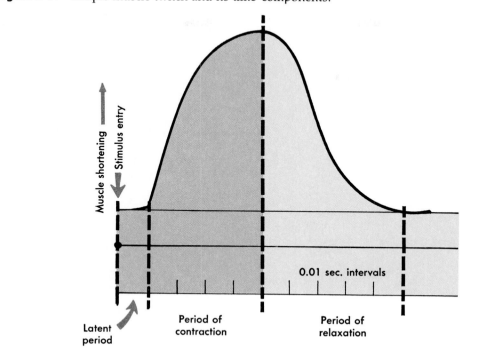

Figure 6-12   Single twitch, incomplete tetanus, and complete tetanus of muscle.

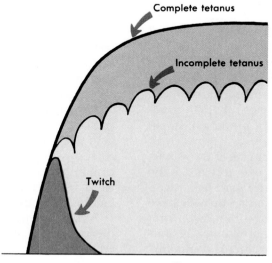

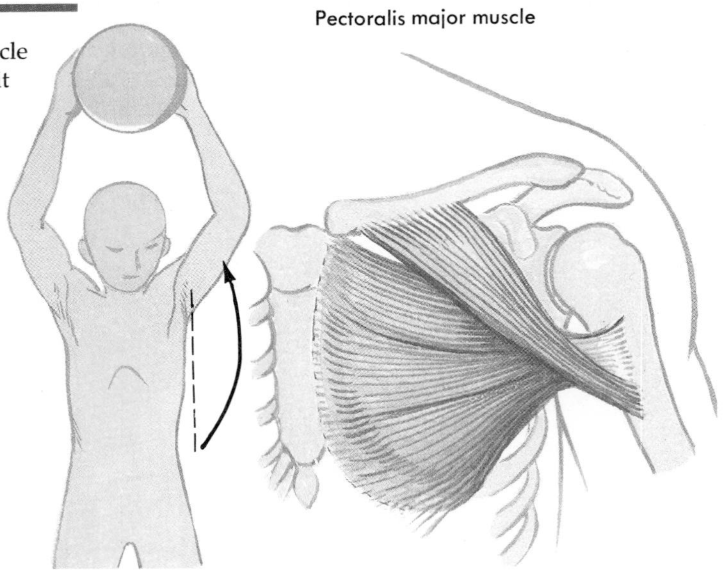

Figure 6-13 When the pectoralis major muscle (shown on the figure at the right) contracts, it flexes the upper arm at the shoulder joint (figure at the left). In what bone must the pectoralis major insert if it moves the upper arm?

Pectoralis major muscle

lay or lessen the damages of diminished activity. Later chapters describe many harmful effects of immobility.

TYPES OF MOVEMENTS PRODUCED BY SKELETAL MUSCLE CONTRACTIONS

Muscles can move some body parts in several directions and others in only two directions. As mentioned on p. 120, this depends largely on the shapes of the bones and the type of freely movable joint. Some of the movements we make most often are flexion, extension, abduction, adduction, rotation, supination, pronation, dorsiflexion, and plantar flexion.

Suppose that you bend an arm at the elbow, bend a leg at the knee, bend over forward at the waist, or bend your head forward in prayer. With each of these movements you will have flexed some part of the body—the lower arm, the lower leg, the trunk, and the head, respectively. Most flexions are movements commonly described as bending. One exception to this rule is flexion of the upper arm. Flexing the upper arm means moving it forward from the chest,

as shown in Figure 6-13. **Flexion** is defined as a movement that makes the angle between two bones at their joint smaller than it was at the beginning of the movement (Figure 6-16).

Extensions are actions opposite, or antagonistic, to flexions. Thus extensions are movements that make joint angles larger rather than smaller. Extensions are straightening or stretching movements rather than bending movements. Extending the lower arm, for example, is straightening it out at the elbow joint, as shown in Figure 6-17, *A*. Extending the upper arm is stretching it out and back from the chest as shown in Figure 6-14. What movement would you make to extend your lower leg? to extend your thigh or upper leg? Look at Figures 6-17, *B*, and 6-18, *A*, to check your answers. If you straighten your back to stand tall or stretch backward from your waist, are you flexing or extending your trunk?

Abduction means moving a part away from the midline of the body such as moving your arms out to the sides (Figure 6-15).

Adduction means moving a part toward the midline, such as bringing your arms down to your sides from an elevated position.

Figure 6-14   When the latissimus dorsi muscle (shown on the figure at the right) contracts, it extends the upper arm at the shoulder joint (figure at the left). In what bone must the latissimus dorsi insert if it moves the upper arm?

Latissimus dorsi muscle

Figure 6-15   When the deltoid muscle (shown on the figure at the right) contracts, it abducts the upper arm at the shoulder joint (figure at the left). In what bone must the deltoid insert to produce this movement?

Deltoid muscle

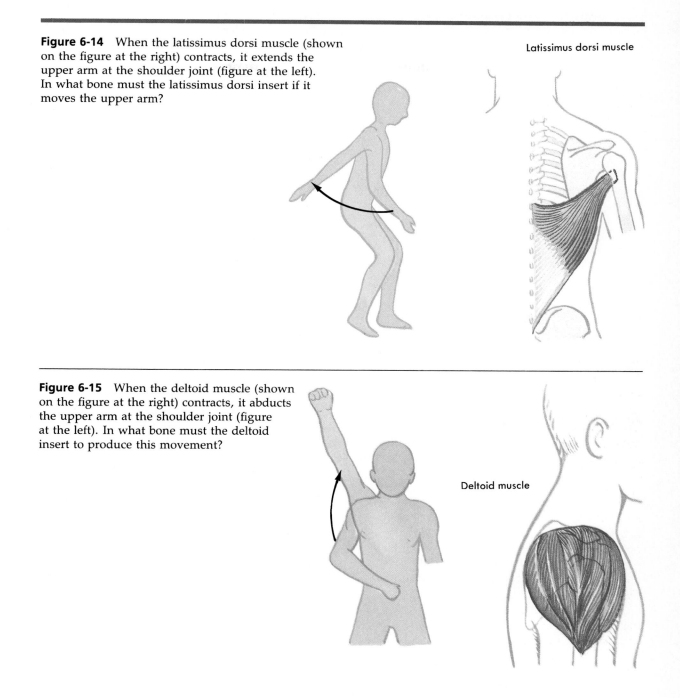

Figure 6-16 Flexion of the lower arm and lower leg. **A,** When the biceps brachii muscle (shown at the right) contracts, it flexes the lower arm at the elbow joint (shown at the left). **B,** When the hamstring muscles (shown at the right) contract, they flex the lower leg at the knee joint (shown at the left).

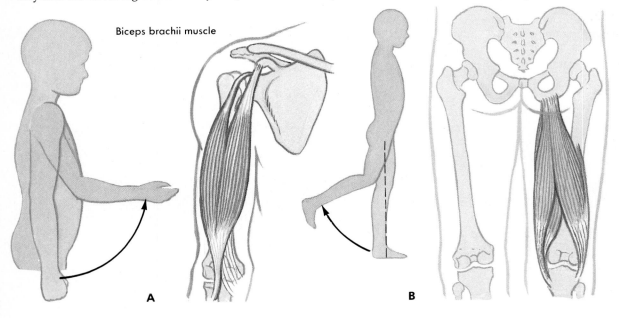

Biceps brachii muscle

Hamstring muscles

A

B

Figure 6-17 Extension of the lower arm and lower leg. **A,** When the triceps brachii muscle (shown at the right) contracts, it extends the lower arm at the elbow joint (shown at the left). **B,** When the rectus femoris muscle (part of the quadriceps femoris muscle group) (shown at the right) contracts, it extends the lower leg at the knee joint (shown at the left).

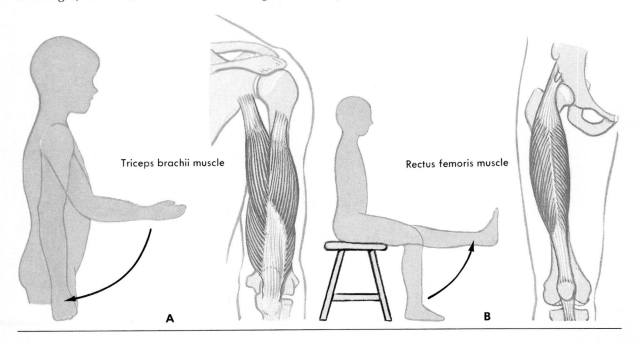

Triceps brachii muscle

Rectus femoris muscle

A

B

Figure 6-18   Flexion and extension of the thigh. **A,** When the gluteus maximus muscle (shown at the right) contracts, it extends the thigh at the hip joint (shown at the left). **B,** When the iliopsoas muscle (shown at the right) contracts and the femur serves as its insertion, it flexes the thigh at the hip joint (shown at the left).

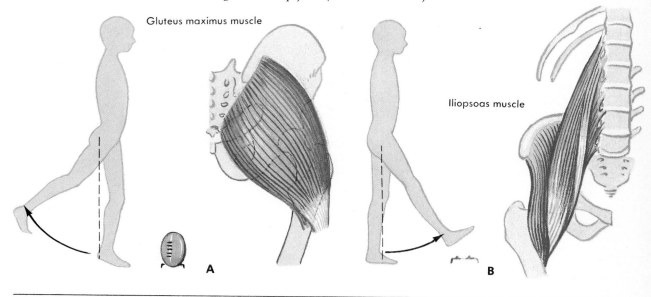

Gluteus maximus muscle

Iliopsoas muscle

A   B

Rotation is movement around a longitudinal axis. You rotate your head by moving your skull from side to side as in shaking your head "no."

Supination occurs when you turn the palm of your hand to the anterior position, as is the case in the anatomical position.

Pronation occurs when you turn the palm of your hand from an anterior to a posterior position.

Dorsiflexion is movement that elevates the top or dorsum of the foot and tilts it up. You dorsiflex your foot by standing on your heels.

Plantar flexion occurs when the bottom of the foot is directed downward so that you are in effect standing on your toes.

SPECIFIC FACTS ABOUT CERTAIN KEY MUSCLES

Flexors (that is, muscles that flex different parts of the body) produce many of the move-ments used in walking, sitting, swimming, typing, and a host of other activities. Extensors also function in these activities but perhaps play their most important role in maintaining upright posture. Study Table 6-1 and Figures 6-6 and 6-7 to learn the names of some of the prime flexors, extensors, abductors, and adductors of the body. Consult Table 6-2 to learn their origins and insertions. Keep in mind that muscles move bones, and the bones they move are their insertion bones.

Table 6-3 lists the names and functions of the muscles of facial expression illustrated in Figure 6-10.

MUSCLE DISORDERS

Perhaps the best known disease of muscles is **muscular dystrophy** (DIS-tro-fe). It is a long-lasting disease whose main characteristics are progressive wasting and weakening of the muscles.

Intramuscular injections

Many drugs are administered by intramuscular injection. If the amount to be injected is less than 5 milliliters, the deltoid muscle is often selected as the site of injection. Note in Figure *A* below that the needle is inserted into the muscle about two-fingers' breadth below the acromion process of the scapula and lateral to the tip of the acromion. If the amount of medication to be injected is more than 5 milliliters, the gluteal area shown in figure *B* below is often used. Injections are made into the gluteus medius muscle near the center of the upper outer quadrant, as shown in the illustration below. Another technique of locating the proper injection site is to draw an imaginary diagonal line from a point of reference on the back of the bony pelvis (posterior superior iliac spine) to the greater trochanter of the femur. The injection is given about three-fingers' breadth above and one third of the way down the line. It is important that the sciatic nerve and the superior gluteal blood vessels be avoided during the injection. Proper technique requires a knowledge of the underlying anatomy.

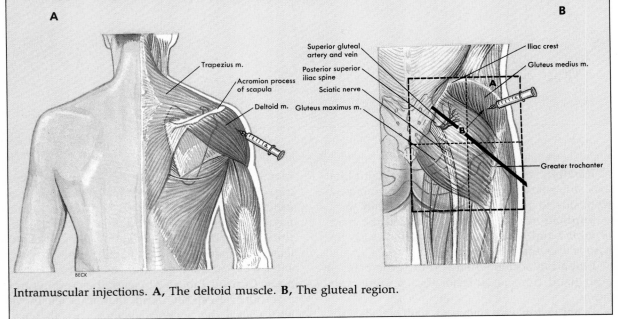

Intramuscular injections. **A,** The deltoid muscle. **B,** The gluteal region.

A person with **paralysis** (pah-RAL-i-sis) cannot contract his muscles when he wants to; his skeletal muscles do not respond to his will. This is not because there is anything wrong with his muscles but because there is disease or injury of his brain or spinal cord or nerves so that they cannot activate his muscles.

Muscle atrophy (AT-ro-fe) is muscle shrinkage, a decrease in muscle size caused by disuse. Many conditions can cause atrophy, such as having a leg in a cast, being a bed patient for a long time, or being paralyzed.

Muscle hypertrophy (hi-PER-tro-fe) is the opposite of atrophy; that is, it is an increase in size resulting from increased use. Skeletal muscles may hypertrophy as a result of exercise. The heart frequently hypertrophies from overwork.

Table 6-1 Muscles grouped according to function

Part moved	Flexors	Extensors	Abductors	Adductors
Upper arm	Pectoralis major	Latissimus dorsi	Deltoid	Pectoralis major and latissimus dorsi contracting together
Lower arm	Biceps brachii	Triceps brachii	None	None
Thigh	Iliopsoas	Gluteus maximus	Gluteus medius and minimus	Adductor group
Lower leg	Hamstrings	Quadriceps femoris group	None	None
Foot	Tibialis anterior	Gastrocnemius and soleus	Peroneus longus	Tibialis anterior
Trunk	Iliopsoas and rectus femoris	Erector spinae (sacrospinalis)	Psoas major and quadratus lumborum	Psoas major and quadratus lumborum

Table 6-2 Muscle functions, origins, and insertions

Muscle	Function	Insertion	Origin
Pectoralis major	Flexes *upper arm* Helps adduct *upper arm*	Humerus	Sternum Clavicle Upper rib cartilages
Latissimus dorsi	Extends *upper arm* Helps adduct *upper arm*	Humerus	Vertebrae Ilium
Deltoid	Abducts *upper arm*	Humerus	Clavicle Scapula
Biceps brachii	Flexes *lower arm*	Radius	Scapula
Triceps brachii	Extends *lower arm*	Ulna	Scapula Humerus
Iliopsoas	Flexes *trunk*	Ilium Vertebrae	Femur
Iliopsoas	Flexes *thigh*	Femur	Ilium Vertebrae
Gluteus maximus	Extends *thigh*	Femur	Ilium Sacrum Coccyx
Gluteus medius	Abducts *thigh*	Femur	Ilium
Gluteus minimus	Abducts *thigh*	Femur	Ilium
Adductors	Adducts *thigh*	Femur	Pubic bone
Hamstring group	Flexes *lower leg* Helps extend *thigh*	Tibia Fibula	Ischium Femur
Quadriceps femoris group, including rectus femoris	Extends *lower leg* Helps flex *thigh*	Tibia	Ilium Femur

Another disease characterized by muscular weakness is **myasthenia gravis.** The cause of this sometimes swiftly fatal disease is still not established. It appears, however, that the disease may involve a defect in the body's immune system, which leads to a lack of communication between nerves and muscles. The result is progressive muscular weakness and paralysis. In severe cases death may occur because of paralysis of the respiratory muscles.

Table 6-3 Muscles of facial expression and of mastication*

Muscle	Origin	Insertion	Function	Innervation
Muscles of facial expression				
Epicranius (occipi-tofrontalis)	Occipital bone	Tissues of eyebrows	Raises eyebrows, wrinkles forehead horizontally	Cranial nerve VII
Corrugator super-cilli	Frontal bone (superciliary ridge)	Skin of eyebrow	Wrinkles forehead vertically	Cranial nerve VII
Orbicularis oculi	Encircles eyelid		Closes eye	Cranial nerve VII
Procerus	Bridge of nose	Skin over epicranius	Narrows eye opening	Cranial nerve VII
Inferior labial depressor	Mandible	Skin of lower lip	Draws lower lip downwards	Cranial nerve VII
Mentalis	Incisive fossa of mandible	Skin of chin	Raises and protrudes lower lip	Cranial nerve VII
Nasalis	Maxilla	Ala of nose	Compresses nasal aperture	Cranial nerve VII
Zygomaticus major	Zygomatic bone	Angle of mouth	Laughing (elevates angle of mouth)	Cranial nerve VII
Zygomaticus minor	Zygomatic bone	Upper lip	Elevates upper lip	Cranial nerve VII
Orbicularis oris	Encircles mouth		Draws lips together	Cranial nerve VII
Platysma	Fascia of upper part of deltoid and pectoralis major	Mandible (lower border) Skin around corners of mouth	Draws corners of mouth down—pouting	Cranial nerve VII
Buccinator	Maxillae	Skin of sides of mouth	Permits smiling Blowing, as in playing a trumpet	Cranial nerve VII
Masseter	Zygomatic arch	Mandible (external surface)	Closes jaw	Cranial nerve V

*When trying to learn the origins and insertion of the muscles listed, refer frequently to illustrations of each muscle and to the skeleton. Also, when possible, feel each muscle on your own body.

OUTLINE SUMMARY

Muscle Tissue

A Structure and types
  1 Striated muscle; also called skeletal muscle or voluntary muscle
  2 Branching muscle; also called cardiac muscle
  3 Nonstriated muscle; also called smooth muscle, visceral muscle, or involuntary muscle

B Function—muscle cells specialize in contraction (shortening)

Skeletal Muscles (Organs)

A Structure
  1 Composed mainly of striated muscle cells (fibers) and connective tissue
  2 Most muscles extend from one bone across movable joint to another bone
  3 Parts of a skeletal muscle
    a Origin—attachment to relatively stationary bone
    b Insertion—attachment to bone that moves
    c Body—main part of muscle
  4 Muscles attach to bones by tendons—strong cords of fibrous connective tissue; some tendons enclosed in synovial-lined tubes, lubricated by synovial fluid; tubes called tendon sheaths; inflammation of tendon sheaths—tenosynovitis
  5 Bursas—small synovial-lined sacs containing small amount of synovial fluid; located between some tendons and underlying bones
  6 Microscopic structure—see Figures 6-3 and 6-4

B Functions
  1 Muscles produce movement—as muscle contracts, it pulls insertion bone nearer origin bone; movement occurs at joint between origin and insertion
    a Groups of muscles usually contract to produce a single movement

(1) Prime mover—muscle whose contraction is mainly responsible for producing a given movement
(2) Synergists—muscles whose contractions help the prime mover produce a given movement

  b Types of movements produced by skeletal muscle contractions
    (1) Flexion—making angle at joint smaller
    (2) Extension—making angle at joint larger
    (3) Abduction—moving a part away from midline
    (4) Adduction—moving a part toward midline
  c Normal muscle functioning depends on normal functioning of various other structures, notably nerves and joints
  d Not all muscle contractions produce movements; isometric contractions increase the tension in muscles without producing movements; isotonic contractions produce movements; tonic contractions, or muscle tone, produce no movement but increase firmness (tension) of muscles that maintain posture, that is, the position of parts of body; a twitch contraction is a jerky contraction in response to a single stimulus; tetanic contraction (tetanus) is a sustained contraction
  e Muscle stimulus—Figure 6-9
    (1) Subminimal stimulus—will not cause contraction
    (2) Minimal stimulus—will initiate contraction
    (3) Maximal stimulus—all fibers in muscle will contract
    (4) Supramaximal stimulus—strength of the stimulus above maximal; no additional effect on strength of contraction

    f Mobility (exercise) absolutely essential for healthy survival—necessary for such functions as maintaining muscle tone, normal functioning of heart and lungs, and normal structure and functioning of bones and joints

2 Posture
    a Posture means position of body parts
    b Good posture important for many reasons—for example, to prevent fatigue and bone and joint deformities
    c Skeletal muscles maintain posture by counteracting the pull of gravity—by maintaining tone (partial contraction)

3 Heat production
    a Body temperature must be maintained within a narrow range of normal (98.6° F)
    b Muscle cells produce body heat by catabolism of foodstuffs as they shorten; because muscle cells are highly active and numerous, they produce largest percent of body heat and constitute one of the most important parts of the mechanism that helps us maintain constant body temperature

Specific facts about certain key muscles
See Tables 6-1, 6-2, and 6-3

Muscle disorders
A Muscular dystrophy—progressive wasting and weakening of muscles
B Paralysis—loss of ability to produce voluntary movements
C Atrophy—decrease in muscle size
D Hypertrophy—increase in muscle size
E Myasthenia gravis—progressive muscular weakness and paralysis

NEW WORDS

A band	fatigue	muscular dystrophy	pronation
abduction	flexion	myasthenia gravis	rotation
actin	graded response	myofilaments	sarcomere
adduction	hypertrophy	myosin	sliding-filament theory
all or none	hypothermia	neuromuscular junction	stimulus
antagonist	I band	origin	supination
atrophy	insertion	oxygen debt	synergist
bursa	isometric	paralysis	tendon
bursitis	isotonic	plantar flexion	tenosynovitis
dorsiflexion	motor neuron	posture	tetanic contraction
extension	motor unit	prime mover	tonic contraction

CHAPTER TEST

1. Cardiac muscle:
    a. Is also called striated muscle
    b. Is composed of cells that branch frequently
    c. Lines many hollow internal organs
    d. All of the above are correct

2. The term *origin* refers to:
   a. The attachment of a muscle to a bone that does not move when contraction occurs
   b. Attachment of a muscle to a bone that moves when contraction occurs
   c. The body of a muscle
   d. None of the above are correct
3. A bursa:
   a. Is a small fluid-filled sac
   b. Is lined with a synovial membrane
   c. Serves as a flexible cushion
   d. All of the above are correct
4. A sarcomere:
   a. Is the basic functional or contractile unit of skeletal muscle
   b. Contains only actin myofilaments
   c. Contains only myosin myofilaments
   d. Is found only in smooth or involuntary muscle
5. The primary functions of the muscular system include:
   a. Generation of action potentials
   b. Support
   c. Heat production
   d. Hemopoiesis
6. The muscle mainly responsible for producing a particular movement is called a:
   a. Synergist   c. Antagonist
   b. Prime mover   d. Fixator
7. A motor unit consists of:
   a. Only a motor neuron
   b. A motor neuron together with the muscle cells it innervates
   c. Only contracting muscle cells
   d. None of the above are correct
8. The fact that muscle cells contract completely when subjected to an adequate stimulus is called the principle of:
   a. Graded response
   b. Threshold response
   c. All or none
   d. Maximal stimulation
9. Which of the following types of muscle contraction does not produce movement?
   a. Isotonic contraction
   b. Tetanic contraction
   c. Twitch contraction
   d. Isometric contraction
10. Which of the following terms refers to moving a part away from the midline of the body?
    a. Abduction   c. Supination
    b. Adduction   d. Flexion

11. Skeletal muscle is also called _____ muscle.
12. Many hollow internal organs and blood vessel walls contain _____ muscle.
13. Muscles are anchored firmly to bones by _____ .
14. Thick myofilaments are formed from a protein called _____ .
15. The basic functional or contractile unit of skeletal muscle is called a _____ .
16. The explanation of muscle contraction resulting from movement of thick and thin myofibrils toward one another is called the _____ _____ theory.
17. The muscles that assist prime movers in producing a particular movement are called _____ .
18. The biceps brachii and triceps brachii muscles produce opposite movements and are called _____ .
19. Skeletal muscle tone maintains _____ by counteracting the pull of gravity.
20. Depletion of oxygen in a muscle during vigorous and prolonged exercise is called _____ _____ .
21. A minimal or threshold stimulus will cause a muscle to contract _____ .
22. Walking and running are examples of _____ muscle contraction.
23. Standing on your toes is an example of _____ flexion.
24. The term _____ _____ is used to describe shrinkage or decrease in the size of a *muscle*.
25. The major flexor muscles of the lower leg are the _____ .

Select the most correct answer from Column B for each statement in Column A. (Only one answer is correct.)

Column A	Column B
26. _____ Contractile unit of muscle	a. Muscle tone
27. _____ Contains only thin myofilaments	b. Biceps brachii
	c. Masseter
28. _____ Maintains posture	d. Hypothermia
29. _____ Flexes lower arm	e. Sarcomere
30. _____ Extends lower leg	f. Quadriceps group
31. _____ Closes jaw	g. Pectoralis major
32. _____ Movable point of attachment	h. I band
	i. Gluteus maximus
33. _____ Fall in body temperature	j. Insertion
34. _____ Flexes upper arm	
35. _____ Extends thigh	

REVIEW QUESTIONS

1 Compare the three kinds of muscle tissue as to location, microscopic appearance, and nerve control.

2 Explain why skeletal muscle functions are so important. What are the general functions of skeletal muscle?

3 Explain the following terms and give an example of each:
   flexion
   extension
   abduction
   adduction

4 Explain how skeletal muscles, bones, and joints work together to produce movements.

5 Why can a spinal cord injury be followed by muscle paralysis?

6 Can a muscle contract very long if its blood supply is shut off? Give a reason for your answer.

7 What two kinds of muscle contractions do not produce movement?

8 The correct term to substitute in the expression "bending your knee" is (extending? flexing?) your lower leg.

9 What is the name of the main muscle that
   a Flexes the upper arm?
   b Flexes the lower arm?
   c Flexes the thigh?
   d Flexes the lower leg?
   e Extends the upper arm?
   f Extends the lower arm?
   g Extends the thigh?
   h Extends the lower leg?

10 Give the approximate location of each of the following muscles and tell what movement is produces.
   biceps brachii
   hamstrings
   deltoid
   pectoralis major
   quadriceps femoris group
   latissimus dorsi

11 Briefly describe changes that gradually take place in bones, joints, and muscles if a person habitually gets too little exercise.

12 Explain the relationship between strength of stimulus and strength of contraction.

13 What is meant by the term *staircase phenomenon?*

14 Discuss the microscopic structure of skeletal muscle tissue.

15 Explain the following terms:
   atrophy
   posture
   good posture
   hypertrophy
   latent period
   sarcomere
   muscle tone
   isometric contractions
   isotonic contractions
   tonic contractions
   treppe
   tetanus
   maximal stimulus

UNIT THREE

Systems That Provide Communication and Control

The Nervous System and Special Senses

CHAPTER OUTLINE

BOXED ESSAYS

Multiple sclerosis
Lumbar puncture
Herpes zoster or shingles

OBJECTIVES

After you have completed this chapter, you should be able to:

1 List the organs and divisions of the nervous system and describe the generalized functions of the system as a whole.

2 Identify and discuss the coverings and fluid spaces of the brain and spinal cord.

3 Identify the major types of cells in the nervous system and discuss the function of each.

4 Identify the anatomical and functional components of a three-neuron reflex arc. Compare and contrast the propagation of an action potential along a nerve fiber and across a synaptic cleft.

5 Identify the major anatomical components of the brain and spinal cord and briefly comment on the function of each.

6 Compare and contrast spinal and cranial nerves.

7 Discuss the anatomy and physiology of the major sense organs.

The normal body must accomplish a gigantic and enormously complex job—that of keeping itself alive and healthy. Each one of its billions of individual cells performs some activity that is a part of this big function. Control of the body's billions of cells is accomplished mainly by two communication systems, namely, the nervous system and the endocrine system. Both systems transmit information from one part of the body to another, but they do it in different ways. The nervous system transmits information very rapidly by means of nerve impulses conducted from one body area to another. The endocrine system transmits information more slowly by means of chemicals that are secreted by ductless glands into the bloodstream and are circulated from the glands to other parts of the body. Nerve impulses and hormones communicate information to body structures—increasing or decreasing their activities as needed for healthy survival. In other words, the communication systems of the body are also its control and integrating systems. They weld the body's hundreds of different functions into its one overall function of keeping itself alive and healthy.

Homeostasis refers to the balanced and controlled internal environment of the body that is basic to life itself. Homeostasis is possible only if our physiological control and integration systems are functioning properly. Our plan for this chapter is to name the organs and divisions of the nervous system, describe the coverings and fluid spaces of the brain and cord, relate basic information about the nervous system's special kinds of cells, and then discuss the various organs of the nervous system and the sense organs of the body. Chapter 8 considers the autonomic nervous system and Chapter 9 discusses the endocrine system.

ORGANS AND DIVISIONS OF NERVOUS SYSTEM

The organs of the nervous system as a whole include the brain and spinal cord, the numerous nerves of the body (Figure 7-1), the specialized sense organs such as the eyes and ears, and the microscopic sense organs found in the skin. The system as a whole consists of two principal divisions called the central nervous system and the peripheral nervous system. Because the brain and spinal cord occupy a midline, or central, location in the body, together they are called the **central nervous system** or **CNS**. Similarly, the usual designation for the nerves of the body is the **peripheral nervous system** or **PNS**. Use of the term *peripheral* is appropriate because nerves extend to outlying or peripheral parts of the body. A subdivision of the peripheral division of the nervous system is the so-called **autonomic nervous system,** which will be discussed in Chapter 8.

COVERINGS AND FLUID SPACES OF BRAIN AND SPINAL CORD

Nervous tissue is not a sturdy tissue. Even moderate pressure can kill nerve cells, so nature safeguards the chief organs made of this tissue—the spinal cord and the brain—by surrounding them with a tough, fluid-containing membrane called the **meninges** (me-NIN-jez). The meninges are then surrounded by bone. The *spinal meninges* form a tubelike covering around the spinal cord and line the bony vertebral foramen of those vertebrae that surround the cord. Look at Figure 7-2 and you can identify the three layers of the spinal meninges. They are the **dura mater** (DU-rah MA-ter), which is the tough outer layer that lines the vertebral canal, the **pia** (PI-ah) **mater,** which is the innermost layer covering the spinal cord itself, and the **arachnoid** (ah-RAK-noid) layer, which is the middle layer situated between the dura and the pia mater. This middle layer of the meninges, the arachnoid membrane, resembles a cobweb with fluid filling in its spaces. The word *arachnoid* means "cobweblike." It comes from *arachne,* the Greek word for spider. Arachne is the name of the girl who was changed into a spider by Athena because she boasted of the fineness of her weaving—at least, so an ancient Greek myth tells us.

The meninges that form the protective covering around the spinal cord also extend up and around the brain to enclose it completely (Figure

Figure 7-1 The central and peripheral divisions of the nervous systems. The central nervous system (CNS) consists of the brain and spinal cord. The peripheral nervous system (PNS) is composed of the cranial and spinal nerves.

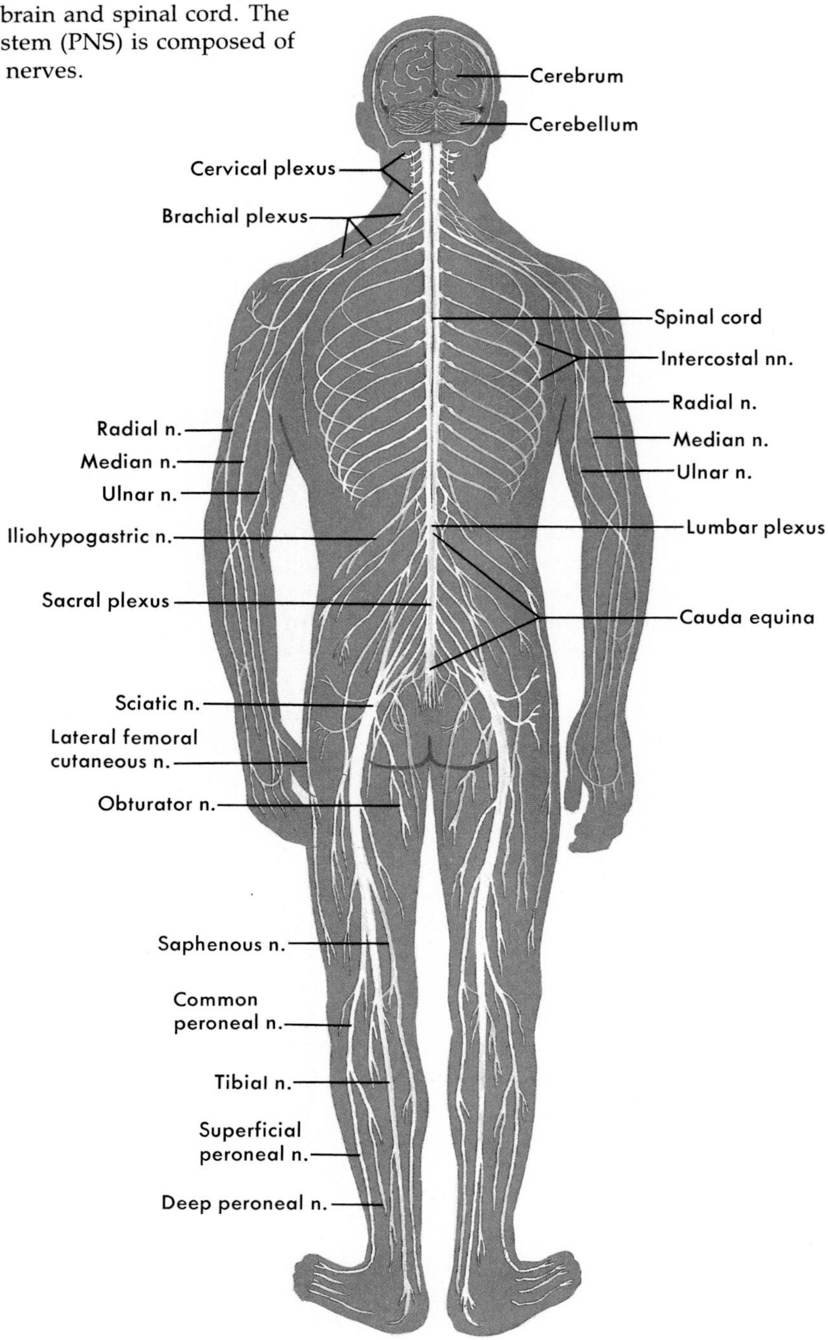

Figure 7-2   Spinal cord showing the meninges, the spinal nerves and their roots and ganglia, and the sympathetic trunk and ganglia.

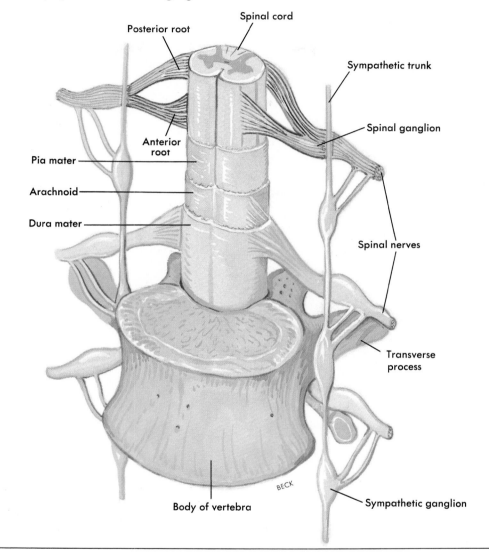

7-3). Fluid fills the arachnoid spaces of the brain meninges as well as those of the spinal cord. Also, as the next paragraph explains, there are fluid-filled spaces inside the brain called the cerebral ventricles.

In Figures 7-4 and 7-5 you can see what ir-regular shapes the ventricles of the brain have. These illustrations can also help you visualize the location of the ventricles if you remember two things—that these large spaces lie deep in-side the brain and that there are two lateral ven-tricles. One lies inside the right half of the ce-

Figure 7-3 Schematic drawing showing the structure of the meninges around the brain.

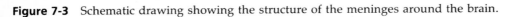

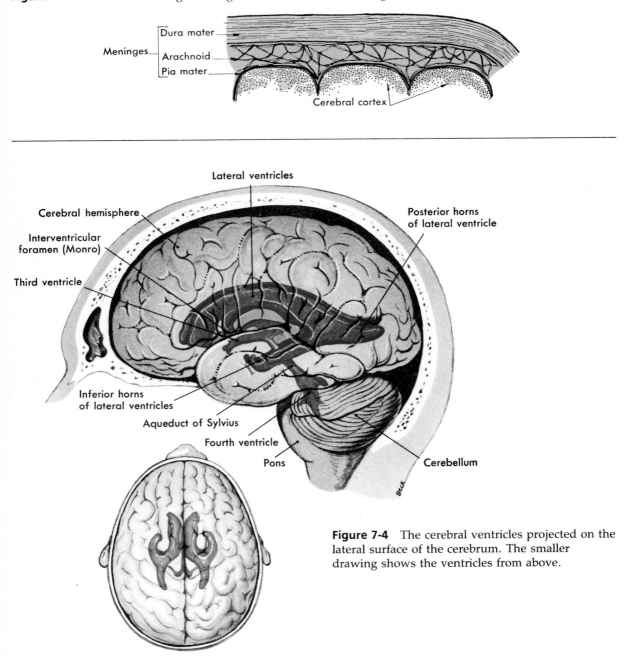

Figure 7-4 The cerebral ventricles projected on the lateral surface of the cerebrum. The smaller drawing shows the ventricles from above.

Figure 7-5 Hydrocephalus. Treatment by draining excess cerebrospinal fluid from the enlarged ventricles.

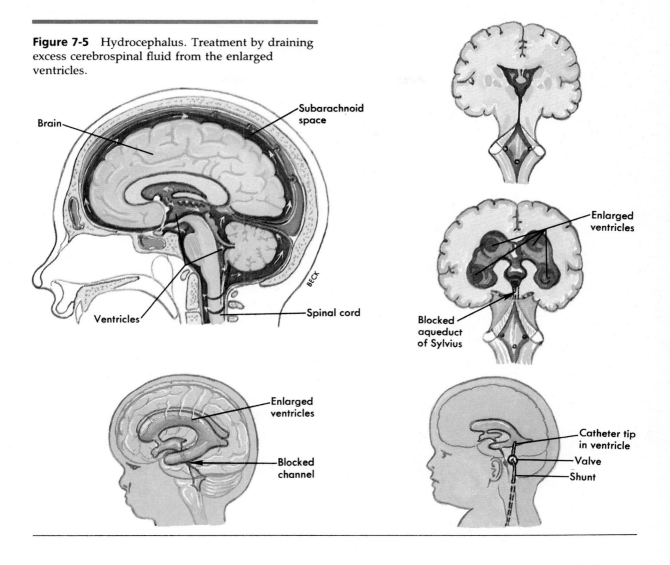

rebrum (the largest part of the human brain), and one lies inside the left half of the cerebrum.

Cerebrospinal (ser-e-bro-SPI-nal) fluid is one of the body's circulating fluids. It forms continually from fluid filtering out of the blood in a network of brain capillaries known as the choroid plexus (KO-roid PLEK-sus) and into the ventricles. From the lateral ventricles cerebrospinal fluid seeps into the third ventricle and flows down through the aqueduct of Sylvius (find this in Figures 7-4 and 7-5) into the fourth ventricle. From the fourth ventricle it moves into the small tubelike central canal of the cord and out into the subarachnoid spaces. Then it moves leisurely down and around the cord and up and around the brain (in the subarachnoid spaces of their meninges) and returns to the blood (in the veins of the brain).

Remembering that this fluid forms continually from blood, circulates, and is resorbed into blood can be useful. It can help you understand certain abnormalities you may see. Suppose a

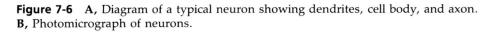

Figure 7-6 **A,** Diagram of a typical neuron showing dendrites, cell body, and axon. **B,** Photomicrograph of neurons.

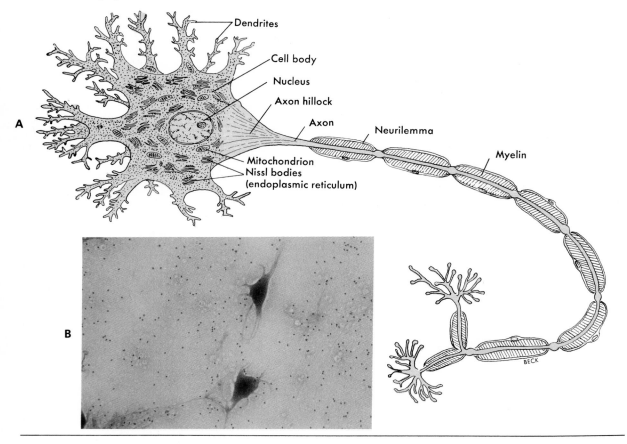

CELLS OF NERVOUS SYSTEM

The two general types of cells found in the nervous system (Figures 7-6 and 7-7) are called **neurons** (NU-rons) or nerve cells and **neuroglia** (nu-ROG-le-ah), which are specialized connective tissue cells.

NEURONS

Each neuron consists of three main parts: a main part called the **neuron cell body,** one or more branching projections called **dendrites** (DEN-drits), and one elongated projection known as an **axon.** Identify each of these parts on the neuron shown in Figure 7-6. Dendrites are the processes or projections that transmit

person has a brain tumor that presses on the aqueduct of Sylvius. This blocks the way for the return of cerebrospinal fluid to the blood. Since the fluid continues forming but cannot drain away, it accumulates in the ventricles or in the meninges (subarachnoid spaces around the brain). Other conditions besides brain tumors can cause an accumulation of cerebrospinal fluid in the ventricles. An example is **hydrocephalus** (hi-dro-SEF-ah-lus), or "water on the brain." One form of treatment involves surgical placement of a hollow tube or catheter through the blocked channel so that cerebrospinal fluid can drain into another location in the body (Figure 7-5).

Figure 7-7 Diagram of a nerve fiber and its coverings. This myelinated axon is located outside the central nervous system. Myelin is produced by the concentric layers of the Schwann cell. The neurilemma is the outer sheath of the Schwann cell and is indented by successive nodes of Ranvier.

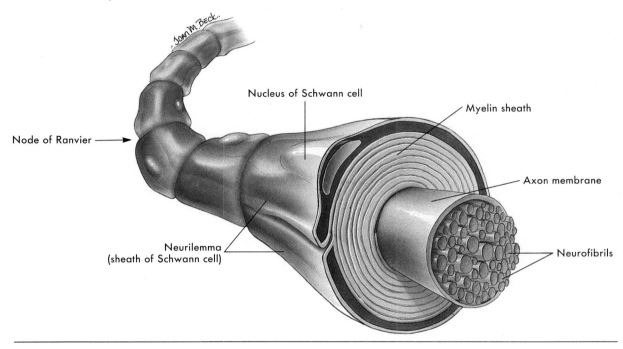

impulses to the neuron cell bodies, and axons are the processes that transmit impulses away from the neuron cell bodies.

There are three types of neurons classified according to the direction in which they transmit impulses: **sensory neurons, motor neurons,** and **interneurons.** *Sensory neurons* transmit impulses to the spinal cord and brain from all parts of the body. *Motor neurons* transmit impulses in the opposite direction—away from the brain and cord. Motor neurons do not conduct impulses to all parts of the body but only to two kinds of tissue—muscle and glandular epithelial tissue. *Interneurons* conduct impulses from sensory neurons to motor neurons. Sensory neurons are also called afferent neurons; motor neurons are called efferent neurons; interneurons are called central, or connecting, neurons.

The axon shown in Figure 7-6 is surrounded

by a segmented wrapping of a material called **myelin** (MI-e-lin). Myelin is a white, fatty substance formed by **Schwann cells** that wrap around some axons outside the central nervous system. Such fibers are called **myelinated fibers,** and they make up the so-called *white matter* of the nervous system. In Figure 7-7 one such axon has been enlarged to show additional detail. Note that the axon is filled with tiny **neurofibrils** (nu-ro-FI-brils) that extend through the axon from the cell body. The multilayered myelin-producing Schwann cells only cover axons outside the central nervous system. The outer cell membrane of a Schwann cell is called the **neurilemma** (nu-ri-LEM-mah). **Nodes of Ranvier** (rahn-ve-A) are the indentations that exist between adjacent Schwann cells. In a type of rapid nerve conduction along a myelinated axon, the impulse jumps from node to node by means of an impulse transmission called **saltatory** (SAL-tah-to-re) **conduction.** The transmission is rapid because the myelin acts as a good "insulator," covering the axon between nodes.

The fact that axons in the brain and cord have no neurilemma is of great clinical significance. The neurilemma plays an essential part in the regeneration of cut and injured axons. Therefore axons in the brain and cord do not regenerate, but those in nerves do.

NEUROGLIA

Neuroglias do not specialize in the function of transmitting impulses. Instead, they are special types of connective tissue cells. Their name is appropriate because it is derived from the Greek word *glia* meaning "glue." One function of neuroglial cells is literally to hold the functioning neurons together and protect them. An important reason for discussing neuroglias is that one of the commonest types of brain tumor—called the **glioma** (gli-O-mah)—develops from them. Neuroglias vary in size and shape (Figure 7-8). Some are relatively large cells that look somewhat like stars because of the many threadlike extensions that jut out from their sur-faces. These neuroglia cells are called **astrocytes** (AS-tro-cytes), a word that means "star cells"; see Figure 7-8, *B.* Their threadlike branches attach both to neurons and small blood vessels, holding these structures close to each other. **Microglia** (mi-KROG-le-ah) are smaller cells than astrocytes. Usually they remain stationary, but in inflamed or degenerating brain tissue they enlarge, move about, and act as microbe-eating scavengers. They surround the microbes, draw them into their cytoplasm, and digest them. **Phagocytosis** (fag-o-si-TO-sis) is the scientific name for this important cellular process. **Oligodendroglia** (ol-i-go-den-DROG-le-ah) help to hold nerve fibers together and also serve another and probably more important function—they produce the fatty myelin sheath that envelops nerve fibers located in the brain and cord. It is the oligodendroglia, not the Schwann cells, that form the myelin sheaths around central nerve fibers, that is, axons located in the brain and cord.

Because of the high fat content of myelin, bundles of myelinated fibers have a creamy white color, and they make up the white matter of the nervous system. White matter in the

---

Multiple sclerosis (MS)

A number of diseases are associated with disorders of the oligodendroglia. Because these neuroglia cells are involved in myelin formation, the diseases as a group are called **myelin disorders.** The most common primary disease of the central nervous system is a myelin disorder called **multiple sclerosis** or **MS.** It is characterized by myelin loss and destruction accompanied by varying degrees of oligodendroglial cell injury and death. The end result is demyelination throughout the white matter of the central nervous system. Hard plaquelike lesions replace the destroyed myelin and affected areas are invaded by inflammatory cells. As the myelin surrounding axons is lost, nerve conduction is impaired and weakness, incoordination, visual impairment, and speech disturbances occur. Although the disease occurs in both sexes and all age groups, it is most common in women between 20 and 40 years.

The cause of multiple sclerosis is thought to be related to autoimmunity and to viral-type infections in some individuals. MS is characteristically relapsing and chronic in nature, but some cases of acute and unremitting disease have been reported. In most instances the disease is prolonged, with remissions and relapses occurring over a period of many years. There is no known cure.

Figure 7-8 **A,** Neuroglia, the special connecting and supporting cells of the brain and spinal cord, are composed of astrocytes, microglia, and oligodendroglia. **B,** Photomicrograph of an astrocyte.

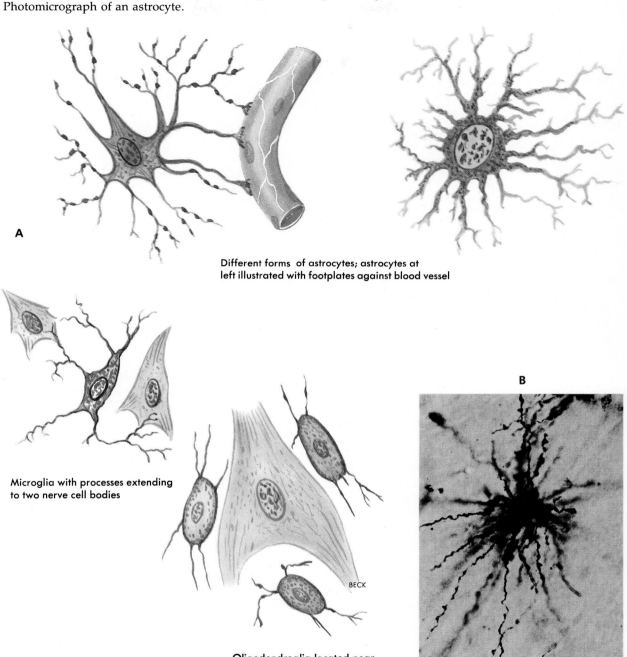

A

Different forms of astrocytes; astrocytes at left illustrated with footplates against blood vessel

Microglia with processes extending to two nerve cell bodies

B

BECK

Oligodendroglia located near a nerve cell body

Figure 7-9 **A,** Cross section of nerve trunk. **B,** Photomicrograph showing cross section of a nerve.

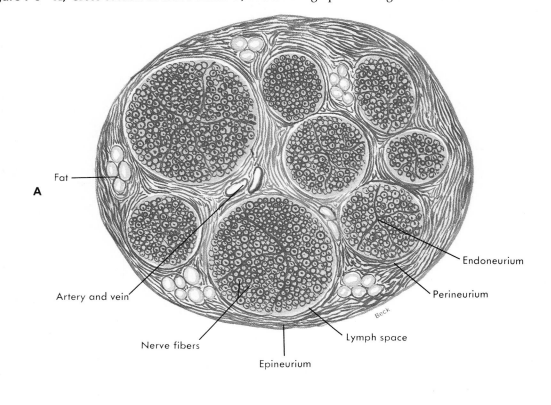

Fat

A

Artery and vein

Nerve fibers

Epineurium

Lymph space

Endoneurium

Perineurium

Beck

REFLEX ARCS

brain and spinal cord is composed of **tracts.** A *tract* is a bundle of myelinated axons. White matter found outside the brain and cord consists of **nerves.** Nerves are cordlike structures composed of bundles of myelinated axons. A number of the important tracts of myelinated axons in the white matter of the spinal cord are shown in Figure 7-14. In addition to "tract," the terms *nerve trunk* and *nerve* may also cause confusion. A nerve trunk (Figure 7-9, *A*) is a grouping of several bundles of nerves, each containing many axons. A nerve (Figure 7-9, *B*) is a single group of nerve fibers (axons in this illustration) surrounded by a connective tissue sheath.

REFLEX ARCS

Every moment of our lives, nerve impulses speed over neurons to and from our spinal cords and brains. If all impulse conduction ceases, life itself ceases. Only neurons can provide the rapid communication between the body's billions of cells that is necessary for maintaining life. Chemical "messages" are the only other kind of communications the body can send, and these travel much more slowly than impulses. They can move from one part of the body to another only by way of the circulating blood. Compared with impulse conduction, circulation is a very slow process indeed.

Nerve impulses can travel over literally tril-

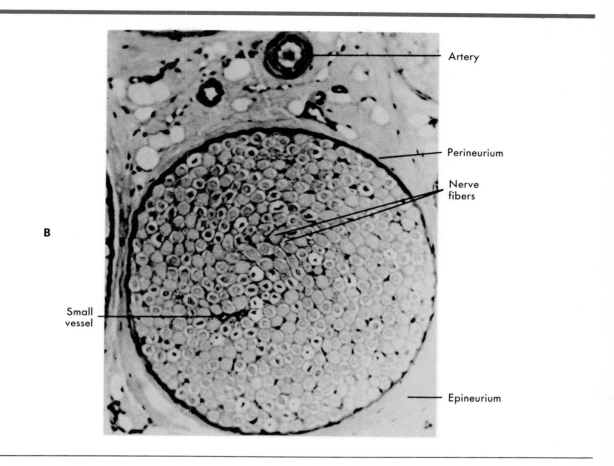

B

Artery

Perineurium

Nerve
fibers

Small
vessel

Epineurium

lions of routes—routes made up of neurons, because neurons are the cells that conduct impulses. Hence the routes traveled by nerve impulses are sometimes spoken of as neuron pathways. Their scientific name, however, is **reflex arcs.** The simplest kind of reflex arc is a *two-neuron arc,* so called because it consists of only two types of neurons—sensory neurons and motor neurons. *Three-neuron arcs* are the next simplest kind. They, of course, consist of all three kinds of neurons—sensory neurons, interneurons, and motor neurons. Reflex arcs are like one-way streets: they allow impulse conduction in only one direction. The next paragraph describes this direction in detail. Look frequently at Figure 7-10 as you read it.

Impulse conduction normally starts in receptors. **Receptors** are the beginnings of dendrites of sensory neurons. In Figure 7-10 the sensory receptors are located in the patellar tendon of the knee joint. Receptors are often located at some distance from the cord—in other tendons, in skin, and in mucous membranes, for example. In the example illustrated in Figure 7-10 receptors in the patellar tendon are being stimulated by a blow from a neurological hammer used by physicians to elicit a reflex during a physical examination. From receptors, impulses travel the full length of the sensory neuron's dendrite, through its cell body and axon, to the branching brushlike ends of the axon. The ends of the sensory neuron's axon contact interneurons. Ac-

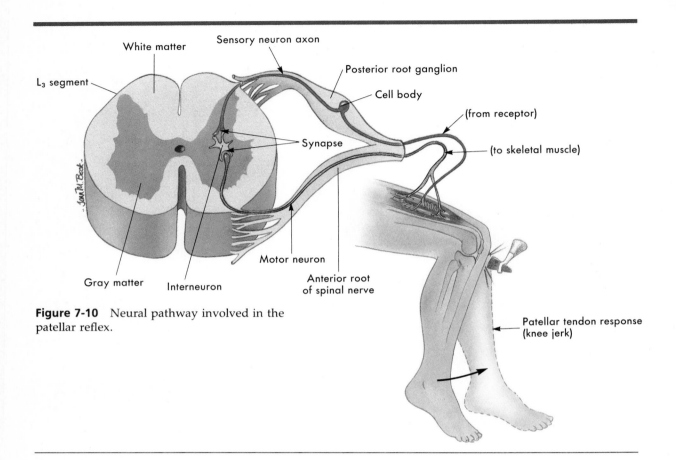

Figure 7-10 Neural pathway involved in the patellar reflex.

tually, a microscopic space separates the axon endings of one neuron from the dendrites of another neuron. This space is called a **synapse.** After crossing the synapse the impulses continue along the dendrites, cell bodies, and axons of the interneurons. Then they cross another synapse. Finally, they travel over the dendrites, cell bodies, and axons of motor neurons to a structure called an *effector* (because it "puts into effect" the message brought to it by nerve impulses). Effectors are either muscles or glands. When impulses reach skeletal muscle cells, they cause the cells to contract. This brings us to another definition. The response to impulse conduction over reflex arcs—in this instance muscle contraction—is called a **reflex.** In short, impulse conduction by a reflex arc causes a reflex to occur. Muscle contractions and gland secretion are the only two kinds of reflexes.

The term *reflex center* means the center of a reflex arc. The first part of a reflex arc consists of sensory neurons, and the last part consists of motor neurons. In spinal cord arcs the sensory neurons conduct impulses to the cord, and the motor neurons conduct them away from the cord and out to muscles. The reflex centers of all spinal cord arcs, then, lie in spinal cord gray matter. In three-neuron arcs the reflex centers consist of interneurons. In two-neuron arcs they are simply the synapses between sensory and motor neurons. We might define a reflex center as the place in a reflex arc where incoming impulses become outgoing impulses.

Figure 7-10 reveals a number of important facts. Note, for example, that the dendrite of the sensory neuron in a reflex arc can be quite long. In the example shown, it extends from the patellar tendon and ends at its neuron cell body

Figure 7-11 **A,** Mechanism of transmission of a nerve impulse. **B,** Saltatory conduction.

A

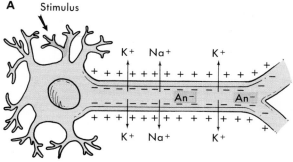

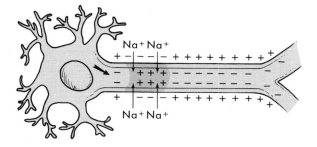

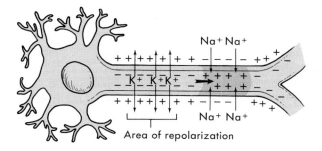

Area of repolarization

located in a structure called the **posterior root ganglion** (GANG-le-on). This ganglion is located near the spinal cord. Note that it is a small swelling on the posterior root of a spinal nerve. (Because of this location, spinal ganglia are also called posterior root ganglia.) Each spinal ganglion contains not one sensory neuron cell body as shown in Figure 7-10 but hundreds of them. Now turn your attention to the interneuron in this figure. All interneurons lie entirely within the gray matter of the central nervous system, that is, the gray matter of either the brain or spinal cord. Gray matter forms the H-shaped inner core of the spinal cord.

Identify the motor neuron in Figure 7-10. Observe that its dendrites and cell body are located in the cord's gray matter. The axon of this motor neuron runs through the anterior root of the spinal nerve and terminates in the quadriceps muscle group of the thigh. Contraction of this muscle group following stimulation of receptors in the patellar tendon produces the familiar "knee jerk."

NERVE IMPULSES

What are nerve impulses? Here is one widely accepted definition: a nerve impulse is a self-propagating wave of electrical negativity that travels along the surface of a neuron's cytoplasmic membrane (Figure 7-11). You might visualize this as a tiny spark sizzling its way along a fuse. Nerve impulses do not continually race along every nerve cell's surface. First they have to be initiated by a stimulus, a change in the neuron's environment. Pressure, temperature, and chemical changes are the usual stimuli.

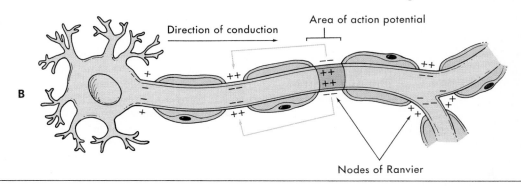

Direction of conduction

Area of action potential

B

Nodes of Ranvier

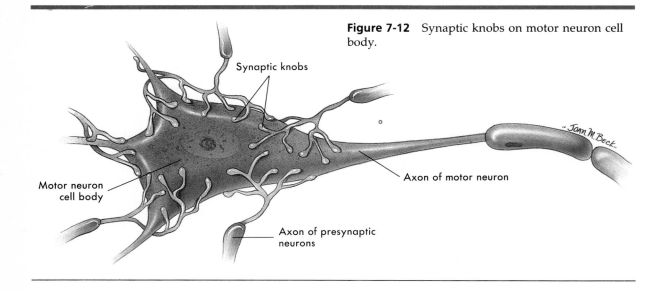

Figure 7-12 Synaptic knobs on motor neuron cell body.

Synaptic knobs

Motor neuron cell body

Axon of motor neuron

Axon of presynaptic neurons

When an adequate stimulus acts on a neuron, it greatly increases the permeability of the stimulated point of its membrane to sodium ions. These positively charged ions therefore rush through the stimulated point into the interior of the neuron. The inward movement of positive ions leaves a slight excess of negative ions outside. A point of electrical negativity has been created on the neuron's surface. A nerve impulse has begun—a self-propagating wave of electrical negativity that speeds point by point along the entire length of the neuron's surface. A nerve impulse that "jumps" from one node of Ranvier to the next along a myelinated axon constitutes a type of propagation called **saltatory conduction** (Figure 7-11, *B*).

Look again at Figure 7-10 and note the two labels for synapse. The first synapse lies between the sensory neuron's axon terminals and the interneuron's dendrites. The second synapse lies between the interneuron's axon terminals and the motor neuron's dendrites. Conduction across synapses is an important part of the nerve conduction process. By definition, a synapse is the place where impulses are transmitted from one neuron, called the **presynaptic neuron,** to another neuron, called the **postsynaptic neuron.** Three structures make up a synapse: a synaptic knob, a synaptic cleft, and the

plasma membrane of a postsynaptic neuron. A **synaptic knob** is a tiny bulge at the end of a terminal branch of a presynaptic neuron's axon (Figures 7-12 and 7-13). Each synaptic knob contains numerous small sacs or vesicles. Each vesicle contains a very small quantity of a chemical compound called a **neurotransmitter.** A **synaptic cleft** is the space between a synaptic knob and the plasma membrane of a postsynaptic neuron. It is an incredibly narrow space—only about one millionth of an inch in width. Identify the synaptic cleft in Figure 7-13. The plasma membrane of a *postsynaptic neuron* has protein molecules embedded in it opposite each synaptic knob. These serve as receptors to which neurotransmitter molecules bind.

Once an impulse is generated and conduction by postsynaptic neurons is initiated, neurotransmitter activity is rapidly terminated. Either one or both of two mechanisms cause this. Some neurotransmitter molecules diffuse out of the synaptic cleft back into synaptic knobs. Other neurotransmitter molecules are metabolized into inactive compounds by specific enzymes.

Neurotransmitters are chemicals by which neurons talk to one another. As noted above, at trillions of synapses in the central nervous system, presynaptic neurons release neurotransmitters that act to assist, stimulate, or inhibit

Figure 7-13   Diagram of a synapse showing its components: a synaptic knob or axon terminal of a presynaptic neuron, the plasma membrane of a postsynaptic neuron, and a synaptic cleft. On the arrival of an action potential at a synaptic knob, neurotransmitter molecules are released from vesicles in the knob into the synaptic cleft. The combining of neurotransmitter molecules with receptor molecules in the plasma membrane of the postsynaptic neuron initiates impulse conduction by it.

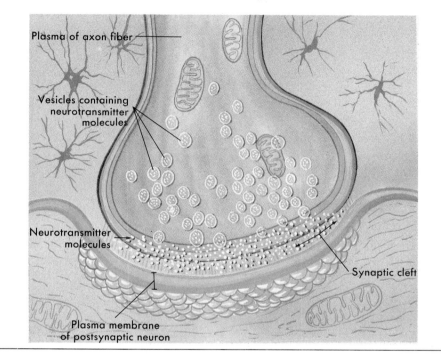

Plasma of axon fiber

Vesicles containing neurotransmitter molecules

Neurotransmitter molecules

Synaptic cleft

Plasma membrane of postsynaptic neuron

postsynaptic neurons. Established as neurotransmitters are at least 30 different compounds. They are not distributed diffusely or at random through the cord and brain. Instead, specific neurotransmitters are localized in discrete groups of neurons and released in specific pathways.

For example, the substance named **acetylcholine** (as-e-til-KO-lean) is known to be released at some of the synapses in the spinal cord and at myoneural (muscle-nerve) junctions. Other well-known neurotransmitters include **norepinephrine** (nor-ep-i-NEF-rin), **dopamine** (DO-pah-men), and **serotonin** (ser-o-TO-nin). They belong to a group of compounds called **catecholamines** (kat-e-kol-AM-ines), which may play a role in sleep, motor function, mood, and pleasure recognition.

Two morphinelike neurotransmitters called **endorphins** (en-DOR-fins) and **enkephalins** (en-KEF-a-lins) are released at various cord and brain synapses in the pain conduction pathway. These neurotransmitters inhibit conduction of pain impulses. They serve the body as its own natural pain killers.

SPINAL CORD
STRUCTURE

If you are of average height, your spinal cord is about 17 or 18 inches long. It lies inside the spinal column in the spinal cavity and extends from the occipital bone down to the bottom of the first lumbar vertebra. Place your hands on your hips, and they will line up with your fourth lumbar vertebra. Your spinal cord ends just

Figure 7-14 Cross section of the spinal cord showing a number of the tracts that carry impulses to or from the brain.

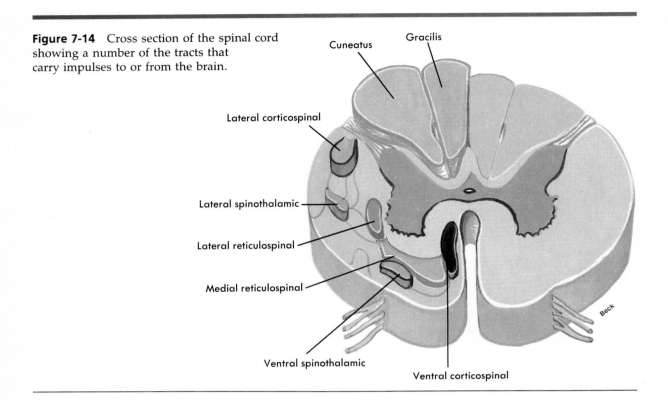

Cuneatus

Gracilis

Lateral corticospinal

Lateral spinothalamic

Lateral reticulospinal

Medial reticulospinal

Beck

Ventral spinothalamic

Ventral corticospinal

Lumbar puncture

The extension of the meninges beyond the cord is convenient for performing lumbar punctures without danger of injuring the spinal cord. A **lumbar puncture** is the withdrawal of some of the cerebrospinal fluid from the subarachnoid space in the lumbar region of the spinal cord. The physician inserts a needle just above or below the fourth lumbar vertebra, knowing that the spinal cord ends an inch or more above that level. The fourth lumbar vertebra can be easily located because it lies on a line with the iliac crest. Placing a patient on the side and arching the back by drawing the knees and chest together separates the vertebrae sufficiently to introduce the needle.

above this level. The spinal meninges, however, do not end here. They continue down almost to the end of the spinal column—a useful fact for a physician to know.

Look now at Figure 7-14. Notice the H-shaped core of the spinal cord. It consists of gray matter and so is composed mainly of dendrites and cell bodies of neurons. Columns of white matter form the outer portion of the cord, and bundles of myelinated nerve fibers—the **spinal tracts**—make up the white columns.

Spinal cord tracts provide two-way conduction paths to and from the brain. **Ascending tracts** conduct impulses up the cord to the brain. **Descending tracts** conduct impulses down the cord from the brain. Tracts are functional organizations in that all the axons that compose one tract serve one general function. For instance, fibers of the spinothalamic tracts serve a sensory function. They transmit impulses that produce our sensations of crude touch, pain, and temperature. In addition to the lateral and ventral spinothalamic tracts, other ascending tracts shown in Figure 7-14 include the gracilis and cuneate tracts, which transmit sensations of touch and pressure up to the brain. Descending tracts are represented by the lateral and ventral corticospinal tracts, which transmit impulses controlling many voluntary movements, and the lateral and medial reticulospinal tracts, which also control skeletal muscle-type activity.

FUNCTIONS

To try to understand spinal cord functions, let us start by thinking about a hotel telephone switchboard. Suppose a guest in Room 108 calls the switchboard operator and asks for Room 520, and in a second or so someone in that room answers. Very briefly, three events took place: a message traveled into the switchboard, a connection was made in the switchboard, and a message traveled out from the switchboard. The telephone switchboard provided the connection that made possible the completion of this call. Or we might say that it transferred the incoming call to an outgoing line. The spinal cord functions similarly. It contains the centers for thousands and thousands of reflex arcs. Look back at Figure 7-10. The interneuron shown there is an example of a spinal cord reflex center. It switches, or transfers, incoming sensory impulses to outgoing motor impulses, thereby making it possible for a reflex to occur. Reflexes that result from conduction over arcs whose centers lie in the cord are called spinal cord reflexes. Two common kinds of spinal cord reflexes are the withdrawal reflexes and the jerk reflexes. An example of a withdrawal reflex is pulling one's hand away from a hot surface. The familiar knee jerk is an example of a jerk reflex.

In addition to functioning as the primary reflex center of the body, the spinal cord tracts, as noted above, function to carry impulses to and from the brain. Sensory impulses travel up to the brain in ascending tracts and motor impulses travel down from the brain in descending tracts. Therefore, if an injury cuts the cord all the way across, impulses can no longer travel to the brain from any part of the body located below the injury, nor can they travel from the brain down

Figure 7-15 Dissection of the cervical segment of the spinal cord showing emerging cervical nerves. The cord is viewed from behind (posterior aspect).

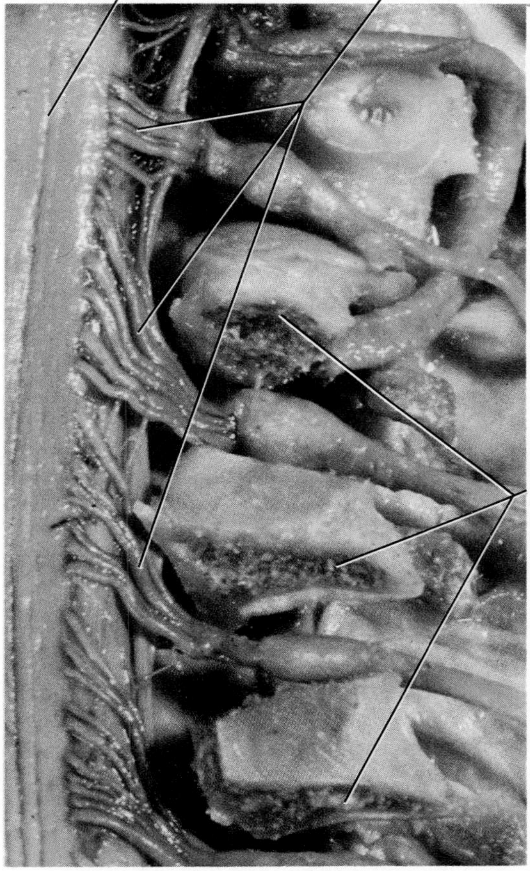

to these parts. In short, this kind of spinal cord injury produces both a loss of sensation, which is called **anesthesia** (an-es-THEE-ze-ah), and a loss of the ability to make voluntary movements, which is called **paralysis** (pah-RAL-i-sis).

SPINAL NERVES
STRUCTURE

Thirty-one pairs of nerves attach to the spinal cord in the following order: eight pairs attach to the cervical segments, twelve pairs attach to the thoracic segments, five pairs attach to the lumbar segments, five pairs attach to the sacrospinal segments, and one pair attaches to the coccygeal segment. Unlike cranial nerves, spinal nerves have no special names; instead, a letter and number identify each one. C1, for example, indicates the pair of spinal nerves attached to the first segment of the cervical part of the cord, T8 indicates those attached to the eighth segment of the thoracic part of the cord, and so on. In Figure 7-15 the cervical area of the spine has been dissected to show the emerging spinal nerves in that area.

FUNCTIONS

Spinal nerves conduct impulses between the spinal cord and the parts of the body not supplied by cranial nerves. The spinal nerve shown in Figure 7-10 contains, as do all spinal nerves, both sensory and motor fibers. Spinal nerves therefore function to make possible both sensations and movements. A disease or injury that prevents conduction by a spinal nerve therefore results in both a loss of feeling and a loss of movement in the part supplied by that nerve.

Detailed mapping of the skin surface reveals a close relationship between the source on the cord of each spinal nerve and the level of the body it innervates (Figures 7-16 and 7-17). Knowledge of the segmental arrangement of spinal nerves has proved useful to physicians. For instance, a neurologist can identify the site of spinal cord or nerve abnormality from the area of the body that is insensitive to a pinprick. Skin surface areas supplied by a single spinal nerve are called **dermatomes** (DER-mah-tomes). A dermatome ''map'' of the body is shown in Figures 7-16 and 7-17.

You may have known someone who suffered with **sciatica.** This is an inflammation of the spinal nerve branch called the sciatic nerve—the largest nerve in the body, incidentally. **Shingles** is another painful and fairly common spinal nerve inflammation. It is a viral infection that almost always affects the skin of a single dermatome (Figure 7-18.)

Herpes zoster or shingles

Herpes zoster or **shingles** is a unique viral infection that almost always affects the skin of a single dermatome. It is caused by a varicella zoster virus of chicken pox. About 3% of the population will suffer from shingles at some time in their lives. In most cases the disease results from reactivation of the varicella virus. The virus most likely traveled through a cutaneous nerve and remained dormant in a dorsal root ganglion for years after an episode of chicken pox. If the body's immunological protective mechanism becomes diminished in the elderly, or following stress, or in individuals undergoing radiation therapy or taking immunosuppressive drugs, the virus may reactivate. If this occurs, the virus will travel over the sensory nerve to the skin of a single dermatome. The result is a painful eruption of red swollen plaques or vesicles that eventually rupture and crust before clearing in 2 to 3 weeks. In severe cases extensive inflammation, hemorrhagic blisters, and secondary bacterial infection may lead to permanent scarring. In most cases of shingles, the eruption of vesicles is preceded by 4 to 5 days of preeruptive pain, burning, and itching in the affected dermatome. Unfortunately, an attack of herpes zoster does not confer lasting immunity. Many individuals suffer three or four episodes in a lifetime.

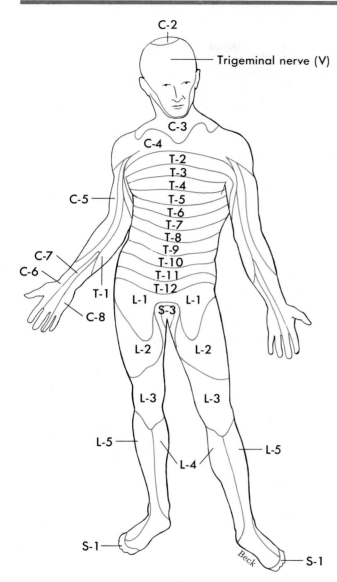

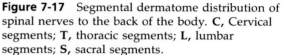

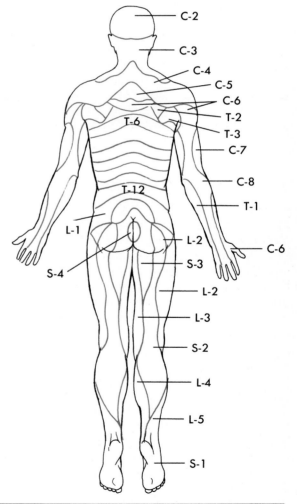

Figure 7-17   Segmental dermatome distribution of spinal nerves to the back of the body. **C,** Cervical segments; **T,** thoracic segments; **L,** lumbar segments; **S,** sacral segments.

Figure 7-16   Segmental dermatome distribution of spinal nerves to the front of the body. **C,** Cervical segments; **T,** thoracic segments; **L,** lumbar segments; **S,** sacral segments.

Figure 7-18 Herpes zoster or shingles. Dermatome involvement in a 13-year-old boy.

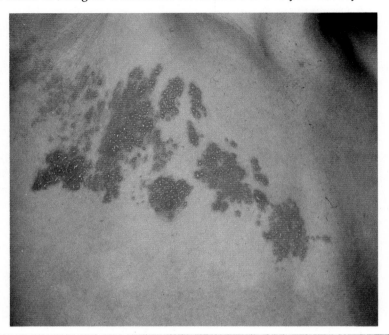

DIVISIONS OF THE BRAIN

The brain, one of our largest organs, consists of the following major divisions, named in ascending order beginning with the lowest part: brain stem (medulla, pons, and midbrain), cerebellum, diencephalon (hypothalamus and thalamus), and cerebrum. Observe in Figure 7-19 the location and relative sizes of the medulla, pons, cerebellum, and cerebrum. Identify the midbrain in Figure 7-20.

BRAIN STEM

The lowest part of the brain stem is formed by the medulla. Immediately above the medulla lies the pons and above that the midbrain. Together these three structures are called the brain stem. Look at Figure 7-20 and you can see why.

The **medulla** is a bulb-shaped extension of the spinal cord. It lies just inside the cranial cavity above the large hole in the occipital bone called the foramen magnum. Like the spinal cord, the medulla consists of both gray and white matter. But their arrangement differs in the two organs. In the medulla, bits of gray matter mix closely and intricately with white matter to form what is called the reticular formation (reticular means netlike). In the cord, gray and white matter do not intermingle; gray matter forms the interior core of the cord and white matter surrounds it. The **pons** and **midbrain,** like the medulla, consist of white matter and scattered bits of gray matter.

All three parts of the brain stem function as two-way conduction paths. Sensory fibers conduct impulses up from the cord to other parts of the brain, and motor fibers conduct impulses down from the brain to the cord. In addition, many important reflex centers lie in the brain stem. The cardiac, respiratory, and vasomotor centers (collectively called the vital centers), for example, are located in the medulla. Impulses from these centers control the heartbeat, respirations, and blood vessel diameter.

Figure 7-19 Lateral view of the brain.

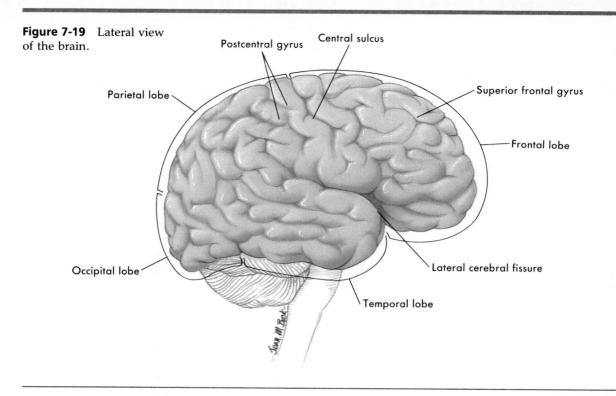

Postcentral gyrus

Central sulcus

Parietal lobe

Superior frontal gyrus

Frontal lobe

Occipital lobe

Lateral cerebral fissure

Temporal lobe

Figure 7-20 Sagittal section of the brain. (Note position of midbrain.)

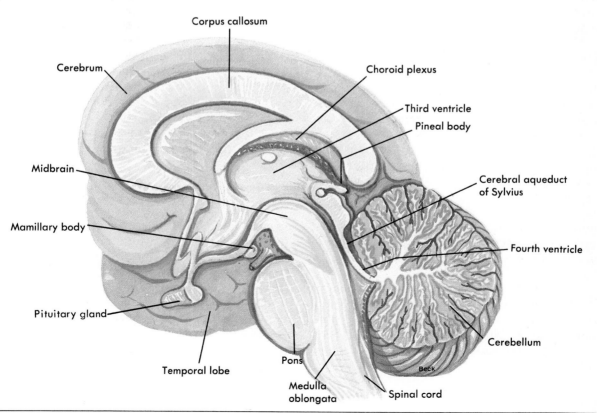

Corpus callosum

Cerebrum

Choroid plexus

Third ventricle

Pineal body

Midbrain

Cerebral aqueduct of Sylvius

Mamillary body

Fourth ventricle

Pituitary gland

Cerebellum

Temporal lobe

Pons

Medulla oblongata

Spinal cord

DIENCEPHALON

The **diencephalon** (di-en-SEF-ah-lon) is the small but important part of the brain located between the midbrain below and the cerebrum above. It consists of two major structures, namely, the hypothalamus and the thalamus.

The **hypothalamus** (hi-po-THAL-ah-mus), as its name suggests, is located below the thalamus. Major parts of the hypothalamus are the posterior pituitary gland, the stalk that attaches it to the undersurface of the brain, and two pairs of clusters of neuron cell bodies (named the paraventricular and supraoptic nuclei) located in the side walls of the third ventricle. Identify the pituitary gland, its stalk, and the third ventricle in Figure 7-20. The mamillary body shown in this figure is also part of the hypothalamus.

The old adage, "Don't judge by appearances," applies well to appraising the importance of the hypothalamus. Measured by size, it is one of the least significant parts of the brain, but measured by its contribution to healthy survival, the hypothalamus is one of the most important of all brain structures. Impulses from neurons whose dendrites and cell bodies lie in the hypothalamus are conducted by their axons to neurons located in the spinal cord, and many of them are then relayed to muscles and glands all over the body. Thus the hypothalamus exerts a major control over virtually all internal organs. Among the vital functions it helps control are the beating of the heart, the constriction and dilatation of blood vessels, and the contractions of the stomach and intestines.

Neurons in the supraoptic and paraventricular nuclei of the hypothalamus function in a surprising way: they make the hormones that the posterior pituitary gland secretes into the blood. Because one of these hormones (called antidiuretic hormone or ADH) affects the volume of urine excreted, the hypothalamus plays an essential role in maintaining the body's water balance.

Some of the neurons in the hypothalamus function as endocrine (ductless) glands. Their axons secrete chemicals called releasing hormones into the blood, which then carries them to the anterior pituitary gland. Releasing hormones, as their name suggests, control the release of certain anterior pituitary hormones. These in turn influence hormone secretion by various other endocrine glands. Thus the hypothalamus indirectly helps control the functioning of every cell in the body.

The hypothalamus performs several other important functions. It helps regulate our appetites and therefore the amount of food we eat. It also plays a central role in maintaining normal body temperature.

Located deep inside each half of the cerebrum, lateral to the third ventricle, lies a rounded mass of gray matter. This is the **thalamus** (THAL-ah-mus), the other major part of the diencephalon. Each thalamus consists chiefly of dendrites and cell bodies of neurons whose axons extend to various sensory areas of the cerebral cortex.

The thalamus performs the following functions:

1 It helps produce sensations. Its neurons relay impulses to the cerebral cortex from the various sense organs of the body (except possibly those responsible for the sense of smell).

2 It associates sensations with emotions. Almost all sensations are accompanied by a feeling of some degree of pleasantness or unpleasantness. Just how these pleasant and unpleasant feelings are produced is not known except that they seem to be associated with the arrival of sensory impulses in the thalamus.

3 It plays a part in the so-called arousal or alerting mechanism.

CEREBRUM

The **cerebrum** (SER-e-brum) is the largest and uppermost part of the brain. If you were to look at the outer surface of the cerebrum, perhaps the first features you would notice are its many ridges and grooves. The ridges are called convolutions or *gyri* (JI-ri) and the grooves are called *fissures.* The deepest fissure, namely the longitudinal fissure, divides the cerebrum into right and left halves or hemispheres. They are almost separate structures except for their lower midportions, which are connected by a structure called the **corpus callosum** (COR-pus kal-LO-

Figure 7-21 Localization of function in the cerebral cortex.

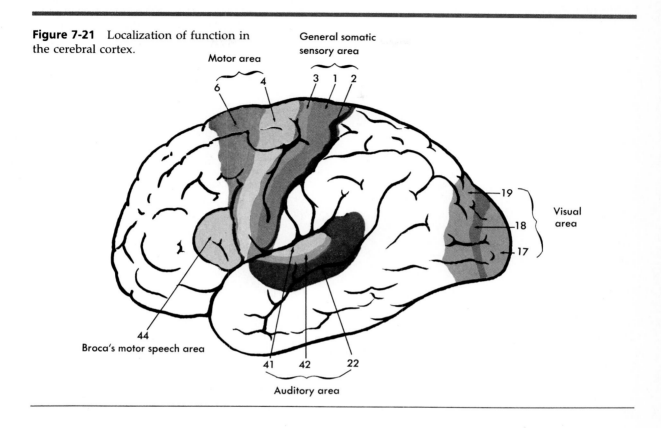

sum) (Figure 7-20). Shallower fissures subdivide each cerebral hemisphere into four major lobes and each lobe into numerous convolutions. The lobes are named for the bones that lie over them: the frontal lobe, the parietal lobe, the temporal lobe, and the occipital lobe. Identify these in Figures 7-19 and 7-21.

A thin layer of gray matter, made up of neuron dendrites and cell bodies, composes the surface of the cerebrum. Its name is the cerebral cortex. White matter, made up of bundles of nerve fibers (tracts), composes most of the interior of the cerebrum. Within this white matter, however, are a few islands of gray matter known as the **basal ganglia,** whose functioning is essential for producing our automatic movements and postures. Parkinson's disease is a disease of the basal ganglia. Because shaking or tremors are common symptoms of Parkinson's disease, it is also called "shaking palsy."

What functions does the cerebrum perform? This is a hard question to answer briefly, because the neurons of the cerebrum do not function alone. They function with many other neurons located in many other parts of the brain and in the spinal cord. Neurons of these structures are continually bringing impulses to cerebral neurons and continually transmitting impulses away from them. If all other neurons were functioning normally and only cerebral neurons were not functioning, here are some of the things that you could not do. You could not think or use your will. You could not decide to make the smallest movement nor could you make it. You could not see or hear. You could not experience any of the sensations that make life so rich and varied. Nothing would anger or frighten you, and nothing would bring you joy or sorrow. You would, in short, be unconscious. These five terms, then, sum up cerebral func-

Figure 7-22 Undersurface of the brain showing attachments of the cranial nerves.

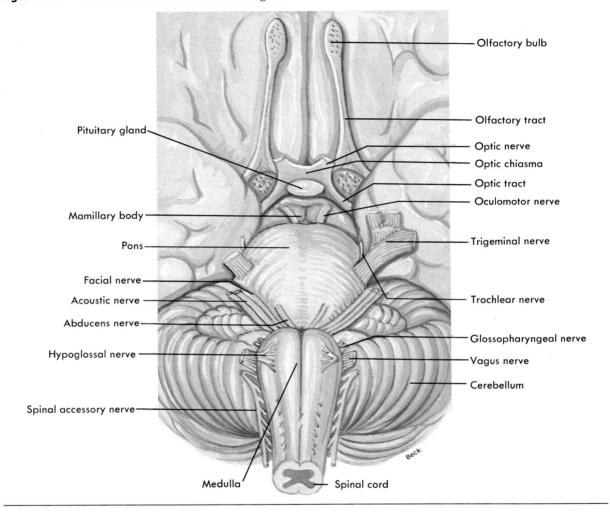

tions: consciousness, mental processes, sensations, emotions, and willed movements. Figure 7-21 shows the areas of the cerebral cortex essential for willed movements, general sensations, vision, hearing, and normal speech.

Injury or disease can destroy neurons. An all too common example is the destruction of neurons of the motor area of the cerebrum that results from a **cerebrovascular accident (CVA)**—medical parlance for a hemorrhage from cerebral blood vessels. When this happens, the victim can no longer voluntarily move the parts of his body on the side opposite to the side on which the cerebral hemorrhage occurred. In nontechnical language, he has suffered a stroke. Note in Figure 7-21 the location of the motor area in the frontal lobe of the cerebrum.

CEREBELLUM
Structure

Look at Figures 7-4, 7-19, 7-20 and 7-22 to find the location, appearance, and size of the cerebellum. The cerebellum is the second largest part of the human brain. It lies under the occipital lobe of the cerebrum. Another fact about the cerebellum, one visible in Figure 7-20, is that gray matter composes its outer layer and white matter composes the bulk of its interior.

Function

Most of our knowledge about cerebellar functions has come from observing patients who have some sort of disease of the cerebellum and from animals who have had the cerebellum removed. From such observations we know that the cerebellum plays an essential part in the production of normal movements. Perhaps a few examples will make this clear. A patient who has a tumor of the cerebellum frequently loses his balance and topples over; he may reel like a drunken man when he walks. He probably cannot coordinate his muscles normally. He may complain, for instance, that he is clumsy about everything he does—that he cannot even drive a nail or draw a straight line. With the loss of normal cerebellar functioning, he has lost the ability to make precise movements. The general functions of the cerebellum, then, are to produce smooth coordinated movements, maintain equilibrium, and sustain normal postures.

CRANIAL NERVES

Twelve pairs of cranial nerves attach to the undersurface of the brain, mostly from the brain stem. Figure 7-22 shows the attachments of these nerves. Their fibers conduct impulses between the brain and various structures in the head and neck and in the thoracic and abdominal cavities. For instance, the second cranial nerve (optic nerve) conducts impulses from the eye to the brain, where these impulses produce the sensation of vision. The third cranial nerve (oculomotor nerve) conducts impulses from the brain to certain muscles of the eye, where they cause contractions that move the eye. The tenth cranial nerve (vagus nerve) conducts impulses between the medulla and various structures in the neck and thoracic and abdominal cavities. The names of each of the cranial nerves and a brief description of their functions are listed in Table 7-1.

SENSE ORGANS

If you were asked to name the sense organs, what organs would you name? Can you think of any besides the eyes, ears, nose, and taste buds? Actually there are millions of other sense organs—all receptors are microscopic-sized sense organs. Receptors, you will recall, are the beginnings of dendrites of sensory neurons.

Receptors are generously scattered about in almost every part of the body. To demonstrate this fact, try pricking any point of your skin with a fine needle. You can hardly miss stimulating at least one receptor and almost instantaneously experiencing a sensation of mild pain. Stimulation of some receptors leads to the sensation of heat, stimulation of other receptors gives the sensation of cold, and stimulation of still others gives the sensation of touch or pressure. Receptors classified by structure are listed in Table 7-2 and illustrated in Figures 7-23 to 7-27. When special receptors in the muscles and joints are stimulated, you sense the position of the different parts of the body and know whether they are moving and in which direction they are moving without even looking at them. Perhaps you have never realized that you have this sense of position and movement—a sense called **proprioception** (pro-pre-o-SEP-shun) or **kinesthesia** (kin-es-THE-ze-ah). Let us turn our attention now to two complex and remarkable sense organs—the eyes and ears.

EYE

When you look at a person's eye, you see only a small part of the whole eye. Three layers of tissue form the eyeball: the **sclera** (SKLE-rah), the **choroid** (KO-roid), and the **retina** (RET-i-nah). The outer layer of *sclera* consists of tough fibrous tissue. What we call the "white" of the eye is part of the front surface of the sclera. The other part of the front surface of the sclera is called the cornea and is sometimes spoken of as the "window" of the eye because of its transparency. At a casual glance, however, it does not look transparent but appears blue or brown or gray or green because it lies over the **iris,** the colored part of the eye. Mucous membrane known as the **conjunctiva** (kon-junk-TI-vah)

Table 7-1 Cranial nerves

Nerve*	Conducts impulses	Functions
I Olfactory	From nose to brain	Sense of smell
II Optic	From eye to brain	Vision
III Oculomotor	From brain to eye muscles	Eye movements
IV Trochlear	From brain to external eye muscles	Eye movements
V Trigeminal (or trifacial)	From skin and mucous membrane of head and from teeth to brain; also from brain to chewing muscles	Sensations of face, scalp, and teeth, chewing movements
VI Abducens	From brain to external eye muscles	Turning eyes outward
VII Facial	From taste buds of tongue to brain; from brain to face muscles	Sense of taste; contraction of muscles of facial expression
VIII Acoustic	From ear to brain	Hearing; sense of balance
IX Glossopharyngeal	From throat and taste buds of tongue to brain; also from brain to throat muscles and salivary glands	Sensations of throat, taste, swallowing movements, secretion of saliva
X Vagus	From throat, larynx, and organs in thoracic and abdominal cavities to brain; also from brain to muscles of throat and to organs in thoracic and abdominal cavities	Sensations of throat, larynx, and of thoracic and abdominal organs; swallowing, voice production, slowing of heartbeat, acceleration of peristalsis
XI Spinal accessory	From brain to certain shoulder and neck muscles	Shoulder movements; turning movements of head
XII Hypoglossal	From brain to muscles of tongue	Tongue movements

*The first letter of the words of the following sentence at the first letters of the names of cranial nerves: "On Old Olympus' Tiny Tops A Finn and German Viewed Some Hops." Many generations of students have used this or a similar sentence to help them remember the names of cranial nerves

covers the entire front surface of the eyeball and lines both the upper and lower lids (Figure 7-28).

Two involuntary muscles make up the front part of the middle, or *choroid* coat, of the eyeball. One is the *iris*, the colored structure seen through the cornea, and the other is the *ciliary muscle* (part of the ciliary body shown in Figure 7-29). What appears to be a black center in the iris is really a hole in this doughnut-shaped muscle; it is the **pupil** of the eye. Some of the fibers of the iris are arranged like spokes in a wheel. When they contract the pupils dilate, letting in more light rays. Other fibers are circular. When they contract the pupils constrict, letting in fewer light rays. Normally, the pupils constrict in bright light and dilate in dim light.

The **lens** of the eye lies directly behind the pupil. It is held in place by a ligament attached to the ciliary muscle. When we look at distant objects, the ciliary muscle is relaxed, and the lens has only a slightly curved shape. To focus on near objects, however, the ciliary muscle must contract. As it contracts, it pulls the choroid coat forward toward the lens, thus causing the lens to become more bulging and more curved. Most of us become more farsighted as we grow older. The reason we do is that our lenses lose their

Table 7-2   Receptors classified by structure

Types	Main locations	General senses
Free nerve endings (naked nerve endings)	Skin, mucosa (epithelial layers)	Pain, crude touch, possibly temperature
Encapsulated nerve endings		
Meissner's corpuscles	Skin (in papillae of dermis); numerous in finger-tips and lips	Fine touch, vibration
Ruffini's corpuscles	Skin (dermal layer) and subcutaneous tissue of fingers	Touch, pressure
Pacinian corpuscles	Subcutaneous, submucous, and subserous tissues, around joints, in mammary glands and external genitals of both sexes	Pressure, vibration
Krause's end-bulbs	Skin (dermal layer), subcutaneous tissue, mucosa of lips and eyelids, external genitals	Touch
Golgi tendon receptors	Near junction of tendons and muscles	Subconscious muscle sense
Muscle spindles	Skeletal muscles	Subconscious muscle sense

Figure 7-23   Free nerve endings. **A,** In dermis of skin. **B,** Surrounding, in linear and circular fashion, the root of a hair follicle. **C,** In the cornea.

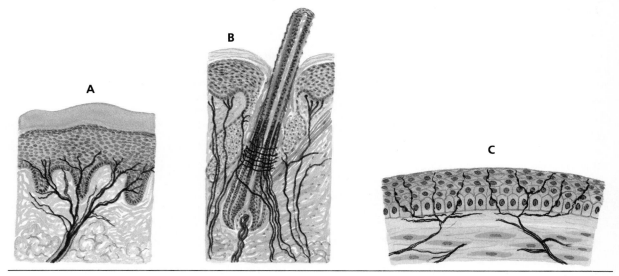

Figure 7-24 Meissner's corpuscle (tactile corpuscle) in skin papilla, an encapsulated nerve ending found in hairless portions of skin. Meissner's corpuscles mediate sensation of touch.

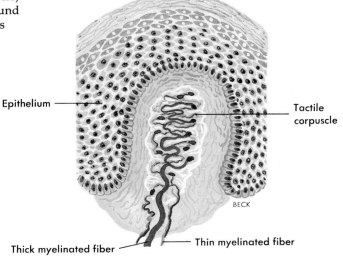

Epithelium

Tactile corpuscle

BECK

Thick myelinated fiber

Thin myelinated fiber

Figure 7-25 Ruffini's corpuscle, a skin receptor that probably mediates touch, rather than heat, as formerly thought.

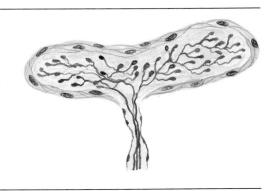

Figure 7-26 Pacinian corpuscle, an encapsulated nerve ending widely distributed in subcutaneous tissue; mediates sensation of pressure.

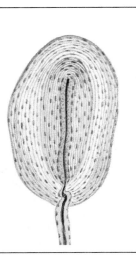

Figure 7-27 Krause's end-bulb, an encapsulated nerve ending; may mediate sensation of cold, but evidence indicates that it is not the only type of receptor for cold.

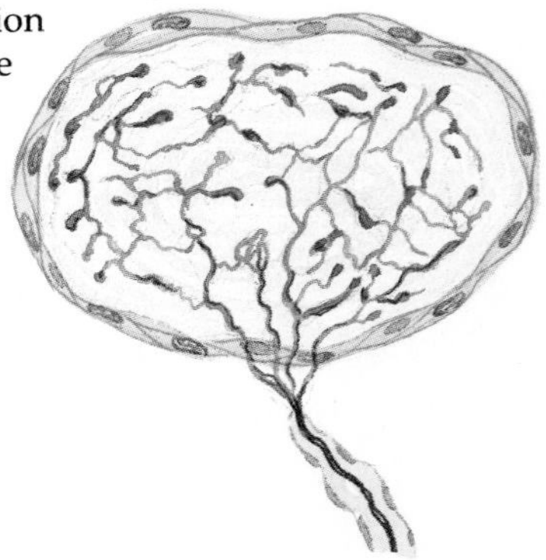

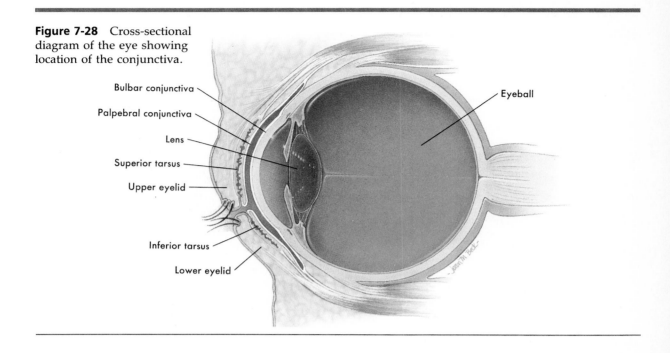

Figure 7-28 Cross-sectional diagram of the eye showing location of the conjunctiva.

Bulbar conjunctiva

Palpebral conjunctiva

Lens

Superior tarsus

Upper eyelid

Inferior tarsus

Lower eyelid

Eyeball

elasticity and can no longer bulge enough to bring near objects into focus. **Presbyopia** or "oldsightedness," is the name for this condition.

The *retina*, or innermost coat of the eyeball, contains microscopic structures called rods and cones because of their shapes. Both are receptors for vision. Dim light can stimulate the rods, but fairly bright light is necessary to stimulate the cones. In other words, **rods** are the receptors for night vision and **cones** for daytime vision. (Cones are also the receptors for color vision.)

Fluids fill the hollow inside of the eyeball. They maintain the normal shape of the eyeball and help refract light rays; that is, the fluids bend light rays so as to bring them to a focus on the retina. **Aqueous humor** is the name of the fluid in front of the lens (in the anterior cavity of the eye) and **vitreous humor** is the name of the jellylike fluid behind the lens (in the posterior cavity). The aqueous humor is constantly being formed, drained, and replaced in the anterior cavity. If drainage is blocked for any reason, the internal pressure within the eye will

increase and damage that could lead to blindness will occur. The condition is called **glaucoma** (glaw-KO-mah).

EAR

The ear is much more than a mere appendage on the side of the head. A large part of the ear—and by far its most important part—lies hidden from view deep inside the temporal bone. Part of the external ear, all of the middle ear, and all of the internal ear are located here.

The **external ear** has two parts: the pinna (or auricle) and the ear canal (or external acoustic meatus). The *pinna* is the appendage on the side of the head. The *ear canal* is a curving tube in the temporal bone that leads from the pinna to the middle ear.

The **middle ear** (or tympanic cavity) is a tiny cavity hollowed out of the temporal bone. This cavity is lined with mucous membrane and contains three very small bones. The names of these ear bones (Figure 7-30) are Latin words that describe their shapes—*malleus* (hammer), *incus*

Figure 7-29   Horizontal section through the right eyeball.

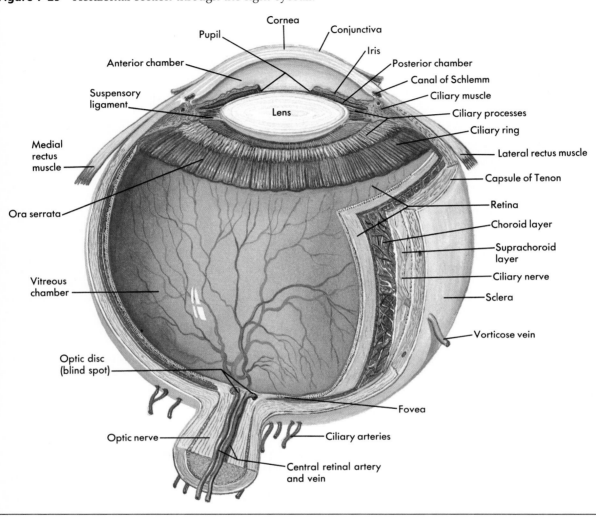

(anvil), and *stapes* (stirrup). The *tympanic membrane* (commonly called the eardrum) separates the middle ear from the external ear canal. The "handle" of the malleus attaches to the inside of the tympanic membrane, and the "head" attaches to the incus. The incus attaches to the stapes, and the stapes fits into a small opening, the *oval window,* that opens into the internal ear. A point worth mentioning, because it explains the frequent spread of infection from the throat to the ear, is the fact that a tube—the *auditory*

or *eustachian tube*—connects the throat with the middle ear. The mucous lining of the middle ears, eustachian tubes, and throat are extensions of one continuous membrane. Consequently a sore throat may spread to produce a middle ear infection *(otitis media).* It can even cause a mastoid infection *(mastoiditis).* Mastoid spaces (sinuses) also open into the middle ear cavity. Because the mucous lining of the mastoid sinuses is also continuous with the mucous lining of the middle ear, it provides a direct route for a mid-

Figure 7-30   Components of the ear. External ear consists of auricle (pinna), external acoustic meatus (ear canal), and tympanic membrane (eardrum). Middle ear (tympanic cavity) includes malleus (hammer), incus (anvil), and stapes (stirrup). Internal ear contains semicircular canals, vestibule, and cochlea.

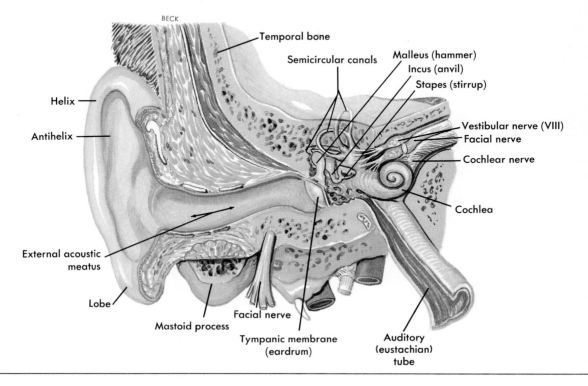

dle-ear infection to spread to produce *mastoiditis.*

The **internal ear** is made of bone and a membrane inside the bone. Because of its complicated shape, the internal ear is called a *labyrinth.* It has three parts: *vestibule, semicircular canals,* and *cochlea* (Figure 7-30). The semicircular canals are three half circles; the cochlea is shaped like a snail shell, which is what the word "cochlea" means.

Two special sense organs for two different kinds of sensations—hearing and balance—are located in the internal ear. The hearing sense organ, which lies inside the cochlea, is called the **organ of Corti.** There are two balance, or equilibrium, sense organs—one called the *macula,* located in the vestibule of the inner ear, and the other called the *crista,* located in the semicircular canals.

OLFACTORY SENSE ORGANS

The receptors for the fibers of the olfactory or first cranial nerves lie in the mucosa of the upper part of the nasal cavity (Figure 7-31). Their location here explains the necessity for sniffing or drawing air forcefully up into the nose to smell delicate odors. The olfactory sense organ consists of hair cells and is relatively simple compared with the complex visual and auditory organs. Whereas the olfactory receptors are extremely sensitive, that is, stimulated by even very slight odors, they are also easily fatigued—a fact that explains why odors that are at first very noticeable are not sensed at all after a short time. Once the olfactory cells are stimulated by odor-causing chemicals, the resulting nerve impulse travels through the olfactory nerves in the

Figure 7-31 Stimulation of olfactory cells in the nasal epithelium.

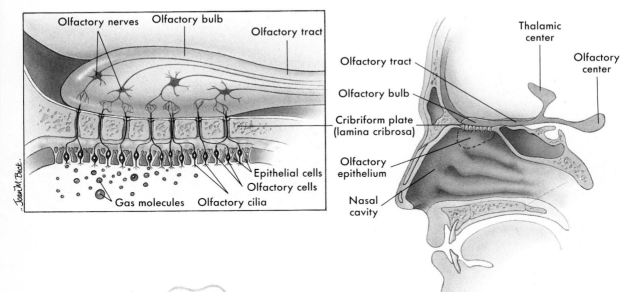

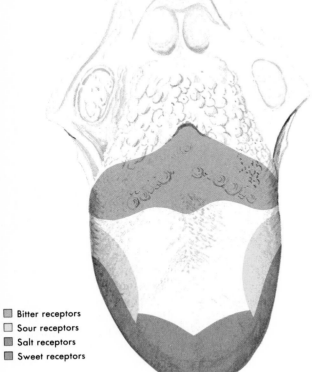

Figure 7-32 Dorsal surface of tongue showing location of taste buds and specific location of the four kinds of taste sensations.

■ Bitter receptors
□ Sour receptors
■ Salt receptors
■ Sweet receptors

olfactory bulb and tract and then enters the thalamic and olfactory centers of the brain where the nervous impulses are interpreted as specific odors.

GUSTATORY SENSE ORGANS

The receptors for the taste nerve fibers in branches of the seventh and ninth cranial nerves are known as **taste buds** (Figure 7-32). Taste buds are called *chemoreceptors* for the obvious reason that chemicals stimulate them to generate nervous impulses. Only four kinds of taste sensations—sweet, sour, bitter, and salty—result from stimulation of taste buds. All the other flavors we sense result from a combination of taste bud and olfactory receptor stimulation. In other words, the myriads of tastes recognized are not tastes alone but tastes plus odors. For this reason a cold that interferes with the stimulation of the olfactory receptors by odors from foods in the mouth markedly dulls one's taste sensations.

OUTLINE SUMMARY

Organs and Divisions of Nervous System
A Central nervous system (CNS)—brain and spinal cord.
B Peripheral nervous system (PNS)—all nerves
C Autonomic nervous system

Coverings and Fluid Spaces of Brain and Spinal Cord
A Coverings
   1 Cranial bones and vertebrae
   2 Cerebral and spinal meninges
B Fluid spaces—subarachnoid spaces of meninges, central canal inside cord, and ventricles in brain

Cells of Nervous System
A Neurons
   1 Consist of three main parts: dendrites—conduct impulses to cell body of neuron; cell body of neuron; and axon—conduct impulses away from cell body of neuron
   2 Neurons classified according to function as sensory—conduct impulses to spinal cord and brain; motor neurons—conduct impulses away from brain and cord out to muscles and glands; and interneurons—conduct impulses from sensory neurons to motor neurons
B Neuroglias—connective tissue cells of two main types
   1 Astrocytes—star-shaped cells that anchor small blood vessels to neurons
   2 Microglias—small cells that move about in inflamed brain tissue carrying on phagocytosis: that is, they engulf and destroy microorganisms and other injurious particles
   3 Oligodendroglia

Reflex Arcs
A Nerve impulses are conducted from receptors to effectors over neuron pathways or reflex arcs; conduction by reflex arc results in a reflex, that is, contraction by a muscle or secretion by a gland.
B Simplest reflex arcs called two-neuron arcs—consist of sensory neurons synapsing in spinal cord with motor neurons; three-neuron arcs consist of sensory neurons synapsing in spinal cord with interneurons that synapse with motor neurons

Nerve Impulses
A Definition—self-propagating wave of electrical negativity that travels along surface of neuron membrane
B Mechanism
   1 Stimulus increases permeability of neuron membrane to positive sodium ions
   2 Inward movement of positive sodium ions leaves slight excess of negative ions outside at stimulated point; marks beginning of nerve impulses

Neurotransmitters

A Definition—chemical compounds released from axon terminals (of presynaptic neuron) into synaptic cleft

B Neurotransmitters bind to specific receptor molecules in membrane of postsynaptic neuron, thereby stimulating impulse conduction by it

C Names of neurotransmitters—acetylcholine, catecholamines (norepinephrine, dopamine, serotonin), and other compounds

Spinal Cord

A Outer part composed of white matter made up of many bundles of axons called tracts; interior composed of gray matter made up mainly of neuron dendrites and cell bodies

B Functions as center for all spinal cord reflexes; sensory tracts conduct impulses to brain and motor tracts conduct impulses from brain

Spinal Nerves

A Structure—contain dendrites of sensory neurons and axons of motor neurons

B Functions—conduct impulses necessary for sensations and voluntary movements

Divisions of the Brain

A Brain stem

  1 Consists of three parts of brain; named in ascending order, they are the medulla, pons, and midbrain

  2 Structure—medulla, pons, and midbrain; consist of white matter with bits of gray matter scattered through them

  3 Function—gray matter in brain stem functions as reflex centers, for example, for heartbeat, respirations, and blood vessel diameter; sensory tracts in brain stem conduct impulses to higher parts of brain; motor tracts conduct from higher parts of brain to cord

B Diencephalon

  1 Structure and function of hypothalamus

    a Consists mainly of posterior pituitary gland, pituitary stalk, and the paraventricular and supraoptic nuclei

    b Acts as the major center for controlling autonomic nervous system; therefore helps control the functioning of most internal organs

    c Controls hormone secretion by both anterior and posterior pituitary glands; therefore indirectly helps control hormone secretion by most other endocrine glands

    d Acts as center for controlling appetite; therefore helps regulate amount of food eaten and body weight

    e Functions in some way to maintain waking state

    f Probably contains reward and punishment centers

  2 Structure and function of thalamus

    a Rounded mass of gray matter in each cerebral hemisphere; located lateral to each side of third ventricle

    b Relays sensory impulses to cerebral cortex sensory areas

    c Functions in some way to produce emotions of pleasantness or unpleasantness associated with sensations

C Cerebrum

  1 Largest part of human brain

  2 Outer layer of gray matter called cerebral cortex; made up of lobes, which are made up of convolutions; cortex composed mainly of neuron dendrite and cell bodies

  3 Interior of cerebrum composed mainly of nerve fibers arranged in bundles called tracts

  4 Functions of cerebrum—mental processes of all types, including sensations, consciousness, and voluntary control of movements

D Cerebellum

  1 Second largest part of human brain

  2 Helps control muscle contractions so that they produce coordinated movements so we can maintain balance, move smoothly, and sustain normal postures

Cranial Nerves

See Table 7-1.

Sense Organs

A All receptors (beginning of dendrites of sensory neurons) are sense organs

B Special sense organs—eyes and ears

C Kinds of senses—many more than the familiar five senses: for example, several kinds of touch senses; proprioception—sense of position and movement

D Eye

   1 Coats of eyeball

      a Sclera—tough outer coat; whites of eye; cornea is transparent part of sclera over iris

      b Choroid—front part of this coat made up of ciliary muscle and iris, the colored part of eye; pupil is hole in center of iris; contraction of iris muscle dilates or constricts pupil

      c Retina—innermost coat of eye; contains rods (receptors for night vision) and cones (receptors for day vision and color vision)

   2 Conjunctiva—mucous membrane covering front surface of eyeball lining and lines lid

   3 Lens—transparent body behind pupil; focuses light rays on retina

   4 Eye fluids

      a Aqueous humor—in anterior cavity in front of lens

      b Vitreous humor—in posterior cavity behind lens

E Ear

   1 External ear—consists of pinna and auditory canal

   2 Middle ear—contains auditory bones (malleus, incus, and stapes); lined with mucous membrane; eustachian tubes and mastoid sinuses open into middle ear

   3 Internal ear, or labyrinth—vestibule, semicircular canals, and cochlea are three divisions; organ of Corti, sense organ of hearing, lies in cochlea; sense organs of equilibrium are the macula and the crista; macula lies in vestibule and crista lies in semicircular canals

F Olfactory sense organs

   1 Receptors for fibers of olfactory or first cranial nerve lie in olfactory mucosa of nasal cavity (Figure 7-31)

   2 Olfactory receptors extremely sensitive but easily fatigued

   3 Odor-causing chemicals initiate an action potential that is interpreted as a specific odor by the brain

G Gustatory sense organs (Figure 7-32)

   1 Receptors called taste buds

   2 Cranial nerves seven and nine carry gustatory impulses

   3 Taste buds called chemoreceptors

   4 Only four kinds of taste sensations—sweet, sour, bitter, and salty

   5 Gustatory and olfactory senses work together

NEW WORDS

acetylcholine
anesthesia
arachnoid
astrocytes
autonomic nervous system
axon
catecholamines
central nervous system (CNS)
dendrite
dopamine
dura mater

effectors
endorphins
enkephalins
ganglion
glaucoma
glioma
hydrocephalus
interneuron
meninges
microglia
motor neuron
multiple sclerosis
myelin

neuroglia
neurons
neurotransmitter
node of Ranvier
norepinephrine
oligodendroglia
paralysis
peripheral nervous system
phagocytosis
pia mater
presynaptic neuron
postsynaptic neuron

receptor molecules
receptors
reflex
reflex arc
saltatory conduction
sensory neuron
serotonin
shingles
synapse
synaptic cleft
tracts

CHAPTER TEST

1. The nervous system as a whole is divided into two principal divisions: the _____ and _____ nervous systems.

2. The tough, fluid-containing membrane surrounding the brain and spinal cord is called the _____ .

3. Cerebrospinal fluid is formed from fluid filtered out of a network of brain capillaries known as the _____ _____ .

4. The medical name describing an accumulation of cerebrospinal fluid in the ventricles of the brain is _____ .

5. The two principal types of specialized cells found in the nervous system are _____ and _____ .

6. Nervous system cells that form the myelin sheaths around nerve fibers in the brain and spinal cord are called _____ .

7. The segmented wrapping of a fatty substance around an axon is called the _____ _____ .

8. Nervous impulses travel _____ from a nerve cell body in a single process called the _____ .

9. The microscopic space that separates the axon endings of one neuron from the dendrites of another neuron is called a _____ .

10. The type of nerve impulse conduction that results in an action potential leaping from one node of Ranvier to the next is called _____ _____ .

11. Chemical compounds released from axon terminals into synaptic clefts are called _____ .

12. An area of skin supplied by a single spinal nerve is called a _____ .

13. Two natural morphinelike painkillers produced in the brain are called _____ and _____ .

14. The medulla, pons, and midbrain are all components of the _____ _____ .

15. The two major structures that make up the diencephalon are called the _____ and _____ .

16. The thin layer of gray matter that covers the cerebrum is called the _____ _____ .

17. There are _____ pairs of cranial nerves and _____ pairs of spinal nerves.

18. Loss of balance and muscle coordination may be a sign of a problem in the _____ .

19. A stimulus-response pathway involving a sensory interneuron and motor neuron is called a _____ _____ .

20. Endoneurium, perineurium, and epineurium are all _____ _____ elements associated with nerves and nerve trunks.

Select the most correct answer from Column B for each statement in Column A. (Only one answer is correct.)

	Column A	Column B
21.	_____ Meninges	a. Microglia
22.	_____ Transmit impulses to neuron	b. Schwann cells
23.	_____ Capable of phagocytosis	c. Presbyopia
24.	_____ Produce myelin	d. Organ of Corti
25.	_____ "White" of the eye	e. Pia mater
26.	_____ "Oldsightedness"	f. Olfaction
27.	_____ Eardrum	g. Taste
28.	_____ Hearing sense organ	h. Tympanic membrane
29.	_____ Sense of smell	i. Dendrites
30.	_____ Gustatory sense	j. Sclera

REVIEW QUESTIONS

1 What general function does the nervous system perform?

2 What other system performs the same general function as the nervous system?

3 What general functions does the spinal cord perform?

4 What does "CNS" mean? "PNS"?

5 What are the meninges?

6 Why is the medulla considered the most vital part of the brain?

7 What general functions does the cerebellum perform?

8 What general functions does the cerebrum perform?

9 What general functions do spinal nerves perform?

10 What are some of the functions performed by cranial nerves?

11 Which pair or pairs of cranial nerves would you nickname "seeing nerves," "hearing nerves," "smelling nerves," and "tasting nerves"?

12 Would a person be blind, deaf, or neither, if both of his eighth cranial nerves atrophied?

13 Describe as fully as you can the structure of the eye.

14 Describe as fully as you can the structure of the ear.

15 What is each of the following?
conjunctiva   organ of Corti
cornea   retina
iris

16 Explain briefly why most old people are farsighted.

17 Define briefly each of the terms listed under "New words."

18 Identify each of the following:
interneuron   somatic motor neuron
motor neuron   spinal ganglion
reflex center   synapse
sensory neuron   visceral motor neuron

CHAPTER 8

The Autonomic Nervous System

OBJECTIVES

After you have completed this chapter, you should be able to:

1 Discuss the function of the autonomic nervous system as a whole.

2 Identify the two major subdivisions of the autonomic nervous system.

3 Compare the locations of the sympathetic and parasympathetic neuron cell bodies, dendrites, and axons.

4 Compare the structure and function of adrenergic and cholinergic fibers in the autonomic nervous system.

5 Compare the specific functions of the sympathetic and parasympathetic divisions of the autonomic nervous system.

The titles of this chapter and the preceding one are somewhat misleading. They may make you think that the body has two nervous systems. This is not so. It has only one nervous system, which consists of two major subdivisions called the central nervous system and the peripheral nervous system. Many structures make up these two divisions, but they function together as a single unit. The autonomic nervous system is a subdivision of the peripheral nervous system.

DEFINITIONS

If you are to understand the material presented in this chapter, you will need to know the meanings of the following terms.

1 The **autonomic nervous system,** according to one definition, consists of certain motor neurons that conduct impulses from the spinal cord or brain stem to three kinds of tissues, namely, cardiac muscle tissue, smooth muscle tissue, and glandular epithelial tissue. According to another definition, the autonomic nervous system consists of the parts of the nervous system that regulate the body's automatic or involuntary functions, for example, the beating of the heart, the contracting of the stomach and intestines, and the secreting of chemical compounds by glands.

2 Autonomic neurons are the motor neurons that make up the autonomic nervous system. The dendrites and cell bodies of some autonomic neurons are located in the gray matter of the spinal cord or brain stem. Their axons extend out from these structures and terminate in ganglia. These autonomic neurons are called **preganglionic neurons** because they conduct impulses before they reach a ganglion. In the ganglia the axon endings of preganglionic neurons synapse with the dendrites or cell bodies of postganglionic neurons. **Postganglionic neurons,** as their name suggests, conduct impulses after they reach a ganglion, that is, they conduct from the ganglion to cardiac muscle, or smooth muscle, or glandular epithelial tissue.

3 Autonomic or visceral effectors are the tissues to which autonomic neurons conduct impulses. Specifically, visceral effectors are cardiac muscle that makes up the wall of the heart, smooth muscle that partially makes up the walls of blood vessels and other hollow internal organs, and glandular epithelial tissue that makes up the secreting part of glands. Do you recall what kind of tissue makes up somatic effectors, the structures to which somatic motor neurons conduct impulses?

NAMES OF DIVISIONS

The autonomic nervous system consists of two divisions called the **sympathetic system** and the **parasympathetic system.** Both of these divisions, in turn, consist of various autonomic ganglia and nerves. Another name for the sympathetic system is the *thoracolumbar system,* an appropriate name because sympathetic ganglia are connected to the thoracic and lumbar regions of the spinal cord. Parasympathetic ganglia are connected to the brain stem and to the sacral segments of the spinal cord. Another name for the parasympathetic system, therefore, is the *craniosacral system.*

SYMPATHETIC NERVOUS SYSTEM
STRUCTURE

Sympathetic preganglionic neurons have their dendrites and cell bodies in the gray matter of the thoracic and upper lumbar segments of the spinal cord. Look now at the right side of Figure 8-1. Follow the course of the axon of the sympathetic preganglionic neuron shown there. Note that it leaves the cord in an anterior root of a spinal nerve. It next enters the spinal nerve but soon leaves it to extend to and through a sympathetic ganglion and terminate in a collateral ganglion. Here it synapses with several postganglionic neurons whose axons extend out to terminate in visceral effectors. Although not shown in Figure 8-1, branches of the preganglionic axon ascend and descend to terminate in ganglia above and below their point of origin. All sympathetic preganglionic axons therefore synapse with many postganglionic neurons, and these frequently terminate in widely separated

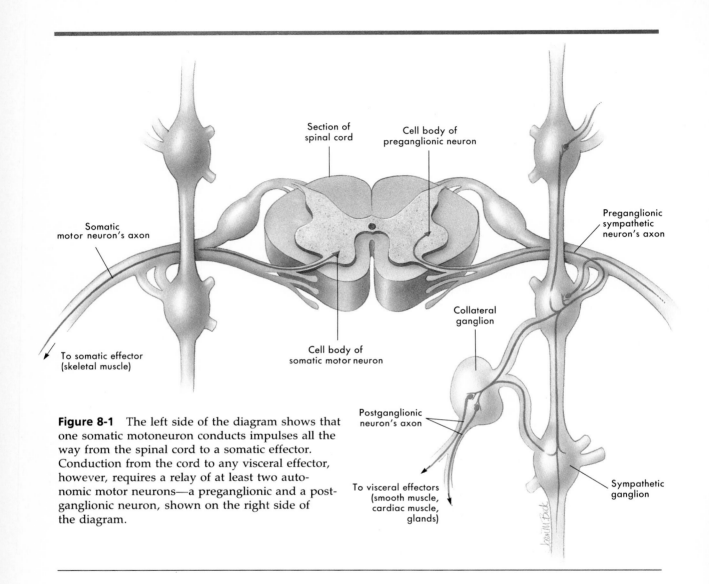

Somatic
motor neuron's axon

Section of
spinal cord

Cell body of
preganglionic neuron

Preganglionic
sympathetic
neuron's axon

To somatic effector
(skeletal muscle)

Cell body of
somatic motor neuron

Collateral
ganglion

Postganglionic
neuron's axon

To visceral effectors
(smooth muscle,
cardiac muscle,
glands)

Sympathetic
ganglion

Figure 8-1   The left side of the diagram shows that one somatic motoneuron conducts impulses all the way from the spinal cord to a somatic effector. Conduction from the cord to any visceral effector, however, requires a relay of at least two autonomic motor neurons—a preganglionic and a postganglionic neuron, shown on the right side of the diagram.

organs. Hence sympathetic responses are usually widespread, involving many organs and not just one.

Sympathetic postganglionic neurons have their dendrites and cell bodies in sympathetic ganglia or in collateral glands. Sympathetic ganglia are located in front of and at each side of the spinal column. Because short fibers extend between the sympathetic ganglia, they look a little like two chains of beads and are often referred to as the "sympathetic chain ganglia."

Axons of sympathetic postganglionic neurons travel in spinal nerves to blood vessels, sweat glands, and arrector hair muscles all over the body. Separate autonomic nerves distribute many sympathetic postganglionic axons to various internal organs.

FUNCTIONS

The sympathetic nervous system functions as an emergency system. Impulses over sympathetic fibers take control of many of our internal

Table 8-1   Autonomic functions

Visceral effectors	Parasympathetic control	Sympathetic control
Heart muscle		
	Slows heartbeat	Accelerates heartbeat
Smooth muscle		
Of most blood vessels	None	Constricts blood vessels
Of blood vessels in skeletal muscles	None	Dilates blood vessels
Of digestive tract	Increases peristalsis	Decreases peristalsis; inhibits defecation
Of anal sphincter	Inhibits—opens sphincter for defecation	Stimulates—closes sphincter
Of urinary bladder	Stimulates—contracts bladder	Inhibits—relaxes bladder
Of urinary sphincters	Inhibits—opens sphincter for urination	Stimulates—closes sphincter
Of eye		
Iris	Stimulates circular fibers—constriction of pupil	Stimulates radial fibers—dilation of pupil
Ciliary	Stimulates—accommodation for near vision (bulging of lens)	Inhibits—accommodation for far vision (flattening of lens)
Of hairs (pilomotor muscles)	No parasympathetic fibers	Stimulates—"goose pimples"
Glands		
Adrenal medulla	None	Increases epinephrine secretion
Sweat glands	None	Increases sweat secretion
Digestive glands	Increases secretion of digestive juices	Decreases secretion of digestive juices

organs when we exercise strenuously and when strong emotions—anger, fear, hate, anxiety—buffet us. In short, whenever we must cope with stress of any kind, sympathetic impulses increase to many visceral effectors and rapidly produce widespread changes within our bodies. The righthand column of Table 8-1 indicates many of these sympathetic responses. The heart beats faster. Most blood vessels constrict, causing blood pressure to shoot up. Blood vessels in skeletal muscles dilate, supplying the muscles with more blood. Sweat glands and adrenal glands secrete more abundantly. Salivary and other digestive glands secrete more sparingly. Digestive tract contractions (peristalsis) become sluggish, hampering digestion. Together these

sympathetic responses make us ready for strenuous muscular work, or, as the famous physiologist Walter Cannon vividly stated, they prepare us for *fight or flight*. The group of changes induced by sympathetic control is known as the **fight-or-flight syndrome.**

PARASYMPATHETIC NERVOUS SYSTEM
STRUCTURE

The dendrites and cell bodies of parasympathetic preganglionic neurons are located in the gray matter of the brain stem and the sacral segments of the spinal cord. Their axons leave the brain stem by way of cranial nerves III, VII, IX,

X, and XI and leave the cord by some pelvic nerves. They extend some distance before terminating in the parasympathetic ganglia located in the head and in the thoracic and abdominal cavities close to visceral effectors. The dendrites and cell bodies of parasympathetic postganglionic neurons lie in these outlying parasympathetic ganglia and their short axons extend into the nearby structures. Each parasympathetic preganglionic neuron synapses, therefore, only with postganglionic neurons to a single effector. For this reason parasympathetic stimulation frequently involves response by only one organ. This is not true of sympathetic responses; as we have noted, sympathetic stimulation usually results in responses by numerous organs.

FUNCTION

The parasympathetic system dominates control of many visceral effectors under normal, everyday conditions. Impulses over parasympathetic fibers, for example, tend to slow the heartbeat, increase peristalsis, and increase secretion of digestive juices and insulin.

AUTONOMIC CONDUCTION PATHS

Conduction paths to visceral and somatic effectors from the central nervous system (spinal cord or brain stem) differ somewhat. Autonomic paths to visceral effectors, as the right side of Figure 8-1 shows, consist of two-neuron relays. Impulses travel over preganglionic neurons from the cord or brain stem to autonomic ganglia. Here they are relayed across synapses to postganglionic neurons, which then conduct the impulses from the ganglia to visceral effectors. Compare the autonomic conduction path with the somatic conduction path illustrated on the left side of Figure 8-1. Somatic motor neurons, like the ones shown here, conduct all the way from cord or brain stem to somatic effectors with no intervening synapses.

AUTONOMIC NEUROTRANSMITTERS

Turn your attention now to Figure 8-2. It reveals information about autonomic neurotransmitters, the chemical compounds released from the axon terminals of autonomic neurons. Observe that three of the axons shown in Figure 8-2—namely, the sympathetic preganglionic axon, the parasympathetic preganglionic axon, and the parasympathetic postganglionic axon—release acetylcholine. These axons are therefore classified as **cholinergic fibers.** Only one type of autonomic axons releases the neurotransmitter norepinephrine. These are the axons of sympathetic postganglionic neurons, and they are classified as **adrenergic fibers.** See p. 171 for other neurons whose axons release acetylcholine and norepinephrine.

AUTONOMIC NERVOUS SYSTEM AS A WHOLE

The function of the **autonomic nervous system** is to regulate the body's automatic, involuntary functions in ways that tend to maintain or quickly restore homeostasis. Many internal organs are doubly innervated by the autonomic nervous system. In other words, they receive fibers from parasympathetic and sympathetic divisions. Both parasympathetic and sympathetic impulses continually bombard them and, as Table 8-1 indicates, influence their functioning in opposite or antagonistic ways. For example, the heart continually receives sympathetic impulses that tend to make it beat faster and parasympathetic impulses that tend to slow it down. The ratio between these two antagonistic forces determines the actual heart rate.

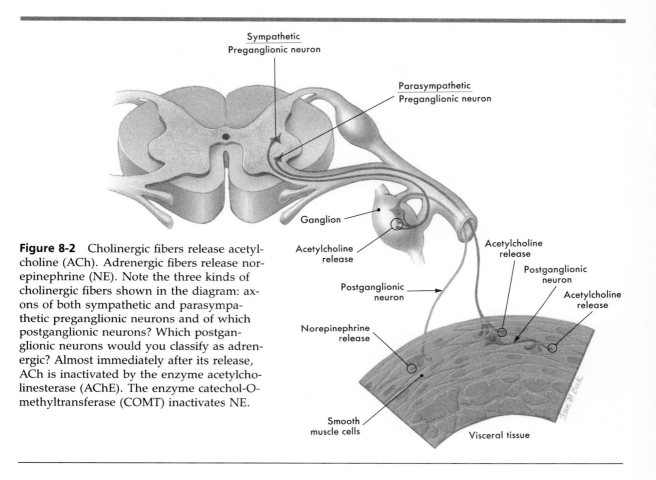

Figure 8-2 Cholinergic fibers release acetylcholine (ACh). Adrenergic fibers release norepinephrine (NE). Note the three kinds of cholinergic fibers shown in the diagram: axons of both sympathetic and parasympathetic preganglionic neurons and of which postganglionic neurons? Which postganglionic neurons would you classify as adrenergic? Almost immediately after its release, ACh is inactivated by the enzyme acetylcholinesterase (AChE). The enzyme catechol-O-methyltransferase (COMT) inactivates NE.

The name *autonomic nervous system* is something of a misnomer. It seems to imply that this part of the nervous system is independent from other parts. This is not true. Dendrites and cell bodies of preganglionic neurons are located, as we have observed, in the spinal cord and the brain stem. They are continually influenced directly or indirectly by impulses from neurons located above them, notably by some in the hypothalamus and in the parts of the cerebral cortex called the "emotional brain." Through conduction paths from these areas, emotions can produce widespread changes in the automatic functions of our bodies, in cardiac and smooth muscle contractions, and in secretion by glands. Anger and fear, for example, lead to increased sympathetic activity and the fight-or-flight syndrome. According to some physiologists, the altered state of consciousness known as meditation or yoga leads to decreased sympathetic activity and a group of changes opposite to those of the fight-or-flight syndrome.

OUTLINE SUMMARY

Definitions

A Autonomic nervous system—motor neurons that conduct impulses from the central nervous system to cardiac muscle, smooth muscle, and glandular epithelial tissue; nervous system structures that regulate the body's automatic or involuntary functions

B Autonomic neurons—motor neurons that conduct from the central nervous system to cardiac muscle, smooth muscle, and glandular epithelial tissue; preganglionic autonomic neurons that conduct from spinal cord or brain stem to an autonomic ganglion; postganglionic neurons that conduct from autonomic ganglia to cardiac muscle, smooth muscle, and glandular epithelial tissue

C Autonomic or visceral effectors—tissues to which autonomic neurons conduct impulses, that is, cardiac and smooth muscle and glandular epithelial tissue

Names of Divisions

A Sympathetic or thoracolumbar system
B Parasympathetic or craniosacral system

Sympathetic Nervous System

A Structure
  1 Dendrites and cell bodies of sympathetic preganglionic neurons located in gray matter of thoracic and upper lumbar segments of cord
  2 Axons leave cord in anterior roots of spinal nerves, extend to sympathetic or collateral ganglia, and synapse with several postganglionic neurons whose axons extend out in spinal nerves or autonomic nerves to terminate in visceral effectors
  3 Chain of sympathetic ganglia in front of and at each side of spinal column

B Functions
  1 Serves as the emergency or stress system, controlling visceral effectors during strenuous exercise and strong emotions (anger, fear, hate, anxiety)
  2 Group of changes induced by sympathetic control is called the fight-or-flight syndrome

Parasympathetic Nervous System

A Structure
  1 Parasympathetic preganglionic neurons have their dendrites and cell bodies in gray matter of the brain stem and of the sacral segments of cord
  2 Their axons leave brain stem by way of cranial nerves III, VII, IX, X, and XI and leave the cord by some pelvic nerves
  3 They terminate in parasympathetic ganglia located in head and thoracic and abdominal cavities close to visceral effectors
  4 Each parasympathetic preganglionic neuron synapses with postganglionic neurons to only one effector

B Function—dominates control of many visceral effectors under normal, everyday conditions

Autonomic Conduction Paths

A Consist of two-neuron relays, that is, preganglionic neurons from central nervous system to autonomic ganglia, synapses, postganglionic neurons from ganglia to visceral effectors

B In contrast, somatic motor neurons conduct all the way from central nervous system to somatic effectors with no intervening synapses

Autonomic Neurotransmitters

A Cholinergic fibers—preganglionic axons of both parasympathetic and sympathetic systems and parasympathetic postganglionic axons release acetylcholine

B Adrenergic fibers—axons of sympathetic postganglionic neurons release norepinephrine (noradrenaline)

C Acetylcholine and norepinephrine are not exclusively autonomic neurotransmitters; for example, axons of somatic motor neurons release acetylcholine and axons of various neurons located in central nervous system are now known to release norepinephrine

Autonomic Nervous System as a Whole

A Regulates the body's automatic functions in ways that tend to maintain or quickly restore homeostasis

B Many visceral effectors are doubly innervated, that is, receive fibers from both parasympathetic and sympathetic divisions and are influenced in opposite ways by the two divisions

C Autonomic nervous system is not autonomous or independent from other parts of the nervous system; conduction paths from hypothalamus and other parts of the brain influence conduction by both parasympathetic and sympathetic divisions

NEW WORDS

acetylcholine
adrenergic fibers
autonomic effectors
autonomic nervous system (ANS)

autonomic neurons
cholinergic fibers
fight-or-flight syndrome

norepinephrine
parasympathetic system
postganglionic neurons

preganglionic neurons
sympathetic system
visceral effectors

CHAPTER TEST

1. Motor neurons in the ANS conduct impulses to three kinds of tissues: _____ _____ , _____ _____ , and _____ _____ .

2. The axons of preganglionic neurons terminate in _____ .

3. Autonomic neurons conduct impulses to tissues called autonomic or visceral _____ .

4. The ANS consists of two divisions called the _____ and _____ subdivisions.

5. Another name for the sympathetic system of the ANS is the _____ system.

6. Strong emotions and strenuous exercise activate the _____ subdivision of the ANS.

7. During sympathetic stimulation sweat gland secretion _____ and digestive gland secretion _____ .

8. The group of changes induced by sympathetic stimulation is known as the _____ syndrome.

9. The neurotransmitter released by sympathetic preganglionic axons, parasympathetic preganglionic axons, and parasympathetic postganglionic axons is called _____ .

10. Axons of sympathetic postganglionic neurons are classified as _____ fibers.

Select the most correct answer from Column B for each statement in Column A. (Only one answer is correct.)

	Column A	Column B
11.	_____ Release norepinephrine	a. Craniosacral system
12.	_____ Visceral effector	b. Increased peristalsis
13.	_____ Parasympathetic system	c. Preganglionic parasympathetic neuron
14.	_____ Sympathetic response	d. Acetylcholine
15.	_____ Very long axon	e. COMT
16.	_____ Neurotransmitter	f. Adrenergic fibers
17.	_____ Parasympathetic response	g. Dr. Walter Cannon
18.	_____ Fight-or-flight syndrome	h. Smooth muscle
19.	_____ Sympathetic system	i. Thoracolumbar system
20.	_____ Inactivates norepinephrine	j. Increased heart rate

REVIEW QUESTIONS

1 Contrast the meanings of the terms *visceral effectors* and *somatic effectors*.

2 Differentiate between a preganglionic neuron and a postganglionic neuron.

3 Compare sympathetic and parasympathetic systems as to locations of their preganglionic dendrites, cell bodies, and axons and their postganglionic dendrites, cell bodies, and axons.

4 All preganglionic neurons release the same neurotransmitter. Name it. Only one kind of autonomic axon releases norepinephrine. Which kind?

5 Compare parasympathetic and sympathetic functions.

6 Explain why the name *autonomic nervous system* is misleading.

7 Explain the meaning of the term *fight-or-flight syndrome*.

8 Which division of the autonomic nervous system functions as an emergency or stress system?

9 What general function does the autonomic nervous system perform?

CHAPTER
9 The Endocrine System

CHAPTER OUTLINE

Regulation of hormone activity and secretion

Prostaglandins

Pituitary gland
Anterior pituitary gland hormones
Posterior pituitary gland hormones

Hypothalamus

Thyroid gland

Parathyroid glands

Adrenal glands
Adrenal cortex
Adrenal medulla

Islands of Langerhans

Female sex glands

Male sex glands

Thymus

Placenta

Pineal gland

BOXED ESSAYS

Endocrine diseases
Production of "natural" painkillers
Diabetes insipidus
Growth hormone abnormalities
Thyroid hormone abnormalities
Adrenal hormone abnormalities

OBJECTIVES

After you have completed this chapter, you should be able to:

1 Distinguish between endocrine and exocrine glands and define the terms *hormone* and *prostaglandin*.

2 Identify and locate the primary endocrine glands and list the major hormones produced by each gland.

3 Discuss how hormones act on target organ cells and how hormone secretions are regulated by negative feedback control mechanisms.

4 Identify the hormones produced by the cell layers (zona) of the adrenal cortex and describe the pathological conditions that result from malfunction.

5 Define *diabetes insipidus, diabetes mellitus, gigantism, acromegaly, goiter, cretinism, glycosuria.*

ave you ever seen a giant or a dwarf?
Have you ever known anyone who had
"sugar diabetes" or a goiter? If so, you have had
visible proof of the importance of the endocrine
system for normal development and health.

The **endocrine system** performs the same
general functions as the nervous system,
namely, communication and control. The nervous system provides rapid, brief control by fast-
traveling nerve impulses. The endocrine system
provides slower but longer-lasting control by
hormones (chemicals) secreted into and circulated by the blood.

The organs of the endocrine system are located in widely separated parts of the body—in
the cranial cavity, in the neck, in the thoracic
cavity, in the abdominal cavity, in the pelvic cavity, and outside the body cavities. Note the
names and locations of the endocrine glands
shown in Figure 9-1 and Table 9-1.

All the organs of the endocrine system are
glands, but not all glands are organs of the endocrine system. Those glands that discharge
their secretions into ducts are called **exocrine
glands.** The salivary glands are good examples
of exocrine glands. Saliva produced by these
glands flows into ducts that empty into the
mouth. **Endocrine glands** are ductless glands.
They secrete chemicals known as **hormones** directly into the blood. Cells that are acted on and
respond in some way to a particular hormone
are referred to as **target organ** cells. The list of
endocrine glands and their target organs continues to grow. Only the names and locations of
the main endocrine glands are given in Table 9-
1 and Figure 9-1.

In this chapter you will read about the functions of the main endocrine glands and discover
why their importance is almost impossible to
exaggerate. Hormones are the main regulators
of metabolism, growth and development, reproduction, and many other body activities.
They play roles of the utmost importance in
maintaining homeostasis—fluid and electrolyte
balance, acid-base balance, and energy balance,
for example. Hormones make the difference between normalcy and all sorts of abnormalities
such as dwarfism, gigantism, and sterility. They

Table 9-1 Names and locations of endocrine glands

Name	Location
Pituitary gland Anterior lobe Posterior lobe	
Pineal gland	Cranial cavity
Hypothalamus	
Thyroid gland	Neck
Parathyroid glands	Neck
Adrenal glands Adrenal cortex Adrenal medulla	Abdominal cavity (retroperitoneal)
Islands of Langerhans	Abdominal cavity (pancreas)
Ovaries Ovarian follicle Corpus luteum	Pelvic cavity
Testes (interstitial cells)	Scrotum
Thymus	Mediastinum
Placenta	Pregnant uterus

are important not only for the healthy survival
of each one of us, but also for the survival of the
human species itself.

REGULATION OF HORMONE ACTIVITY AND SECRETION

What a hormone does to its target cells to
cause them to respond in particular ways has
been the subject of intense interest and research.
The most widely accepted theory of hormone
action is called the **second messenger hypothesis.** A **hypothesis** (hi-POTH-e-sis) is simply a
proposed explanation or theory to explain observed phenomena. The second messenger hypothesis is a theory that attempts to explain why
hormones cause specific effects in target organs
but do not "recognize" or act on other organs
of the body. According to this concept a hor-

Figure 9-1 Location of the endocrine glands in the female. Thymus gland is shown at maximum size at puberty.

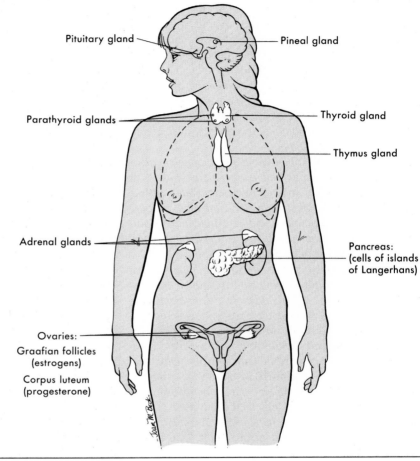

Pituitary gland — — Pineal gland

Parathyroid glands — — Thyroid gland

— Thymus gland

Adrenal glands —

Pancreas: (cells of islands of Langerhans)

Ovaries:
Graafian follicles (estrogens)
Corpus luteum (progesterone)

mone acts as a "first messenger," that is, it delivers its chemical message from the cells of an endocrine gland to highly specific membrane receptor sites on the cells of a target organ. This interaction between a hormone and its specific receptor site on the cell membrane of a target organ cell is often compared to the fitting of a unique key into a lock. After the hormone is attached to its specific receptor site, a number of chemical reactions occur that change energy-rich ATP molecules present inside the cell into a compound called **cyclic AMP** (adenosine monophosphate). Cyclic AMP serves as the "second messenger," delivering information inside the cell that causes the cell to respond by performing its specialized function. For example, cyclic AMP causes thyroid cells to respond by secreting thyroid hormones.

In summary, hormones serve as first messengers, providing communication between endocrine glands and target organs. Cyclic AMP then acts as the second messenger, providing communication within a hormone's target cells. Figure 9-2 summarizes the mechanism of hormone action as explained by the second messenger hypothesis.

The action of small lipid-soluble steroid hormones such as estrogen is not explained by the

Figure 9-2 Mechanism of hormone action. Hormone acts as "first messenger," delivering its message to a membrane receptor in the target organ cell much like a key fits into a lock. Cyclic AMP serves as the "second messenger" that causes the cell to respond by performing its specialized function.

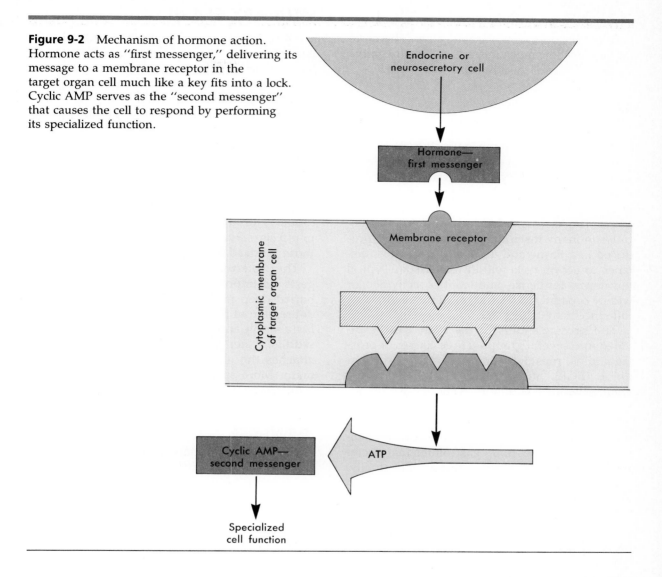

second messenger hypothesis. Because they are lipid soluble, steroid hormones can pass intact directly through the cell membrane of the target organ cell. Once inside the cell they combine with specific proteins and then enter the nucleus to influence cell activity by acting on specific genes.

The regulation of hormone levels in the blood is dependent on a highly specialized homeostatic mechanism called **negative feedback.** The principle of negative feedback can be illustrated

by using the hormone insulin as an example. When released from endocrine cells in the pancreas, insulin acts to lower blood sugar levels. Normally, elevated blood sugar levels will occur after a meal, once the absorption of sugars from the digestive tract has taken place. The elevated blood sugar stimulates the release of insulin from the pancreas. Insulin then assists in the transfer of sugar from the blood into cells, and blood sugar levels will drop. Low blood sugar levels then cause endocrine cells in the pancreas to cease the production and release of insulin. Both of these responses are negative. Therefore the homeostatic mechanism is called a **negative feedback** control mechanism.

PROSTAGLANDINS

Prostaglandins (PGs) or tissue hormones are important and extremely powerful substances that have been found in a wide variety of tissues in the body. They play an important role in communication and control of many body functions but do not meet the definition of a typical hormone. The term *tissue hormone* is appropriate because in many instances a prostaglandin is produced in a tissue and diffuses only a short distance to act on cells within that tissue. Typical hormones influence and control activities of widely separated organs; typical prostaglandins influence activities of neighboring cells.

The prostaglandins in the body can be divided into several groups. Three classes of prostaglandins, prostaglandin A (PGA), prostaglandin E (PGE), and prostaglandin F (PGF) are among the most common. Although we do not know the exact mechanism by which prostaglandins act on cells, they have profound effects on many body functions. They influence respiration, blood pressure, gastrointestinal secretions, and the reproductive system. Eventually, prostaglandins may play an important role in the treatment of such diverse diseases as high blood pressure, asthma, and ulcers.

Endocrine diseases

Diseases of the endocrine glands are numerous, varied, and sometimes spectacular. Tumors or other abnormalities frequently cause the glands to secrete either too much or too little of their hormones. Production of too much hormone by a diseased gland is called **hypersecretion.** If too little hormone is produced, the condition is called **hyposecretion.**

PITUITARY GLAND

The **pituitary** (pih-TOO-ih-tare-e) **gland** truly is a small but mighty structure. Although no larger than a pea, it is really two endocrine glands. One is called the **anterior pituitary gland** or *adenohypophysis* (ad-e-no-hi-POF-i-sis), and the other is called the **posterior pituitary gland** or *neurohypophysis* (nu-ro-hi-POF-i-sis). Differences between the two glands are suggested by their names—*adeno* means "gland" and *neuro* means "nervous." The adenohypophysis has the structure of an endocrine gland, whereas the neurohypophysis has the structure of nervous tissue. Hormones secreted by the adenohypophysis serve very different functions from those released from the neurohypophysis.

The protected location of this dual gland suggests its importance. The pituitary gland lies buried deep in the cranial cavity, in the small depression of the sphenoid bone that is shaped like a saddle and is called the *sella turcica* (Turkish saddle). A stemlike structure, the pituitary stalk, attaches the gland to the undersurface of the brain. More specifically, the stalk attaches the pituitary body to the hypothalamus.

ANTERIOR PITUITARY GLAND HORMONES

The anterior pituitary gland secretes six major hormones. To learn their abbreviated and full names, see Figure 9-3. Each of the four hormones listed as a **trophic** (TROW-pik) **hormone** stimulates another endocrine gland to grow and secrete its hormones. Because the anterior pituitary gland exerts this control over the struc-

---

Production of "natural" painkillers

Research now suggests that both *endorphins* and *enkephalins*, the naturally occurring morphinelike painkillers described in Chapter 7, are formed from a precursor substance called **beta-lipotropin** that is produced in the anterior pituitary gland.

Figure 9-3 Hormones secreted by anterior pituitary gland.

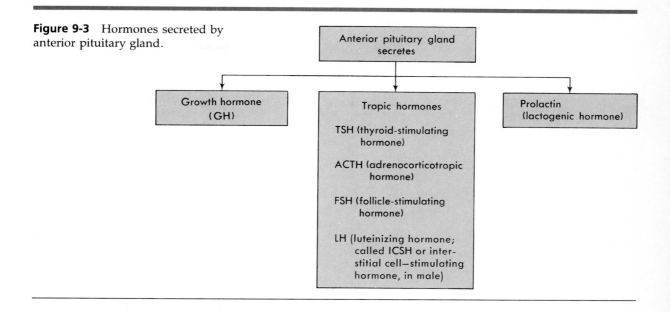

Anterior pituitary gland secretes

Growth hormone (GH)

Tropic hormones

TSH (thyroid-stimulating hormone)

ACTH (adrenocorticotropic hormone)

FSH (follicle-stimulating hormone)

LH (luteinizing hormone; called ICSH or interstitial cell—stimulating hormone, in male)

Prolactin (lactogenic hormone)

ture and function of the thyroid gland, the adrenal cortex, the ovarian follicles, and the corpus luteum, it is often referred to as the *master gland.*

Thyroid-stimulating hormone (TSH) acts on the thyroid gland. As its names suggest, it stimulates the thyroid gland to increase its secretion of thyroid hormone.

The **adrenocorticotropic** (ad-re-no-kor-te-ko-TROP-ik) **hormone (ACTH)** acts on the adrenal cortex. It stimulates the adrenal cortex to increase in size and to secrete larger amounts of its hormones, especially of cortisol (hydrocortisone).

Follicle-stimulating hormone (FSH) stimulates primary ovarian follicles in an ovary to start growing and to continue developing to maturity, that is, to the point of ovulation. FSH also stimulates follicle cells to secrete estrogens. In the male, FSH stimulates the seminiferous tubules to grow and form sperm.

Luteinizing (LU-te-ni-zing) **hormone (LH)** acts with FSH to perform four functions. It stimulates a follicle and ovum to complete their growth to maturity. It stimulates follicle cells to secrete estrogens. It causes ovulation (rupturing of the mature follicle with expulsion of its ripe

ovum). Because of this function, LH is sometimes called the ovulating hormone. Finally, LH stimulates the formation of a golden body, the corpus luteum, in the ruptured follicle; the process is called luteinization. This function, obviously, is the one that earned LH its title of luteinizing hormone. The male pituitary gland also secretes LH; it was formerly called *interstitial cell–stimulating hormone* (ICSH) because it stimulates interstitial cells in the testes to develop and secrete testosterone, the male sex hormone (Figure 9-5).

Melanocyte-stimulating hormone (MSH) causes a rapid increase in the synthesis and dispersion of melanin (pigment) granules in specialized skin cells (Figure 9-4).

The other two important hormones secreted by the anterior pituitary gland are **growth hormone** and **prolactin** (pro-LAK-tin) (lactogenic hormone). Growth hormone (GH) tends to speed up the movement of digested proteins (amino acids) out of the blood and into the cells, and this accelerates the cells' anabolism of amino acids to form tissue proteins; hence this action promotes normal growth. Growth hormone also affects fat and carbohydrate metabolism: it ac-

Figure 9-4 Functions of anterior pituitary hormones.

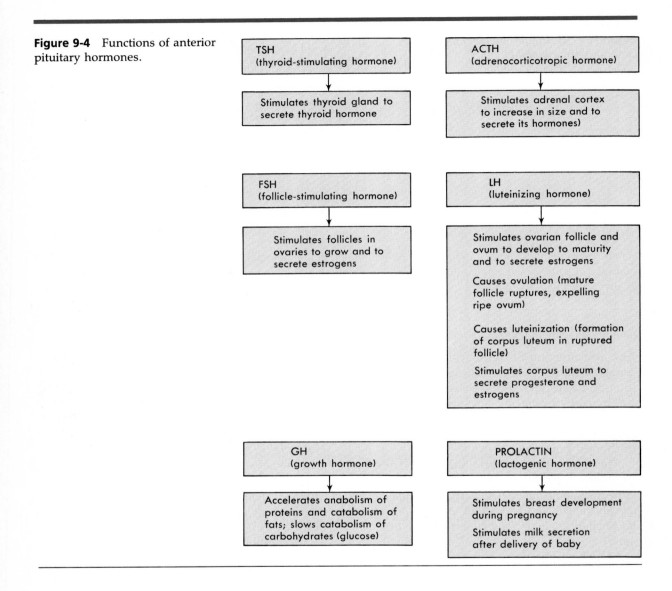

celerates fat catabolism but slows glucose catabolism. This means that less glucose leaves the blood to enter cells and therefore the amount of glucose in the blood tends to increase. Thus growth hormone and insulin have opposite effects on blood glucose. Insulin tends to decrease blood glucose, and growth hormone tends to increase it. Too much insulin in the blood produces **hypoglycemia** (hi-po-gli-SE-me-ah)

(lower than normal blood glucose concentration). Too much growth hormone produces **hyperglycemia** (higher than normal blood glucose concentration). This type of hyperglycemia is appropriately called pituitary diabetes, and growth hormone is referred to as a diabetogenic (diabetes-causing) or hyperglycemic hormone.

During pregnancy, prolactin stimulates the breast development necessary for eventual lac-

Figure 9-5 Anterior pituitary hormones and their target organs: adrenocorticotropic hormone (ACTH); somatotropic (growth) hormone (GH or STH); thyroid-stimulating hormone (TSH); follicle-stimulating hormone (FSH); luteinizing hormone (LH); male analog of LH formerly called (ICSH); melanocyte-stimulating hormone (MSH).

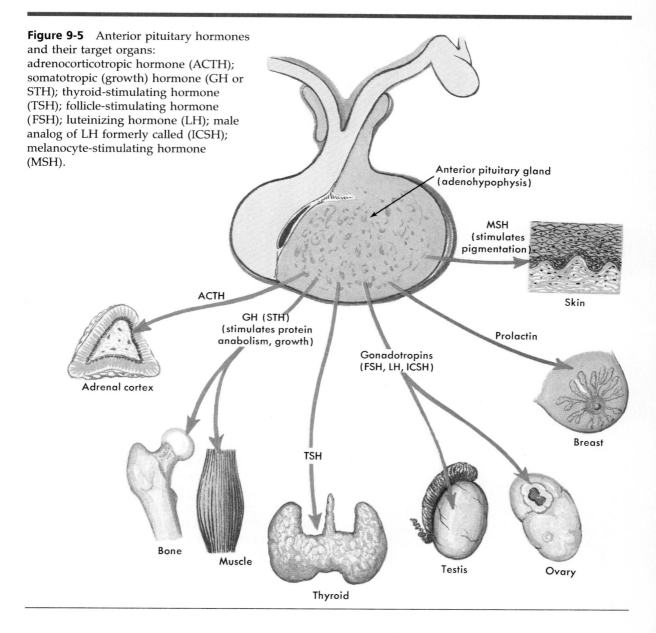

tation (milk secretion). Also, soon after delivery of a baby, prolactin stimulates the breasts to start secreting milk, a function suggested by prolactin's other name, lactogenic hormone.

For a brief summary of anterior pituitary hormone functions, see Figures 9-4 and 9-5.

POSTERIOR PITUITARY GLAND HORMONES

The posterior pituitary gland releases two hormones—**antidiuretic** (an-ti-di-u-RET-ik) **hormone (ADH)** and **oxytocin** (ok-se-TOW-sin). ADH accelerates the reabsorption of water from urine in kidney tubules back into the blood. With

Figure 9-6 Posterior pituitary hormones and their functions.

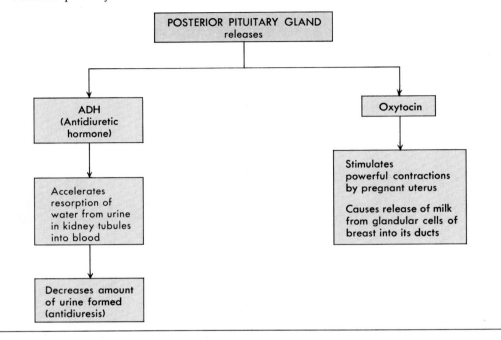

Diabetes insipidus

Diabetes insipidus (di-ah-BE-tes in-SIP-i-dus) is the name of the disease caused by hyposecretion of the antidiuretic hormone by the posterior pituitary gland. This condition is marked by the elimination of extremely large volumes of urine each day. In untreated cases, patients may excrete as much as 25 to 30 *liters* of urine in a 24-hour period. In addition to voiding an abnormally large quantity of urine, these individuals suffer from great thirst, dehydration, and serious electrolyte imbalances. Treatment of diabetes insipidus requires the administration of replacement quantities of antidiuretic hormone, either by injection or by absorption of the hormone through the mucosa of the nose when given as a nasal spray.

more water moving out of the tubules into the blood, less water remains in the tubules, and therefore less urine leaves the body. The reason the name *antidiuretic hormone* is an appropriate one is that *anti-* means "against" and *diuretic* means "increasing the volume of urine excreted." Therefore antidiuretic means "acting against an increase in urine volume"; in other words, the antidiuretic hormone acts to decrease urine volume.

The posterior pituitary hormone oxytocin is secreted by a woman's body before and after she has a baby. Oxytocin stimulates contraction of the smooth muscle of the pregnant uterus and so is believed to initiate and maintain labor. This is why physicians sometimes prescribe oxytocin injections to induce or increase labor. Oxytocin also performs a function that is important to a newborn baby. It causes the glandular cells of the breast to release milk into ducts from which a baby can obtain it by sucking. Figure 9-6 summarizes posterior pituitary functions.

Figure 9-7   A pituitary giant and dwarf contrasted with normal-sized men. Excessive secretion of growth hormone by the anterior lobe of the pituitary gland during the early years of life produces gigantism, whereas deficient secretion of this substance produces well-formed dwarfs.

HYPOTHALAMUS

In discussing the hormones ADH and oxytocin we noted that these hormones were *released* from the posterior lobe of the pituitary. Actual production of these two hormones occurs in the hypothalamus of the brain. Two groups of specialized neurons in the hypothalamus called the **supraoptic** (su-prah-OP-tic) and **paraventricular** (par-ah-ven-TRIK-u-lar) **nuclei** synthesize the posterior pituitary hormones, which then pass down along axons into the pituitary gland. Release of both ADH and oxytocin into the blood is controlled by nervous stimulation. In addition

to ADH (secreted by supraoptic neurons) and oxytocin (secreted by paraventricular neurons), the hypothalamus is now known to play a dominant role in the production and release of many hormonelike substances that have dramatic effects on the control of many body functions related to homeostasis. Examples include the regulation of body temperature, appetite, and thirst.

In addition to oxytocin and ADH, the hypothalamus also produces substances called **releasing** and **inhibiting hormones.** These substances are produced in the hypothalamus and then travel through a specialized blood capillary

Figure 9-8 Acromegaly. Note large head, exaggerated forward projection of jaw, and protrusion of frontal bone.

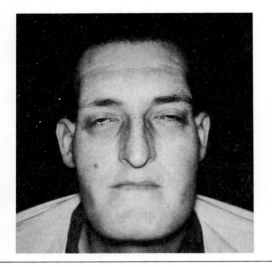

Figure 9-9 Acromegaly. Notice enlarged skin pores and separation of lower teeth.

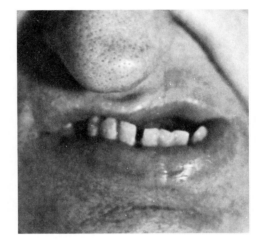

Growth hormone abnormalities

Excessive or hypersecretion of growth hormone during the early years of life produces a condition called **gigantism** (ji-GAN-tizm), illustrated in Figure 9-7. The name suggests the obvious characteristics of this condition. The child grows to giant size. Undersecretion or hyposecretion of the growth hormone produces **dwarfism** (DWARF-izm). If the anterior pituitary gland secretes too much growth hormone after the growth years, then the disease called **acromegaly** (ak-row-MEG-ah-lee) develops. Characteristics of this disease are enlargement of the bones of the hands, feet, jaws, and cheeks. The facial appearance that is typical of acromegaly results from the combination of bone and soft tissue overgrowth. A prominent forehead and large nose are characteristic (Figure 9-8). In addition, Figure 9-9 shows the skin characterized by large, widened pores, and the mandible, which grows in length so that separation of the lower teeth, called "overbite" or "lantern jaw," commonly occurs.

system to the anterior pituitary gland, where they either cause the release of anterior pituitary hormones or, in a number of instances, inhibit both their production and their release into the general circulation.

THYROID GLAND

Earlier in this chapter we mentioned that some endocrine glands are not located in a body cavity. The thyroid is one of these. It lies in the neck just below the larynx (Figure 9-10).

Figure 9-10 Thyroid and parathyroid glands. Note their relations to each other and to the larynx (voice box) and trachea.

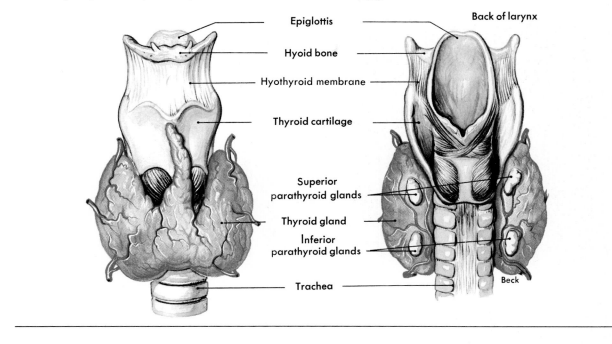

The thyroid gland secretes two thyroid hormones called **thyroxine** (thi-ROK-sin) or **T₄** and **triiodothyronine** (tri-i-o-do-THI-ro-nine) or **T₃**. It also secretes the hormone **calcitonin** (kal-si-TOE-nin). Of the two thyroid hormones thyroxine is the more important and more abundant. One molecule of thyroxine contains four atoms of iodine, and one molecule of triiodothyronine, as its name suggests, contains three iodine atoms. In order for thyroxine to be produced in adequate amounts the diet must contain sufficient iodine.

Most endocrine glands do not store their hormones, but secrete them directly into the blood as they are produced. The thyroid gland is different in that it stores considerable amounts of the thyroid hormones in the form of a *colloid* compound seen in Figure 9-11. The colloid material is stored in the follicles of the gland and, when the thyroid hormones are needed, they

are released from the colloid and secreted into the blood.

The thyroid hormones influence every one of the trillions of cells in our bodies. They make them speed up their release of energy from foods. In other words, the thyroid hormones stimulate cellular metabolism. This has far-reaching effects. Because all body functions depend on a normal supply of energy, they all depend on normal thyroid secretion. Even normal mental and physical growth and development depend on normal thyroid functioning.

Calcitonin tends to decrease the concentration of calcium in the blood, probably by first acting on bone to inhibit its breakdown. With less bone being resorbed, less calcium moves out of bone into blood, and, as a result, the concentration of calcium in blood decreases. An increase in calcitonin secretion quickly follows any increase in blood calcium concentration even if

Figure 9-11 Thyroid gland. Note that each of the follicles is filled with colloid. The colloid serves as a storage medium for the thyroid hormones. (x140.)

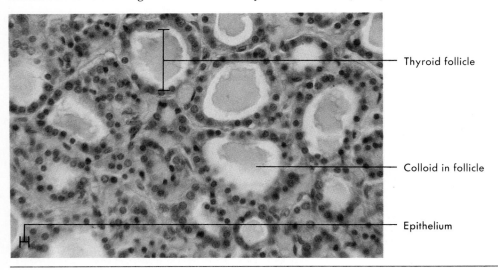

— Thyroid follicle

— Colloid in follicle

— Epithelium

Thyroid hormone abnormalities

Hyperthyroidism (hi-per-THI-roid-izm) or oversecretion of the thyroid hormones dramatically increases the metabolic rate. Food material is burned by the cells at an excessive rate, and these individuals tend to lose weight, have an increased appetite, and show signs of nervous irritability. They appear restless, jumpy, and excessively active. Many patients with hyperthyroidism also have very prominent, almost protruding eyes (Figure 9-12).

Hypothyroidism (hi-po-THI-roid-izm) or undersecretion of thyroid hormones can be caused by and result in a number of different conditions. Low dietary intake of iodine causes a painless enlargement of the thyroid gland called **simple goiter** (GOY-ter), shown in Figure 9-13. This condition was once common in areas of the United States where the iodine content of the soil and water is inad-

equate. The use of iodized salt has dramatically reduced the incidence of simple goiter caused by low iodine intake. In simple goiter the gland enlarges in an attempt to compensate for the lack of iodine in the diet necessary for the synthesis of thyroid hormones.

Hyposecretion of thyroid hormones during the formative years leads to a condition called **cretinism** (KREE-tin-izm). It is characterized by a low metabolic rate, retarded growth and sexual development, and, frequently, mental retardation. Later in life, deficient thyroid hormone secretion produces the disease called **myxedema** (mik-se-DEE-muh). The low metabolic rate that characterizes myxedema leads to lessened mental and physical vigor, weight gain, loss of hair, and an accumulation of fluid in the subcutaneous tissues that is often most noticeable around the eyes (Figure 9-14).

Figure 9-12 Hyperthyroidism. Note the agitated facial expression and the prominent, protruding eyes.

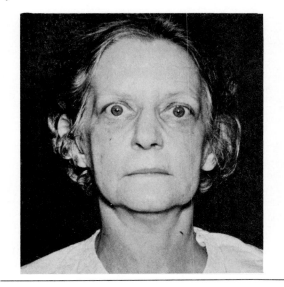

Figure 9-13 Simple goiter. The enlarged thyroid gland is caused by low iodine intake.

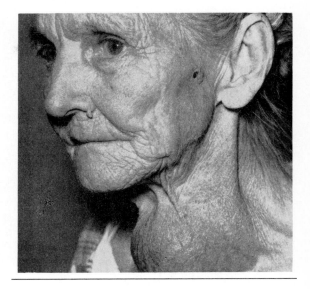

Figure 9-14 Myxedema, a condition produced by hyposecretion of the thyroid gland during the adult years. Note the edema around the eyes and the facial puffiness.

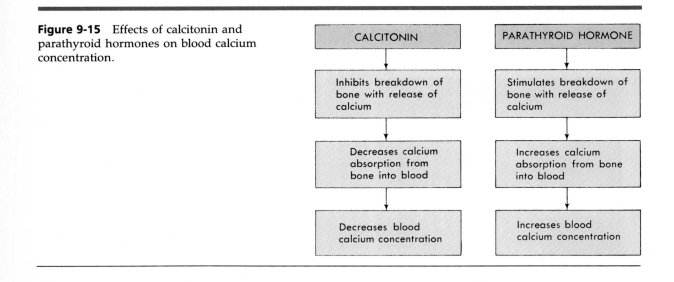

Figure 9-15 Effects of calcitonin and parathyroid hormones on blood calcium concentration.

it is a slight one. This causes blood calcium concentration to decrease back to its normal lower level. Calcitonin thus functions to help maintain homeostasis of blood calcium. It prevents a harmful excess of calcium in the blood, a condition called **hypercalcemia** (hi-per-kal-SE-me-ah), from developing.

PARATHYROID GLANDS

The **parathyroid glands** are small glands. There are usually four of them, and they are found on the back of the thyroid gland (Figure 9-10). The parathyroid glands secrete **parathyroid hormone (PTH).**

Parathyroid hormone tends to increase the concentration of calcium in the blood—the opposite effect of the thyroid gland's calcitonin. Whereas calcitonin acts to decrease the amount of calcium being resorbed from bone, parathyroid hormone acts to increase it. Parathyroid hormone stimulates bone-resorbing cells or osteoclasts to increase their breakdown of bone's hard matrix, a process that frees the calcium stored in the matrix. The released calcium then moves out of bone into blood, and this in turn increases blood's calcium concentration. For a summary of the antagonistic effects of calcitonin and parathyroid hormone, see Figure 9-15. This

is a matter of life-and-death importance because our cells are extremely sensitive to changing amounts of blood calcium. They cannot function normally either with too much or with too little calcium. For example, with too much blood calcium, brain cells and heart cells soon do not function normally; a person becomes mentally disturbed and his heart may stop. However, with too little blood calcium, nerve cells become overactive—sometimes to such an extreme degree that they bombard muscles with so many impulses that the muscles go into spasms. This is called *tetany*.

ADRENAL GLANDS

As you can see in Figures 9-1 and 9-16, an adrenal gland curves over the top of each kidney. Although from the surface an adrenal gland appears to be only one organ, it actually is two separate endocrine glands, namely, the **adrenal cortex** and the **adrenal medulla.** Does this two-glands-in-one structure remind you of another endocrine organ? (See p. 208.) The adrenal cortex is the outer part of an adrenal gland and the medulla is its inner part. Adrenal cortex hormones have different names and quite different actions from adrenal medulla hormones.

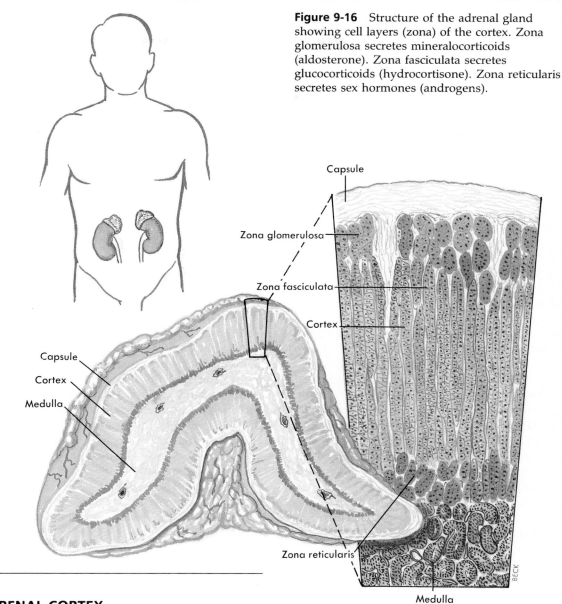

Figure 9-16   Structure of the adrenal gland showing cell layers (zona) of the cortex. Zona glomerulosa secretes mineralocorticoids (aldosterone). Zona fasciculata secretes glucocorticoids (hydrocortisone). Zona reticularis secretes sex hormones (androgens).

Capsule

Zona glomerulosa

Zona fasciculata

Cortex

Zona reticularis

Medulla

Capsule

Cortex

Medulla

ADRENAL CORTEX

Three different zones or layers of cells make up the **adrenal cortex** (Figure 9-16). Starting with the zone or layer directly under the outer capsule of the gland and proceeding toward the center, that is, going from superficial to deep, their names are:

1 Zona glomerulosa (ZO-nah glo-mer-u-LO-sa),

2 Zona fasciculata (ZO-nah fas-cic-u-LA-ta), and

3 Zona reticularis (ZO-nah re-tic-u-LA-ris). Hormones secreted by the three cell layers or zona of the adrenal cortex are called **corticoids** (KOR-ti-koids). The outer zone of adrenal cortex cells or zona glomerulosa secretes hormones called **mineralocorticoids** (min-er-al-o-KOR-ti-koids) or **MCs** for short. The main mineralocorticoid is the hormone **aldosterone** (AL-do-sterone). The middle zone or zona fasciculata secretes **glucocorticoids** (gloo-ko-KOR-ti-koids) or **GCs. Cortisol** or **hydrocortisone** is the chief glucocorticoid. The innermost or deepest zone of the cortex or zona reticularis secretes small amounts of **sex hormones.** Sex hormones secreted by the adrenal cortex resemble testosterone. We shall now discuss briefly the functions of these three kinds of adrenal cortical hormones.

One of the important functions of glucocorticoids is to help maintain normal blood glucose concentration. Glucocorticoids tend to increase **gluconeogenesis** (gloo-ko-ne-o-JEN-e-sis), a process that converts amino acids or fatty acids to glucose and that is carried on mainly by liver cells. Glucocorticoids act in several ways to increase gluconeogenesis. They promote the breakdown to amino acids of tissue proteins, especially those present in muscle cells. The amino acids thus formed move out of the tissue cells into blood and circulate to the liver. Liver cells then change them to glucose by the process of gluconeogenesis. The newly formed glucose leaves the liver cells and enters the blood. This of course increases blood glucose concentration.

In addition to performing these functions that are necessary for maintaining normal blood glucose concentration, glucocorticoids also play an essential part in maintaining normal blood pressure. They act in a complicated way to make it possible for two other hormones secreted by the medulla of the adrenal gland to partially constrict blood vessels, a condition necessary for maintaining normal blood pressure. Also, glucocorticoids act with these hormones from the adrenal medulla to produce an anti-inflammatory effect. They bring about a normal recovery from inflammations produced by many kinds of agents. The use of hydrocortisone to relieve rheumatoid arthritis, for example, is based on the anti-inflammatory effect of glucocorticoids.

Another effect produced by glucocorticoids is called their anti-immunity, antiallergy effect. Glucocorticoids bring about a decrease in the number of certain cells (lymphocytes and plasma cells) that produce antibodies, substances that make us immune to some factors and allergic to others.

When extreme stimuli act on the body, they produce an internal state or condition known as *stress.* Surgery, hemorrhage, infections, severe burns, and intense emotions are examples of extreme stimuli that bring on stress. The normal adrenal cortex responds to the condition of stress by quickly increasing its secretion of glucocorticoids. This fact is well established. What is still not known, however, is whether the increased amount of glucocorticoids helps the body cope successfully with stress. Increased glucocorticoid secretion is only one of many ways in which the body responds to stress, but it is one of the first stress responses and it brings about many of the other stress responses. Examine Figure 9-17 to discover what stress responses are produced by a high concentration of glucocorticoids in the blood.

Mineralocorticoids perform different functions from glucocorticoids. As their name suggests, they help control the amount of certain mineral salts in the blood. Aldosterone is the name of the chief mineralocorticoid. Remember its main functions—to increase the amount of sodium and decrease the amount of potassium in the blood—for these changes lead to other profound changes. Aldosterone increases blood sodium and decreases blood potassium by influencing the kidney tubules. It causes them to speed up their resorption of sodium back into the blood so that less of it will be lost in the urine. At the same time, aldosterone causes the tubules to increase their secretion of potassium so that more of this mineral will be lost in the urine. Aldosterone also tends to speed up kidney resorption of water.

ADRENAL MEDULLA

The **adrenal medulla,** or inner portion of the adrenal gland shown in Figure 9-16, secretes the hormones **epinephrine** (ep-i-NEF-rin) and **norepinephrine** (nor-ep-i-NEF-rin).

Our bodies have many ways to defend themselves against enemies that threaten their wellbeing. A physiologist might say that the body

Figure 9-17   Stress responses induced by high concentration of glucocorticoids in blood.

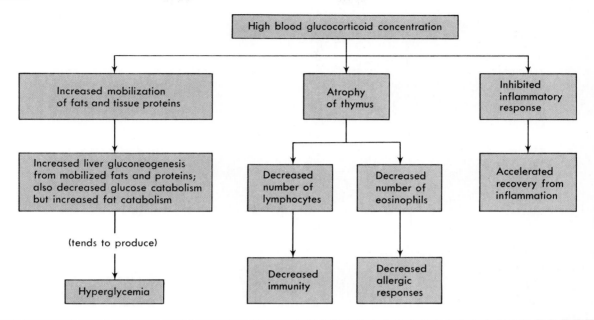

resists stress by making many stress responses. We have just discussed increased glucocorticoid secretion. An even faster-acting stress response is increased secretion by the adrenal medulla. (It occurs very rapidly because nerve impulses conducted by sympathetic nerve fibers stimulate the adrenal medulla.) When stimulated, it literally squirts epinephrine and norepinephrine into the blood. Like glucocorticoids, these hormones may help the body resist stress. Unlike glucocorticoids, they are not essential for maintaining life. Glucocorticoids, the hormones from the adrenal cortex, on the other hand, may help the body resist stress and are essential for life.

Suppose you suddenly faced some threatening situation. Imagine that your doctor told you that you had to have a dangerous operation, or that a gunman threatened to kill you. Almost instantaneously, the medullas of your two adrenal glands would be galvanized into feverish activity. They would quickly secrete large amounts of epinephrine (adrenaline) into your blood. Many of your body functions would seem

to become supercharged. Your heart would beat faster; your blood pressure would rise; more blood would be pumped to your skeletal muscles; your blood would contain more glucose for more energy, and so on. In short, you would be geared for strenuous activity, for "fight or flight." Epinephrine prolongs and intensifies changes in body function brought about by the stimulation of the sympathetic subdivision of the autonomic nervous system. Recall from Chapter 8 that sympathetic or adrenergic nerve fibers release epinephrine and norepinephrine as neurotransmitter substances.

The close functional relationship between the nervous system and the endocrine system is perhaps most noticeable in the body's response to stress. The term **general-adaptation syndrome** or **GAS** is often used to describe how the body mobilizes a number of different defense mechanisms when threatened by harmful stimuli. In generalized stress conditions the hypothalamus acts on the anterior pituitary gland to cause the release of ACTH, which stimulates the adrenal

Figure 9-18 Cushing's syndrome, the result of excess glucocorticoid hormone production by a tumor of the zona fasciculata of the adrenal cortex. **A,** Preoperatively. **B,** Six months postoperatively.

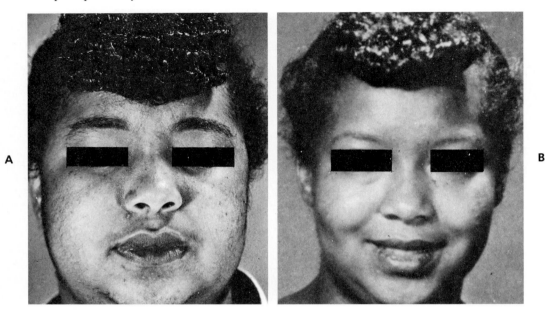

A

B

Figure 9-19 Results of a virilizing tumor of the zona reticularis of the adrenal cortex of a young girl. The tumor secretes testosterone-like sex hormones, thereby producing masculinizing effects that resemble the secondary sex characteristics of the male.

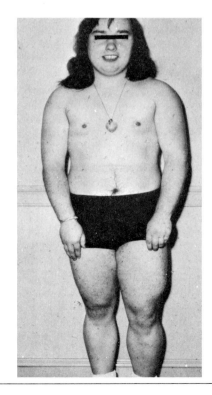

Adrenal hormone abnormalities

Injury, disease states, or malfunction of the adrenal glands can result in either hypersecretion or hyposecretion of hormones.

Tumors of the adrenal cortex located in the zona fasciculata often result in the production of abnormally large amounts of glucocorticoids. The medical name for hypersecretion of glucocorticoids is **Cushing's syndrome.** For some reason many more women than men develop Cushing's syndrome. Its most noticeable features are the so-called moon face (Figure 9-18) and the buffalo hump on the upper back that develop because of the redistribution of body fat. Individuals with Cushing's syndrome also have elevated blood sugar levels and suffer frequent infections. Surgical removal of the glucocorticoid-producing tumor shown in Figure 9-18 resulted in dramatic improvement of the moon face symptom within only 6 months.

Tumors that affect the zona reticularis often produce testosterone-like sex hormones. As a result, the symptoms of hypersecretion often resemble the male secondary sexual characteristics such as beard growth, development of body hair, and increase in muscle mass. If these masculinizing symptoms appear in a female (Figure 9-19), the cause is frequently a **virilizing** (VIR-i-li-zing) **tumor** of the adrenal cortex. The term *virile* is from the Latin word *virilis* meaning "male" or "masculine." Deficiency or hyposecretion of glucocorticoids results in a condition called **Addison's disease.** Reduced cortisone levels result in muscle weakness, reduced blood sugar, nausea, loss of appetite, and weight loss.

cortex to secrete glucocorticoids. In addition, the sympathetic subdivision of the autonomic nervous system is stimulated with the adrenal medulla, so the release of epinephrine and norepinephrine occurs to assist the body in responding to the stressful stimulus. Unfortunately, during periods of prolonged stress glucocorticosteroids may have harmful side effects because they are anti-inflammatory and cause blood vessels to constrict. For example, decreased immune activity in the body may promote the spread of infections, and prolonged blood vessel constriction may lead to increased blood pressure.

ISLANDS OF LANGERHANS

All of the endocrine glands discussed so far are big enough to be seen without a magnifying glass or microscope. The **islands of Langerhans,** in contrast, are too tiny to be seen without a microscope. These glands are merely little clumps of cells scattered like islands in a sea among the pancreatic cells that secrete the pancreatic digestive juice (Figure 9-20). Paul Langerhans discovered these cell islands in the pancreas about 100 years ago—hence their name.

Two kinds of cells in the islands of Langerhans are those called *alpha cells* and *beta cells.* Alpha cells secrete a hormone called **glucagon,** whereas beta cells secrete one of the most famous of all hormones, **insulin.**

Glucagon accelerates a process called **liver glycogenolysis** (gli-ko-je-NOL-i-sis). Glycogenolysis is a chemical process by which the glucose stored in the liver cells in the form of glycogen is converted to glucose. This glucose then leaves the liver cells and enters the blood. Glucagon, therefore, tends to increase blood glucose concentration.

Insulin and glucagon serve as antagonists. In other words, insulin tends to decrease blood glucose concentration; glucagon tends to increase it. Insulin is the only hormone that can decrease blood glucose concentration. Several other hormones, however, tend to increase its concentra-

Figure 9-20 Pancreas. Two islands of Langerhans or hormone-producing areas are evident among the pancreatic cells that produce the pancreatic digestive juice. (x140.)

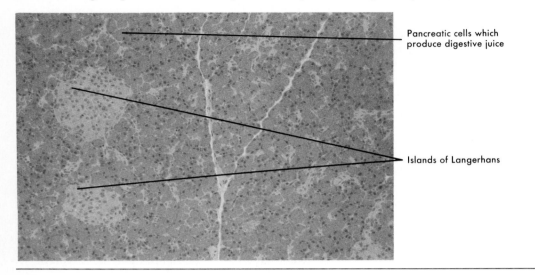

Pancreatic cells which produce digestive juice

Islands of Langerhans

tion. We have already named three of these: glucocorticoids, growth hormone, and glucagon. Insulin decreases blood glucose by accelerating its movement out of the blood, through cell membranes, and into cells. As glucose enters the cells at a faster rate, the cells increase their metabolism of glucose. Briefly then, insulin functions to decrease blood glucose and to increase glucose metabolism.

If the islands of Langerhans secrete a normal amount of insulin, a normal amount of glucose enters the cells, and a normal amount of glucose stays behind in the blood. ("Normal" blood glucose is about 80 to 120 milligrams of glucose in every 100 milliliters of blood.) If the islands of Langerhans secrete too much insulin, as they sometimes do when a person has a tumor of the pancreas, then more glucose than usual leaves the blood to enter the cells and blood glucose decreases. If the islands of Langerhans secrete too little insulin, as they do in **diabetes mellitus,** (di-ah-BE-tez mell-I-tus), less glucose leaves the

blood to enter the cells so that blood glucose increases—sometimes to even three or more times the normal amount.

In diabetes mellitus excess sugar is filtered out of the blood by the kidneys and is lost in the urine. If this condition, called **glycosuria** (gli-ko-SU-re-ah), occurs, it can be detected by a simple test for the presence of sugar in the urine. Such urine sugar tests are frequently used in screening procedures for diabetes mellitus.

FEMALE SEX GLANDS

A woman's sex glands are her two ovaries. Each ovary contains two different kinds of glandular structures, namely, the ovarian follicles and the corpus luteum. **Ovarian follicles** secrete *estrogens,* the "feminizing hormone." The **corpus luteum** chiefly secretes *progesterone* but also some estrogens. We shall save our discussion of the structure of these endocrine glands and the functions of their hormones for Chapter 17.

MALE SEX GLANDS

Some of the cells of the testes produce the male sex cells called **sperm.** Other cells in the testes, male reproductive ducts, and glands produce the liquid portion of the male reproductive fluid called *semen.* The interstitial cells in the testes secrete the male sex hormone called **testosterone** directly into the blood. These cells of the testes therefore are the male endocrine glands. Testosterone is the "masculinizing hormone." Chapter 16 contains more information about the structure of the testes and the functions of testosterone.

THYMUS

The thymus is located in the mediastinum (Figure 9-1), and in infants it may extend up into the neck as far as the lower edge of the thyroid gland. Like the adrenal gland, the thymus has a cortex and medulla. Both portions are composed largely of lymphocytes. There is no longer any doubt that the thymus functions as an endocrine gland. As part of the body's immune system, the endocrine function of the thymus is not only important but essential. This small structure (it weighs at most a little over an ounce) plays a critical part in the body's defenses against infections—in its vital immunity mechanism.

The hormone **thymosin** (THI-mo-sin) has been isolated from thymus tissue and is considered responsible for its endocrine activity. Thymosin is actually a group of several hormonelike substances that together play an important role in the development and functioning of the body's immune system.

Suppression of the immune system sometimes occurs in certain disease states and in patients who are undergoing massive chemotherapy or radiotherapy for the treatment of cancer. Such individuals are said to be "immunosuppressed" and are extremely susceptible to infections. Thymosin may prove useful as an activator of the immune system in such patients.

PLACENTA

The placenta functions as a temporary endocrine gland. During pregnancy, it produces **chorionic gonadotropins** (KO-re-on-ic gon-ah-do-TRO-pins), so called because they are secreted by cells of the **chorion** (KO-re-on), the outermost membrane that surrounds the baby during development in the uterus. In addition to chorionic gonadotropins, the placenta also produces estrogen and progesterone. During pregnancy, the kidneys excrete large amounts of chorionic gonadotropins in the urine. This fact, discovered more than a half century ago, led to the development of the now-familiar pregnancy tests.

PINEAL GLAND

The pineal gland is a small cone-shaped gland that lies near the roof of the third ventricle of the brain. It is easily located in a child but becomes more fibrous as the individual ages. This gland is sometimes called "the third eye" because it is influenced by the amount of light entering the eyes. The pineal gland secretes a hormone called **melatonin,** which has an inhibitory effect on the ovary and may in some as yet unknown way influence the menstrual cycle. Another secretion of the pineal gland is thought to influence aldosterone secretion by stimulating the adrenal cortex.

OUTLINE SUMMARY

General Functions

A Endocrine glands secrete chemicals (hormones) into the blood

B Hormones perform general functions of communication and control—but a slower, longer-lasting type of control than that provided by nerve impulses

C Cells acted on by hormones are called target organ cells

Prostaglandins

A Prostaglandins (PGs) are powerful substances found in a wide variety of body tissues

B PGs are often produced in a tissue and diffuse only a short distance to act on cells in that tissue

C Several classes of prostaglandins, including prostaglandin A (PGA), prostaglandin E (PGE), and prostaglandin F (PGF)

D PGs influence many body functions, including respiration, blood pressure, gastrointestinal secretions, and reproduction

Pituitary Gland

A Anterior pituitary gland (adenohypophysis)

  1 Names of major hormones

    a Thyroid-stimulating hormone (TSH)

    b Adrenocorticotropic hormone (ACTH)

    c Follicle-stimulating hormone (FSH)

    d Luteinizing hormone (LH)

    e Interstitial cell–stimulating hormone (ICSH)

    f Melanocyte-stimulating hormone (MSH)

    g Growth hormone (GH)

    h Prolactin (lactogenic hormone)

  2 Functions of major hormones

    a TSH—stimulates growth of thyroid gland; also stimulates it to secrete thyroid hormone

    b ACTH—stimulates growth of adrenal cortex and stimulates it to secrete glucocorticoids (mainly cortisol)

    c FSH—initiates growth of ovarian follicles each month in ovary and stimulates one or more follicles to develop to stage of maturity and ovulation; FSH also stimulates estrogen secretion by developing follicles

    d LH—acts with FSH to stimulate estrogen secretion and follicle growth to maturity; LH causes ovulation and is "the ovulating hormone"; LH causes luteinization of the ruptured follicle and stimulates progesterone secretion by corpus luteum

    e ICSH—male pituitary secretes LH but it is called interstitial cell–stimulating hormone (ICSH); causes interstitial cells in testes to secrete testosterone

    f MSH—causes rapid increase in synthesis and spread of melanin (pigment) in the skin

    g GH—stimulates growth by accelerating protein anabolism; also, tends to accelerate fat catabolism and slow glucose catabolism; by slowing glucose catabolism, GH tends to increase blood glucose to higher than normal level (hyperglycemia)

    h Prolactin (lactogenic hormone)—stimulates breast development during pregnancy and secretion of milk after delivery of baby

B Posterior pituitary gland (neurohypophysis)

  1 Names of hormones

    a Antidiuretic hormone (ADH)

    b Oxytocin

  2 Functions of hormones

    a ADH—accelerates water resorption from urine in kidney tubules into blood, thereby decreasing urine secretion (antidiuresis)

    b Oxytocin—stimulates pregnant uterus to contract; may initiate labor; causes glandular cells of breast to release milk into ducts

Hypothalamus

A Actual production (secretion) of ADH and oxytocin occurs in hypothalamus
  1 Supraoptic neurons in hypothalamus secrete ADH
  2 Paraventricular neurons in hypothalamus secrete oxytocin
B After production in hypothalamus, hormones pass along axons into pituitary gland
C Secretion and release of posterior pituitary hormones controlled by nervous stimulation
D Hypothalamus controls many body functions related to homeostasis (temperature, appetite, thirst)

Thyroid Gland

A Names of hormones
  1 Thyroid hormone—thyroxine
  2 Calcitonin
B Functions of hormones
  1 Thyroid hormone—accelerates catabolism (tends to increase basal metabolic rate)
  2 Calcitonin—tends to decrease blood calcium concentration, perhaps by inhibiting breakdown of bone with release of calcium into blood

Parathyroid Glands

A Name of hormone—parathyroid hormone
B Function of hormone—tends to increase blood calcium concentration by increasing breakdown of bone with release of calcium into blood

Adrenal Glands

A Adrenal cortex
  1 Names of hormones (corticoids)
    a Glucocorticoids (GCs)—chiefly cortisol (hydrocortisone)
    b Mineralocorticoids (MCs)—chiefly aldosterone
    c Sex hormones—small amounts of male hormones (androgens) and female hormone (estrogens) secreted by adrenal cortex of both sexes
  2 Cell layers (zonae) (Figure 9-16)
    a Zona glomerulosa—outermost layer, secretes mineralocorticoids
    b Zona fasciculata—middle layer, secretes glucocorticoids
    c Zona reticularis—deepest or innermost layer, secretes sex hormones
  3 Functions of glucocorticoids
    a Help maintain normal blood glucose concentration by increasing gluconeogenesis—the formation of "new" glucose from amino acids produced by the breakdown of proteins, mainly those in muscle tissue cells; also the conversion to glucose of fatty acids produced by the breakdown of fats stored in adipose tissue cells
    b Play an essential part in maintaining normal blood pressure—glucocorticoids make it possible for epinephrine and norepinephrine to maintain a normal degree of vasoconstriction, a condition necessary for maintaining normal blood pressure
    c Act with epinephrine and norepinephrine to produce an anti-inflammatory effect, to bring about normal recovery from inflammations of various kinds
    d Produce anti-immunity, antiallergy effect; glucocorticoids bring about a decrease in the number of lymphocytes and plasma cells and therefore a decrease in the amount of antibodies formed
    e Glucocorticoid secretion quickly increases when body is thrown into condition of stress; high blood concentration of glucocorticoids, in turn, brings about many other stress responses (see Figure 9-17)
  4 Mineralocorticoids—tend to increase blood sodium and decrease body potassium concentrations by accelerating kidney tubule reabsorption of sodium and excretion of potassium
B Adrenal medulla
  1 Names of hormones—epinephrine (adrenaline) and norepinephrine
  2 Functions of hormones—epinephrine and norepinephrine help body resist

stress by intensifying and prolonging effects of sympathetic stimulation; increased epinephrine secretion first endocrine response to stress

Islands of Langerhans
A Names of hormones
  1 Glucagon—secreted by alpha cells
  2 Insulin—secreted by beta cells
B Functions of hormones
  1 Glucagon tends to increase blood glucose by accelerating liver glycogenolysis (conversion of glycogen to glucose)
  2 Insulin tends to decrease blood glucose by accelerating movement of glucose out of blood into cells, which tends to increase glucose metabolism by cells

Female Sex Cells
Ovaries contain two kinds of cells that secrete hormones—cells of ovarian follicles and of corpus luteum; see Chapter 17

Male Sex Glands
Interstitial cells of testes secrete male hormone testosterone, which promotes development of male sex organs and male secondary sex characteristics

Thymus
A Name of hormone—thymosin
B Function of hormone—thymosin plays an important role in the development and functioning of the body's immune system

Placenta
A Name of hormones—chorionic gonadotropins
B Function of hormones—maintain the corpus luteum

Pineal Gland (Epiphysis Cerebri)
A Cone-shaped gland near roof of third ventricle of brain
  1 Glandular tissue predominates in children and young adults
  2 Fibrous tissue components increase with age
B Called "third eye" because of influence on secretory activity related to amount of light entering eyes
C Secretes melatonin, which
  1 Inhibits ovarian activity
  2 Influences menstrual cycle
D Other pineal secretions stimulate adrenal cortex and influence aldosterone secretion

NEW WORDS

acromegaly	diabetes mellitus	hypercalcemia	mineralocorticoids
antidiuresis	diuresis	hyperglycemia	myxedema
catecholamines	gigantism	hypoglycemia	prostaglandins
corticoids	glucocorticoids	luteinization	stress
cretinism	gluconeogenesis	melanin	target organ cell
Cushing's syndrome	glycogenolysis	melatonin	tetany
diabetes insipidus	goiter		

CHAPTER TEST

1. Chemicals secreted directly into the blood by endocrine glands are called
   _____ .
2. Glands that discharge their secretions into ducts are called
   _____ glands.
3. Cells acted on by a particular hormone are called _____
   organ cells.
4. Cyclic AMP is said to serve as the _____
   _____ providing communication within the cells acted
   upon by hormones.
5. The regulation of hormone levels in blood is dependent on a specialized
   homeostatic mechanism called _____ feedback.
6. Another name for the group of powerful "tissue hormones" found in
   many body tissues is _____ .
7. Production of too much hormone by a diseased gland is called
   _____ .
8. The hormone that acts on the thyroid gland and is produced by the pituitary gland is called _____ _____ hormone.
9. The anterior pituitary has the structure of an endocrine gland, whereas
   the posterior pituitary has the structure of _____ tissue.
10. Growth hormone and prolactin are both secreted by the
    _____ _____ gland.
11. The letters ADH stand for _____ hormone.
12. Secretion of too much growth hormone in an adult produces the condi-
    tion called _____ .
13. One of the primary thyroid hormones is called T$_4$ or
    _____ .
14. Low dietary intake of iodine may cause an enlargement of the thyroid
    gland called simple _____ .
15. Osteoclasts or bone-resorbing cells are stimulated by the
    _____ hormone.
16. The layer of the adrenal cortex that secretes mineralocorticoids is called
    the zona _____ .
17. Epinephrine and norepinephrine are secreted by the adrenal
    _____ .
18. The medical name for hypersecretion of glucocorticoids is
    _____ syndrome.
19. The alpha cells of the pancreatic islands of Langerhans secrete
    _____ .
20. Inadequate insulin secretion is associated with the disease
    _____ _____ .
21. The corpus luteum of the ovary secretes chiefly _____ .
22. The male sex hormone is called _____ .
23. The hormone of the thymus gland is called _____ .
24. The hormone melatonin is secreted by the _____ gland.
25. In the female the ovarian follicles secrete the hormone
    _____ .

Circle the T before each true statement and the F before each false statement.

T F   26. Both the endocrine and nervous systems perform the same general functions.

T F   27. Endocrine and exocrine glands both discharge their secretions directly into the blood.

T F   28. Hormones act on target organ cells.

T F   29. The placenta is considered an endocrine gland.

T F   30. The second messenger hypothesis is used to explain the action of steroid hormones.

T F   31. Prostaglandins are also called tissue hormones.

T F   32. The term adenohypophysis is used to describe the posterior pituitary gland.

T F   33. TSH, ACTH, FSH, and LH are all trophic hormones of the anterior pituitary gland.

T F   34. The hormone LH causes ovulation.

T F   35. Hyposecretion of antidiuretic hormone causes diabetes insipidus.

T F   36. Oxytocin and ADH are produced in the anterior pituitary gland.

T F   37. Gigantism and dwarfism are both caused by hyposecretion of growth hormone.

T F   38. Calcitonin is secreted by the pineal gland.

T F   39. Parathyroid hormone stimulates breakdown of bone with release of calcium into the blood.

T F   40. The zona fasciculata of the adrenal cortex secretes sex hormones.

T F   41. High blood glucocorticoid concentration inhibits the inflammatory response.

T F   42. Hyposecretion of glucocorticoids results in Addison's disease.

T F   43. Beta cells of the islands of Langerhans secrete insulin.

T F   44. Glycogenolysis converts glycogen to glucose.

T F   45. The thymus does not function as an endocrine gland.

REVIEW QUESTIONS

1 What endocrine glands are located in the following parts of the body:

abdominal cavity	neck
cranial cavity	pelvic cavity
mediastinum	

2 What endocrine gland is known as the "master gland"? Why?

3 How are the prostaglandins similar or dissimilar to regular hormones? Name three classes of prostaglandins.

4 What endocrine gland secretes each of the following:

ACTH	epinephrine
aldosterone	growth hormone
calcitonin	insulin
chorionic	oxytocin
gonadotropins	progesterone

5 Many changes occur in the body when it is in a condition of stress—for example, after a person has had major surgery. Name two endocrine glands that greatly increase their secretion of hormones in times of stress. Name the hormones they secrete.

6 Metabolism changes when the body is in a condition of stress. How does the metabolism of proteins, of fats, and of carbohydrates change, and what hormones cause the changes?

7 What hormones, with a high concentration present in the blood, tend to make us less immune to infectious diseases?

8 What hormone is called the "water-retaining hormone" because it decreases the amount of urine formed?

9 What hormone tends to cause potassium loss from the body through the urine?

10 What hormone is called the "salt-retaining hormone" because it makes the kidneys resorb sodium into the blood more rapidly, so that less sodium is lost in the urine?

11 Name two or more hormones that tend to increase blood glucose.

12 What hormone speeds up the rate of catabolism—that is, makes you burn up your foods faster?

13 What is the main function of each of the following:

ACTH	insulin
ADH	parathyroid hormone
calcitonin	thyroid hormone
epinephrine	testosterone

14 Which hormone is called the ovulating hormone?

15 Does acromegaly result from deficient or excess secretion of what hormone?

16 Does cretinism result from deficient or excess secretion of what hormone?

17 Does Cushing's syndrome result from deficient or excess secretion of what hormone?

18 Does diabetes insipidus result from deficient or excess secretion of ADH or insulin?

19 Does diabetes mellitus result from deficient or excess secretion of ADH or insulin?

20 What endocrine disorder produces gigantism?

21 What hormone tends to decrease blood calcium concentration? What effect does this have on bone breakdown?

22 What hormone produces opposite effects from the hormone referred to in question 21?

23 What hormone or hormones prepare the body for strenuous activity—for "fight or flight," in other words?

24 What hormone is important in the development and functioning of the immune system?

25 Name the hormone found in the urine of pregnant women.

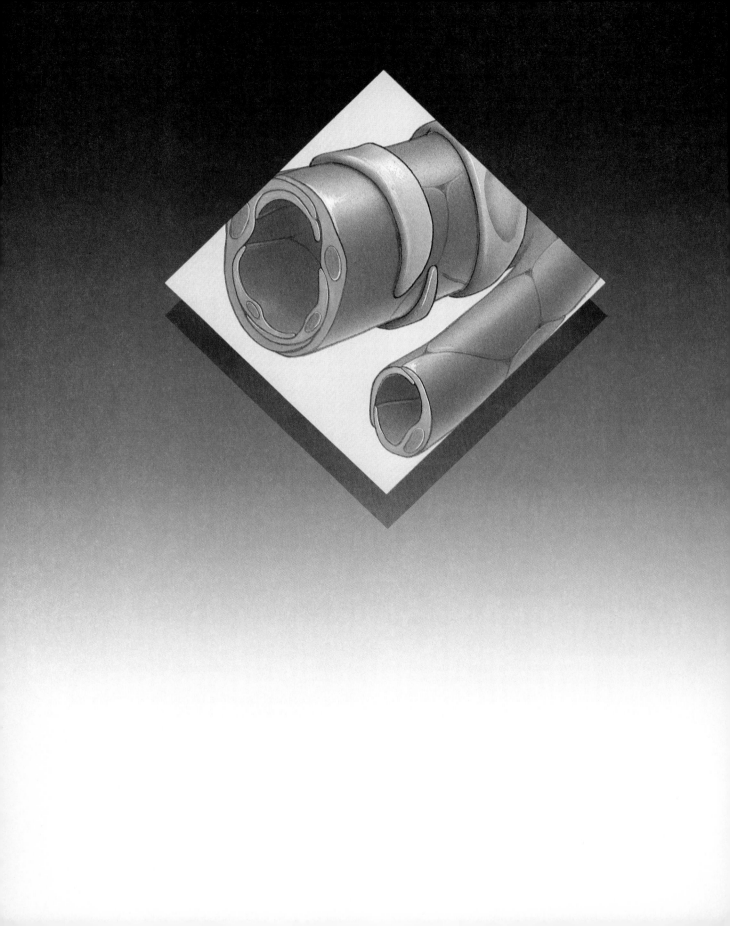

UNIT FOUR

Systems That Provide Transportation and Immunity

CHAPTER
10   Blood

OBJECTIVES

After you have completed this chapter, you should be able to:

1 Describe the generalized function of blood.

2 List the types of cells found in the formed elements of blood and identify the most important function of each.

3 Discuss anemia in terms of red blood cell numbers and hemoglobin content.

4 Explain the steps involved in blood coagulation and describe ABO and Rh blood grouping.

5 Define the following medical terms associated with blood: *hematocrit, leukocytosis, leukopenia, polycythemia, sickle cell, phagocytosis, serum, fibrinogen, Rh factor, anemia.*

Chapters 10 and 11 deal with **transportation,** one of the body's most important functions. Have you ever thought what would happen if all the transportation ceased in your city or town? How soon would you have no food to eat and how soon would rubbish pile up, for instance? Stretch your imagination just a little and you can visualize many disastrous results. Lack of food transportation alone would threaten every individual's life. Similarly, lack of food transportation to cells—the "individuals" of the body—threatens their lives. The system that provides this vital transportation service is the **circulatory system.** In this chapter we shall discuss the complex transportation fluid—blood. It is blood that performs the vital pickup and delivery services so necessary for the survival of the body's cells. The heart, blood vessels, lymphatic vessels, and lymph nodes will be discussed in Chapter 11.

BLOOD STRUCTURE AND FUNCTIONS

CELLS

Blood is a fluid that has many kinds of chemicals dissolved in it and millions upon millions of cells floating in it. The liquid part of blood that has not clotted is called **plasma.** If blood is drawn into a tube and allowed to clot, the liquid that can be separated from the clot is called **serum.** You can remember the major difference between plasma and serum if you think of serum as plasma from which fibrinogen has been removed.

There are three main types and several subtypes of blood cells:
1 Red blood cells or **erythrocytes** (e-RITH-ro-cytes)
2 White blood cells or **leukocytes** (LU-ko-cytes)
  a Granular leukocytes (have granules in their cytoplasm)
    (1) Neutrophils
    (2) Eosinophils
    (3) Basophils

  b Nongranular leukocytes (do not have granules in their cytoplasm)
    (1) Lymphocytes
    (2) Monocytes
3 Platelets or **thrombocytes** (THROM-bo-cytes)

Examine Figure 10-1 to see what each of these different kinds of blood cells looks like under the microscope.

It is difficult to believe how many blood cells there are in the body. For instance, 5,000,000 red blood cells, 7,500 white blood cells, and 300,000 platelets in 1 cubic millimeter of blood (a drop only about $\frac{1}{25}$ of an inch long and wide and high) would be considered normal red blood cell, white blood cell, and platelet counts. The term **formed elements** is often used to describe these nonfluid elements of the blood. Since red blood cells, white blood cells, and platelets are continually being destroyed, the body must continually make new ones to take their place at a really staggering rate—a few million red blood cells are manufactured *each second!*

Two kinds of connective tissue—**myeloid tissue** and **lymphatic tissue**—make blood cells for the body. Myeloid tissue is better known as red bone marrow. In the adult body it is present chiefly in the sternum, ribs, and hipbones. A few other bones such as the vertebrae, clavicles, and cranial bones also contain small amounts of this valuable substance. Red bone marrow forms all types of blood cells except some lymphocytes and monocytes. Most of these are formed by the lymphatic tissue, which is located chiefly in the lymph nodes, thymus, and spleen.

The term **anemia** (ah-NE-me-ah) is used to describe a number of different disease conditions caused by an inability of the blood to carry sufficient oxygen to the body cells. Anemias can result from: (1) inadequate numbers of red blood cells or (2) a deficiency of oxygen-carrying **hemoglobin** (he-mo-GLO-bin). Thus anemia can occur if the hemoglobin in red blood cells is inadequate even if adequate numbers of RBCs are present. Anemias caused by an actual decrease in the number of red cells can occur if: (1) blood is lost by hemorrhage as in the case of accident

Figure 10-1   Human blood cells. There are approximately 30 trillion blood cells in the adult. Each cubic millimeter of blood contains from 4½ to 5½ million red blood cells, 7,500 white blood cells, and an average of 300,000 platelets.

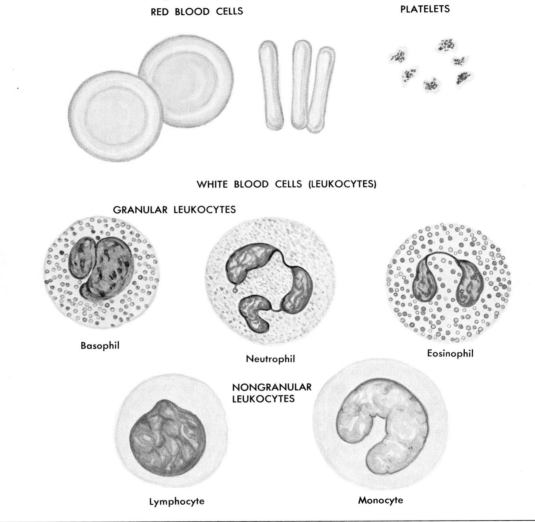

RED BLOOD CELLS

PLATELETS

WHITE BLOOD CELLS (LEUKOCYTES)

GRANULAR LEUKOCYTES

Basophil

Neutrophil

Eosinophil

NONGRANULAR LEUKOCYTES

Lymphocyte

Monocyte

victims or patients with bleeding ulcers or (2) if the blood-forming tissues cannot maintain normal numbers of blood cells. Such failures of blood-forming tissues can occur because of cancer, radiation (x-ray) damage, and certain types of infections. The term **pernicious** (per-NISH-us) **anemia** is used to describe a deficiency of red cells caused by the lack of Vitamin B_{12}. If bone marrow produces an *excess* of red blood cells, the result is a condition called **polycythemia** (pol-e-si-THE-me-ah). The blood in individuals suffering from this condition may contain so many red cells that it may become too thick to flow properly.

A common laboratory test called the **hematocrit** can tell a physician a great deal about the volume of red cells in a blood sample. If whole blood is placed in a special hematocrit tube and

Artificial blood

Transfusion of blood from one individual to another always involves some risk. For years scientists have been engaged in the search for an acceptable blood substitute that would eliminate many of the risks of transfusion and yet provide a critically ill person with one or more of the benefits of additional whole blood. One of the most important functions that transfused blood performs is additional oxygen transport to the body tissues; therefore one of the critical requirements for any artificial substitute is an ability to function in the same way. Extensive clinical trials of an artificial blood capable of carrying significant amounts of oxygen are currently in progress in the United States and Japan. A complex chemical called **Fluosol-DA** is capable of delivering substantial amounts of oxygen to tissues and may eventually prove useful in treating serious anemias, carbon monoxide poisoning, and heart attacks.

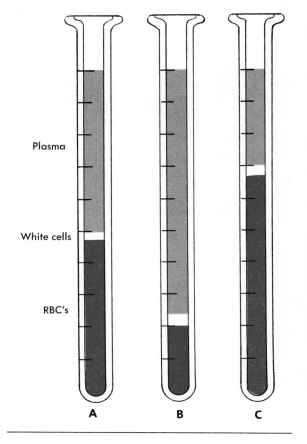

Figure 10-2 Hematocrit tubes showing normal blood, anemia, and polycythemia. Note the buffy coat located between the packed red cells and the plasma. **A,** Normal percent RBC. **B,** Anemia (low percent RBC). **C,** Polycythemia (high percent RBC).

then "spun down" in a centrifuge, the heavier formed elements will quickly settle to the bottom of the tube. During the hematocrit procedure, the red blood cells are forced to the bottom of the tube first. The white blood cells and platelets then settle out in a layer called the **buffy coat.** In Figure 10-2 the buffy coat can be seen between the packed red blood cells on the bottom of the hematocrit tube and the liquid layer of plasma above. Normally about 45% of the blood volume consists of red cells. For a patient with anemia, the percentage of red cells will drop, and for a patient with polycythemia, it will increase dramatically (Figure 10-2).

The term **leukopenia** (lu-ko-PE-ne-ah) refers to an abnormally low white blood cell count (under 5000 WBCs/dl of blood). A number of disease conditions may affect the immune system and decrease the amount of circulating white blood cells. **Acquired immune deficiency syn-**

drome or **AIDS,** which will be discussed in Chapter 12, is one example of a disease that is characterized by marked leukopenia. **Leukocytosis** (lu-ko-SI-toe-sis) refers to an abnormally high white blood cell count (over 10,000 WBCs/dl of blood). It is a much more common problem than leukopenia and almost always accompanies infections. There is also a malignant disease, **leukemia** (lu-KEY-me-ah), in which the number of white blood cells increases tremendously. You may have heard of this disease as "blood cancer." The buffy coat is thicker and more notice-

Figure 10-3 Scanning electron photomicrograph of normal red blood cells. (x3000.)

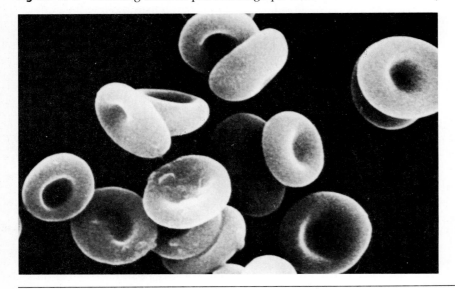

able in the hematocrit of blood from patients with leukemia because of the elevated white blood cell counts in these individuals.

Red Blood Cell Functions

Red blood cells perform several important functions. They transport oxygen to all the other cells in the body. The hemoglobin in them unites with oxygen to form **oxyhemoglobin** (ok-se-he-mo-GLO-bin). It is this combined oxygen-hemoglobin complex that makes possible the efficient transport of large quantities of oxygen to body cells.

Iron is a critical component of the hemoglobin molecule. Without adequate iron in the diet the body cannot manufacture enough hemoglobin. The result is **iron deficiency anemia**—a worldwide medical problem. If hemoglobin falls below the normal level, as it does in anemia, it starts an unhealthy chain reaction: less hemoglobin—less oxygen transported to cells—slower breakdown and use of nutrients by cells—less energy produced by cells—decreased cellular functions. If you understand this relationship between he-moglobin and energy, you can guess correctly that an anemic person's chief complaint will probably be that he or she feels "so tired all the time." Red blood cells perform another essential function: besides transporting oxygen, they also transport carbon dioxide.

As you can see in Figures 10-1 and 10-3, red blood cells have a most unusual shape. The cell is "caved in" on both sides so that each one has a thin center and thicker edges. Figure 10-3 shows red blood cells photographed with a scanning electron microscope. With this instrument, extremely small objects can be enlarged far more than is possible with a standard light microscope, and, as you can see in the illustration, objects appear more three-dimensional. Because of their unique shape and because of the large numbers of red cells, their total surface area is enormous! It provides an area larger than a football field for the exchange of oxygen and carbon dioxide between blood and the body's cells.

Sickle cell anemia is a severe and sometimes fatal hereditary disease caused by the production of an abnormal type of hemoglobin. In af-

Figure 10-4 Sickle cell. This abnormal type of red blood cell is caused by the presence of defective hemoglobin. Sickle cell anemia is a hereditary disease.

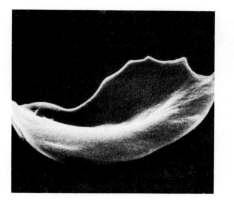

fected individuals the defective hemoglobin is produced because of the inheritance of a single gene. The faulty gene causes production of varying amounts of a less soluble hemoglobin that crystallizes much like the formation of ice in freezing water when oxygen levels in the blood are reduced. This crystallization of hemoglobin causes affected red blood cells to be distorted into a sickle shape (Figure 10-4). These odd-shaped cells not only cannot perform their role of oxygen transport, they also obstruct blood flow.

White Blood Cell Functions

White blood cells carry on a function perhaps slightly less vital than that of red cells but one nevertheless that often saves our lives. They de-

Figure 10-5 Phagocytosis of bacteria by a white blood cell. Phagocytosis can occur in the blood stream, or white cells may squeeze through capillary walls and destroy bacteria in the tissues.

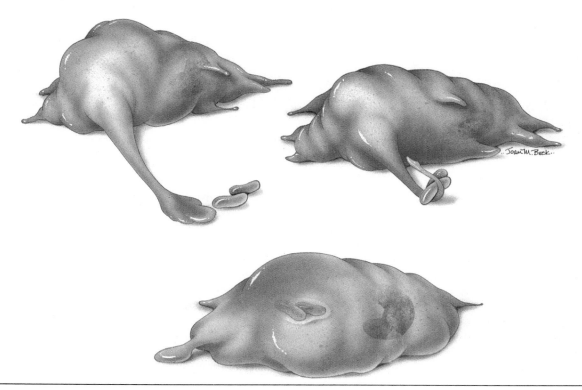

Figure 10-6 Leukocytes in human blood smears. **A,** Neutrophil. **B,** Lymphocyte.

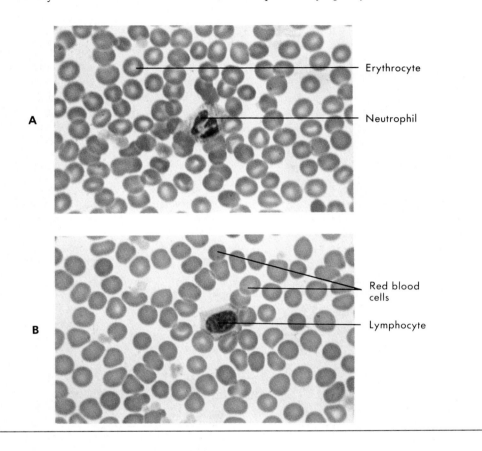

fend the body from microorganisms that have succeeded in invading the tissues or bloodstream. **Neutrophils** (NU-tro-fils) (Figure 10-6, *A*) and **monocytes** (MON-o-cytes) (Figure 10-6, *D*), for example, engulf microbes. They actually take them into their own cell bodies and digest them. The process is called **phagocytosis** (fag-o-si-TOE-sis), and the cells that carry on this process are called **phagocytes** (FAG-o-cytes) (Figure 10-5). Most numerous of the phagocytes are the neutrophils.

White blood cells of the type called **lymphocytes** (LIM-fo-cytes) (Figure 10-6, *B*) also help protect us against infections, but they do it by a process different from phagocytosis. Lympho-cytes function in the immune mechanism, the complex process that makes us immune to infectious diseases. The immune mechanism starts to operate, for example, when microbes invade the body. In some way their presence acts to stimulate lymphocytes to start multiplying and to become transformed into plasma cells. Presumably each kind of microbe can stimulate only one specific kind of lymphocyte to multiply and form one kind of plasma cell that makes a specific antibody. This antibody is specific for destroying the particular microbe that invaded the body in the first place and set the immune mechanism in operation. Details of the immune system will be discussed in Chapter 12.

Figure 10-6, cont'd.   **C,** Basophil. **D,** Monocyte.

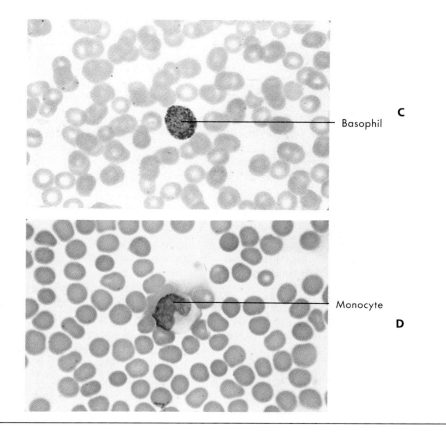

Basophil   C

Monocyte   D

Blood doping

Reports that Olympic athletes have used blood transfusions to improve performance have surfaced repeatedly during the past 20 years. The practice—called **blood doping** or **blood boosting**—is intended to increase oxygen delivery to muscles. Theoretically, infused red blood cells and elevation of hemoglobin levels following transfusion should increase oxygen consumption and muscle performance during exercise. In practice, however, questions of how effective transfusions might be in affording even world-class and professional athletes substantial competitive advantages have not been resolved. Improved performance might result but the advantage appears to be minimal.

All blood transfusions carry some risk and unnecessary or questionably indicated transfusions are medically unacceptable.

Figure 10-7   Diagram showing the main steps in blood clotting—a process far more complex than is indicated here.

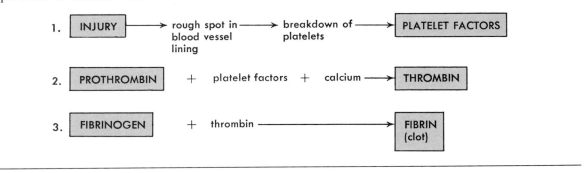

Eosinophils (e-o-SIN-o-fils) are granulocytic white blood cells that help protect the body from numerous irritants that cause allergies. They are also capable of phagocytosis. **Basophils** (BA-so-fils) (Figure 10-6, C) also function in allergic reactions. These rare leukocytes also secrete a number of important substances including the potent chemical **heparin.** Heparin may help to prevent the clotting of blood as it flows through the blood vessels of the body.

Platelet Function (Blood Clotting)

Platelets, the third main type of blood cells, play an essential part in blood clotting. Your life might someday be saved just because your blood can clot. A clot plugs up torn or cut vessels and so stops bleeding that otherwise might prove fatal. The story of how blood clots is the story of a chain of rapid-fire reactions. The first step in the chain is some kind of an injury to a blood vessel that makes a rough spot in its lining. (Normally the lining of blood vessels is extremely smooth.) Almost immediately some of the blood platelets break up as they flow over the rough spot in the vessel's lining and release a substance into the blood that leads to the formation of other substances called **platelet factors.** Platelets become "sticky" at the point of injury and soon accumulate near the opening in the cut blood vessel. As platelet numbers increase, so does the release of platelet factors. Soon the next step in the blood clotting mechanism is triggered. Plate-

let factors combine with **prothrombin** (pro-THROM-bin) (a protein present in normal blood), calcium, and other substances to form **thrombin** (THROM-bin). Then in the last step thrombin reacts with **fibrinogen** (fi-BRIN-o-jen) (another protein present in normal blood) to change it to a gel called **fibrin.** Under the microscope fibrin looks like a tangle of fine threads with red blood cells caught in the tangle. Figures 10-7 and 10-8 illustrate the steps in the blood clotting mechanism.

The clotting mechanism just described contains clues for ways to stop bleeding by speeding up blood clotting. For example, you might simply apply gauze to a bleeding surface. Its slight roughness would cause more platelets to break down and release more platelet factors. The additional platelet factors would then make the blood clot more quickly.

Physicians frequently prescribe vitamin K before surgery. Why? The reason is to make sure that the patient's blood will clot fast enough to prevent hemorrhage. Figure 10-9 shows the somewhat roundabout way in which vitamin K acts to hasten blood clotting.

Unfortunately, clots sometimes form in uncut blood vessels of the heart, brain, lungs, or some other organ—a dreaded thing, because clots may produce sudden death by shutting off the blood supply to a vital organ. When a clot stays in the place where it formed, it is called a **thrombus** (THROM-bus) and the condition is spoken

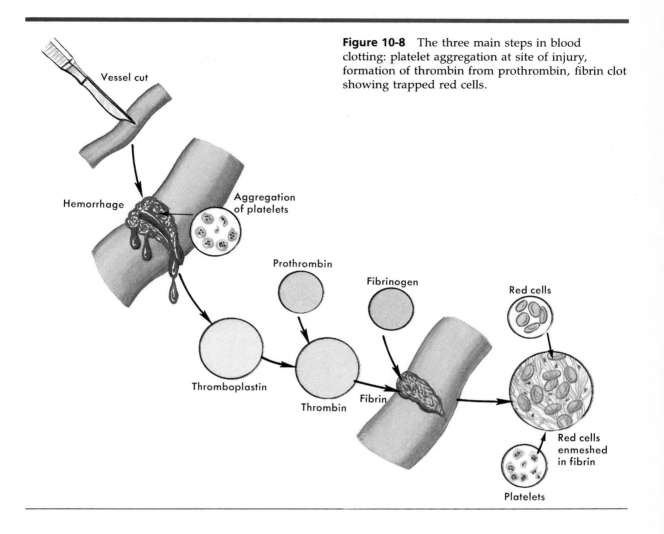

Figure 10-8 The three main steps in blood clotting: platelet aggregation at site of injury, formation of thrombin from prothrombin, fibrin clot showing trapped red cells.

of as **thrombosis** (THROM-bo-sus). If part of the clot dislodges and circulates through the bloodstream, the dislodged part is then called an **embolus** (EM-bo-lus) and the condition is called an **embolism** (EM-bo-lizm). Suppose that your doctor told you that you had a clot in one of your coronary arteries. Which diagnosis would he make—coronary thrombosis or coronary embolism—if he thought that the clot had formed originally in the coronary artery as a result of the accumulation of fatty material in the vessel wall? Doctors now have some drugs that they can use to help prevent thrombosis and embolism. Dicumarol is one. It blocks the stimulating effect of vitamin K on the liver, and consequently the liver cells make less prothrombin. The blood prothrombin content soon falls low enough to prevent clotting.

BLOOD PLASMA AND SERUM

Blood plasma is the liquid part of the blood, or blood minus its cells. It consists of water with many substances dissolved in it. All of the chemicals needed by cells to stay alive—food, oxygen,

Figure 10-9 How vitamin K acts to accelerate blood clotting.

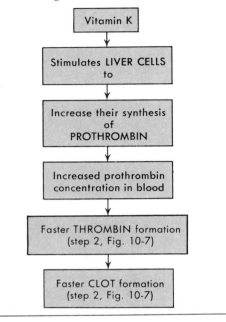

part of whole blood's volume. Examples of normal volumes are the following: plasma volume—2.6 liters; blood cell volume—2.4 liters; total blood volume—5 liters.

If you read many advertisements or watch many television commercials, you may think that almost everyone has "acid blood" at some time or other. Nothing could be farther from the truth. Blood is alkaline; it rarely reaches even the neutral point. If the alkalinity of your blood decreases toward neutral, you are a very sick person; in fact, you have what is called **acidosis.** But even in this condition, blood almost never becomes the least bit acid; it just becomes less alkaline than normal.

BLOOD TYPES (OR BLOOD GROUPS)

Blood types are identified by the presence of certain antigens in the red blood cells. An **antigen** (AN-ti-jen) is a substance that can stimulate the body to make antibodies. Almost all substances that act as antigens are foreign proteins. In other words, they are not the body's own natural proteins but are proteins that have entered the body from the outside—by injection of transfusion or some other method.

The word "antibody" can be defined in terms of what causes its formation or in terms of how it functions. Defined the first way, an **antibody** (an-ti-BOD-e) is a substance made by the body in response to stimulation by an antigen. Defined according to its functions, an antibody is a substance that reacts with the antigen that stimulated its formation. Many antibodies react with their antigens to clump or agglutinate (ah-GLOO-tin-ate) them. In other words, they cause their antigens to stick together in little clusters (Figure 10-10).

Every person's blood belongs to one of the following four blood types:

1 Type A
2 Type B
3 Type AB
4 Type O

Suppose that you have type A blood (as do about 41% of Americans). The letter *A* stands for a certain type of antigen (a protein) present in the cytoplasmic membrane of your red blood cells when you were born. Because you were born

and salts, for example—have to be brought to them by the blood. Food and salts are dissolved in plasma; so, too, is a small amount of oxygen. (Most of the oxygen in the blood is carried in the red blood cells as oxyhemoglobin.) Wastes that cells must get rid of are dissolved in plasma and transported to the excretory organs. And, finally, the hormones that help control our cells' activities and the antibodies that help protect us against microorganisms are dissolved in plasma.

Many people seem curious about how much blood they have. The amount depends on how big they are and whether they are male or female. A big person has more blood than a small person, and a man has more blood than a woman. But as a general rule, most adults probably have between 4 and 6 quarts (4 and 6 liters) of blood. It normally accounts for about 7% to 9% of the total body weight.

The volume of the plasma part of blood is usually a little more than half the volume of whole blood. Blood cells make up the remaining

Figure 10-10   Appearance of agglutination or clumping tests used to type blood. Blood type O red cells do not clump when mixed with either anti-A or anti-B serum; blood type A red cells clump when mixed with anti-A serum; and blood type B red cells clump with anti-B serum. Red cells in type AB blood are agglutinated (clumped) by both anti-A and anti-B serums.

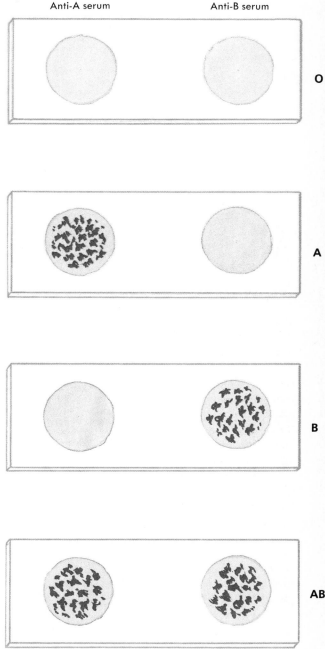

with type A antigen, your body does not form antibodies to react with it. In other words, your blood plasma contains no anti-A antibodies. It does, however, contain anti-B antibodies. For some unknown reason these antibodies are present naturally in type A blood plasma. The body did not form them in response to the presence of their antigen. In summary, then, in type A blood the red blood cells contain type A antigen and the plasma contains anti-B antibodies.

In type AB blood, as its name indicates, the red blood cells contain both type A and type B antigens and the plasma contains neither anti-A nor anti-B antibodies. The opposite is true of type O blood—its red blood cells contain neither type A nor type B antigens and its plasma contains both anti-A and anti-B antibodies.

Harmful effects or even death can result from a blood transfusion if the donor's red blood cells become agglutinated by antibodies in the recipient's plasma. If a donor's red blood cells do not contain any A or B antigen, they of course cannot be clumped by anti-A or anti-B antibodies. For this reason the type of blood that contains neither A nor B antigens—namely, type O blood—can be used as donor blood without the danger of anti-A or anti-B antibodies clumping its red blood cells. Type O blood is therefore called **universal donor** blood. **Universal recipient** blood is type AB, it contains neither anti-A nor anti-B antibodies in its plasma. Therefore it cannot clump any donor's red blood cells containing A or B antigens.

Rh Factor

In recent years the expression **Rh-positive** blood has become a familiar one. It means that the red blood cells of this type of blood contain

Figure 10-11 Mechanism used to identify the Rh factor in blood. Anti-Rh antibodies added to blood containing the Rh antigen (Rh⁺ blood) will cause agglutination. Addition of anti-Rh antibodies to blood from an individual who does not have an Rh antigen (Rh⁻ blood) will not result in agglutination.

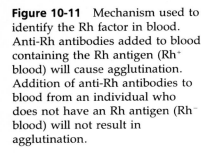
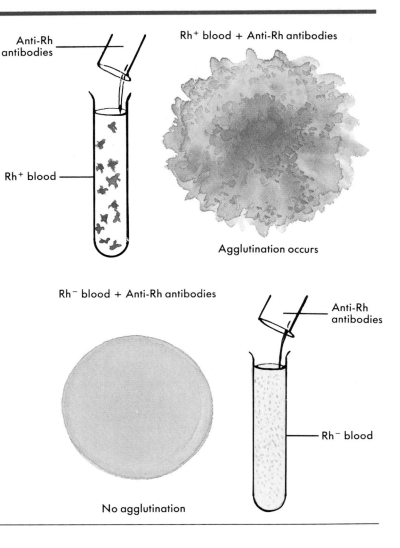

Anti-Rh antibodies

Rh⁺ blood

Rh⁺ blood + Anti-Rh antibodies

Agglutination occurs

Rh⁻ blood + Anti-Rh antibodies

Anti-Rh antibodies

Rh⁻ blood

No agglutination

an antigen called the Rh factor (Figure 10-11). If, for example, a person has type AB, Rh-positive blood, his red blood cells contain type A antigen, type B antigen, and the Rh factor antigen.

In **Rh-negative** blood the red blood cells do not contain the Rh factor. Plasma never naturally contains anti-Rh antibodies. But if Rh-positive blood cells are introduced into an Rh-negative person's body, anti-Rh antibodies soon appear in his blood plasma. In this fact lies the danger for a baby born to an Rh-negative mother and Rh-positive father. If the baby is Rh-positive, the Rh factor on its red blood cells may stimulate the mother's body to form anti-Rh antibodies. Then, if she later carries another Rh-positive fetus, it may develop a disease called **erythroblastosis fetalis** (e-rith-ro-blas-TOE-sis fe-TAL-is), caused by the mother's Rh antibodies reacting with the baby's Rh-positive cells. The term Rh is used because this important blood cell antigen was first discovered in the blood of *Rh*esus monkeys.

OUTLINE SUMMARY

Blood Structure and Functions

A Cells
  1 Kinds
    a Red blood cells (erythrocytes)
    b White blood cells (leukocytes)
      (1) Granular leukocytes—neutrophils, eosinophils, basophils
      (2) Nongranular leukocytes—lymphocytes, monocytes
    c Platelets or thrombocytes
  2 Numbers
    a Red blood cells—4½ to 5 million per cubic millimeter of blood
    b White blood cells—5,000 to 9,000 per cubic millimeter of blood
    c Platelets—300,000 per cubic millimeter of blood
  3 Formation—red bone marrow (myeloid tissue) forms all blood cells except some lymphocytes and monocytes, which are formed by lymphatic tissue in lymph nodes, thymus, and spleen
  4 Anemia—inability of blood to carry adequate oxygen to tissues caused by:
    a Inadequate red blood cell numbers
    b Deficiency of hemoglobin
    c Pernicious anemia—deficiency of Vitamin B_{12}
  5 Hematocrit—medical test in which a centrifuge is used to separate whole blood into formed elements and liquid fraction
    a Buffy coat is WBC and platelet fraction
    b Normal RBC level is 45%
  6 Clinical conditions related to blood:
    a leukopenia—abnormally low WBC count
    b leukocytosis—abnormally high WBC count
    c leukemia—"blood cancer"—elevated WBC count
  7 Red blood cell functions—transport oxygen and carbon dioxide
  8 White blood cell functions—neutrophils and monocytes carry on phagocytosis; lymphocytes function to produce immunity; eosinophils protect against irritants that cause allergies and basophils produce heparin
  9 Platelet function—Blood clotting (summarized in Figures 10-7 and 10-8)

B Blood plasma
  1 Definition—blood minus its cells
  2 Composition—water containing many dissolved substances, for example, foods, salts, hormones
  3 Amount of blood—varies with size and sex; 4 to 6 liters about average; about 7% to 9% of body weight
  4 Reaction—slightly alkaline

C Blood types or blood groups (see Figure 10-10)
  1 Type A blood—type A antigens in red cells; anti-B type antibodies in plasma
  2 Type B blood—type B antigens in red cells; anti-A type antibodies in plasma
  3 Type AB blood—type A and type B antigens in red cells; no anti-A or anti-B antibodies in plasma; therefore type AB blood is called universal recipient blood
  4 Type O blood—no type A or type B antigens in red cells; therefore type O blood is called universal donor blood; both anti-A and anti-B antibodies in plasma
  5 Rh-positive blood—Rh factor antigen present in red blood cells (Figure 10-11)
  6 Rh-negative blood—no Rh factor present in red blood cells; no anti-Rh antibodies present naturally in plasma; anti-Rh antibodies, however, appear in plasma of Rh-negative persons if Rh-positive blood cells have been introduced into their bodies

NEW WORDS

anemia
antibodies
antigens
basophil
buffy coat
embolism
embolus
eosinophil
erythroblastosis fetalis
erythrocyte

fibrin
fibrinogen
fluosol-DA
hematocrit
hemoglobin
heparin
iron deficiency anemia
leukemia
leukocyte
leukocytosis

leukopenia
lymphocyte
monocyte
neutrophil
oxyhemoglobin
pernicious anemia
plasma
phagocyte
phagocytosis
polycythemia

prothrombin
serum
sickle cell anemia
thrombocytes
thrombin
thrombosis
thrombus
universal donor
universal recipient

CHAPTER TEST

1. Erythrocytes is another name for:
   a. White blood cells
   b. Red blood cells
   c. Platelets
   d. Thrombocytes
2. Which of the following is classified as a granular leukocyte?
   a. Monocyte
   b. Neutrophil
   c. Lymphocyte
   d. None of the above are correct

3. Which of the following terms refers to an abnormally low white blood cell count?
   a. Anemia
   b. Leukemia
   c. Leukocytosis
   d. Leukopenia
4. The most numerous of the phagocytes are the:
   a. Neutrophils
   b. Red blood cells
   c. Basophils
   d. Lymphocytes
5. Heparin secretion is a function of:
   a. Monocytes
   b. Neutrophils
   c. Basophils
   d. Eosinophils
6. Vitamin K:
   a. Increases prothrombin levels in blood
   b. Is necessary for normal blood clotting
   c. Stimulates liver cells
   d. All of the above are correct
7. A thrombus:
   a. Is a stationary blood clot
   b. Is a moving blood clot
   c. Is seldom serious
   d. Is a normal component of plasma
8. Which of the following blood types is called "universal donor" blood?
   a. Type A
   b. Type AB
   c. Type O
   d. Type B
9. The liquid fraction of blood that has clotted is called:
   a. Serum
   b. Plasma
   c. Hematocrit
   d. Hemoglobin
10. Erythroblastosis fetalis is caused by:
   a. Anti Rh antibodies in the father's blood
   b. Anti Rh antibodies in the mother's blood
   c. Both of the above are correct
   d. None of the above are correct

11. The liquid part of blood that has not clotted is called
    _____ .

12. White blood cells are also called _____ .

13. A better-known name for myeloid tissue is
    _____ _____ _____ .

14. An excess of red blood cells is called _____ .

15. Sickle cell anemia is caused by the production of an abnormal type of
    _____ .

16. Prothrombin is necessary for blood to _____ .

17. A moving blood clot is called an _____ .

18. A substance that can stimulate the body to produce antibodies is called
    an _____ .

19. Universal recipient blood is type _____ .

20. The blood cell antigen first discovered in the blood of Rhesus monkeys is
    the _____ factor.

REVIEW QUESTIONS

1 What is the normal number of red blood cells per cubic millimeter of blood? White blood cells? Platelets?

2 Name the granular leukocytes.

3 What two kinds of connective tissue make blood cells for the body?

4 Suppose that your doctor told you that your "red count was 3 million." What does "red count" mean? Might the doctor say that you had any of the following conditions—anemia, leukocytosis, leukopenia, polycythemia—with a red blood count of this amount? If so, which one?

5 What does the hematocrit blood test measure? What is the normal value for this test?

6 If you had appendicitis or some other acute infection, would your white blood cell count be more likely to be 2,000, 7,000, or 15,000? Give a reason for your answer.

7 Your circulatory system is the transportation system of your body. Mention some of the substances it transports and tell whether each is carried in blood cells or in the blood plasma.

8 What organs of the body contain large numbers of reticuloendothelial cells? What role do these cells play in the circulatory system?

9 Briefly explain what happens when blood clots, including what makes it start to clot.

10 You hear that a friend has a "coronary thrombosis." What does this mean to you?

11 Define the terms *antibody* and *antigen* as they apply to blood typing.

12 Identify the four blood types.

13 What is meant by the terms "universal donor" and "universal recipient"?

14 What is the difference between blood type and Rh factor?

15 Why would a physician prescribe vitamin K before surgery for a patient with a history of bleeding problems?

The Circulatory System: Cardiovascular and Lymphatic Subdivisions

OBJECTIVES

After you have completed this chapter, you should be able to:

1 Discuss the location, size, and position of the heart in the thoracic cavity and identify the heart chambers, sounds, and valves.

2 Trace blood through the heart and compare the functions of the heart chambers on the right and left sides.

3 List the anatomical components of the heart conduction system and discuss the features of the normal electrocardiogram.

4 Compare the anatomy and function of the blood vessel types and discuss both blood pressure and pulse.

5 Discuss the systemic, pulmonary, hepatic, portal, and fetal circulations.

6 Describe the generalized functions of the lymphatic system and list the primary lymphatic structures.

Billions of cells make up our bodies, and in order to survive each cell must have numerous substances transported to and from it. The system that supplies our cells' transportation needs is the **circulatory system.** Recall from Chapter 3 that the circulatory system is composed of two major subdivisions called the **cardiovascular subdivision** and the **lymphatic subdivision.**

This chapter begins with the study of the *cardiovascular subdivision* of the circulatory system. *Cardiovascular* is an appropriate name for this subdivision because it consists of the heart, a pump that keeps the blood moving around through a closed circle of blood vessels. Not only must the blood be kept moving through the vessels by the heart's pumping action, but the amount of blood flowing to specific regions must also be varied according to their degree of activity. When a structure's activity increases, the volume of blood flow to the structure also increases. For example, during exercise, skeletal muscle activity increases and so does the volume of blood flowing to muscles; following a meal, digestive organ activity increases and so does blood flow to digestive organs.

A discussion of the *lymphatic subdivision* will conclude our coverage of the circulatory system. The lymphatic subdivision includes a specialized fluid called lymph, lymph nodes, lymphatic vessels, and specialized lymphatic organs such as the thymus, the spleen, and the tonsils. The functions of the lymphatic subdivision of the circulatory system involve both transportation and body defense. Therefore the section on the lymphatic system will be followed by a discussion of the immune system in Chapter 12.

HEART
LOCATION, SIZE, AND POSITION

No one needs to be told where the heart is or what it does. Everyone knows that the heart is in the chest, that it beats night and day to keep the blood flowing, and that if it stops, life stops.

Most of us probably think of the heart as located on the left side of the body. As you can see in Figure 11-1, the heart is located between the lungs in the lower portion of the mediastinum. Draw an imaginary line through the middle of the trachea in Figure 11-1 and continue the line down through the thoracic cavity to divide it into right and left halves. Note that about two thirds of the mass of the heart is to the left of this line and one third to the right. The heart is often described as a triangular organ, shaped and sized roughly like a closed fist. In Figure 11-1 you can see that the **apex** or blunt point of the lower edge of the heart lies on the diaphragm, pointing toward the left. To count the apical beat, one must place a stethoscope directly over the apex, that is, in the space between the fifth and sixth ribs on a line with the midpoint of the left clavicle.

The heart is positioned in the thoracic cavity between the sternum in front and the bodies of the thoracic vertebrae behind. Because of this placement, it can be compressed or squeezed by application of pressure to the lower portion of the body of the sternum using the heel of the hand. Rhythmic compression of the heart in this way can maintain blood flow in cases of cardiac arrest and, if combined with effective artificial respiration, the resulting procedure, called **cardiopulmonary resuscitation (CPR),** can be lifesaving.

ANATOMY
Heart Chambers

If you cut open a heart, you can see many of its main structural features (Figure 11-2). This organ is hollow, not solid. A partition called the **septum** divides it into right and left sides. The heart contains four cavities or hollow chambers. The two upper chambers are called **atria** (A-tre-ah), and the two lower chambers are called **ventricles** (VEN-tri-kls). The atria are smaller than the ventricles, and their walls are thinner and less muscular. Atria are often called *receiving chambers* because blood enters the heart through veins than open into these upper cavities. Eventually, blood is pumped from the heart into arteries that exit from the ventricles; therefore the ventricles are sometimes referred to as the *discharging chambers* of the heart. Each heart chamber is named according to its location. Thus there

Figure 11-1 **A,** Anatomical relationship of the heart to other structures in the thoracic cavity.

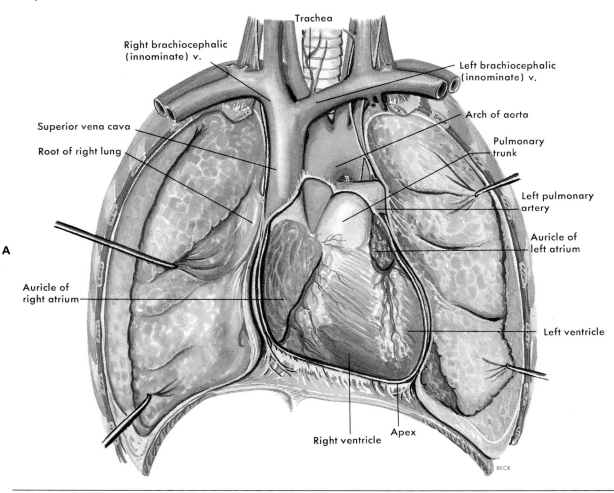

is a right and left atrial chamber above and a right and left ventricular chamber below. The wall of each heart chamber is composed of cardiac muscle tissue usually referred to as the **myocardium** (mi-o-KAR-de-um).

Each chamber of the heart is lined by a thin layer of very smooth tissue called the **endocardium** (en-do-KAR-de-um). Inflammation of this lining is referred to as **endocarditis** (en-do-kar-

DI-tis). If inflamed, the endocardial lining can become rough and abrasive to red blood cells passing over its surface. Blood flowing over a rough surface is subject to clotting, and a **thrombosis** (throm-BO-sis) or clot may form. Unfortunately, rough spots caused by endocarditis or injuries to blood vessel walls often cause the release of platelet factors. The result is often the formation of a fatal blood clot.

Figure 11-1, cont'd.   B, Heart and major blood vessels viewed from the front (anterior).

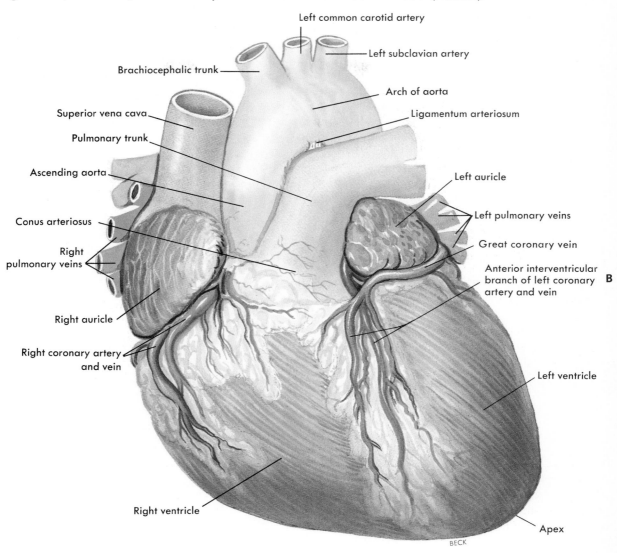

Covering Sac or Pericardium

The heart has a covering as well as a lining. Its covering, called the **pericardium** (per-i-KAR-de-um), consists of two layers of fibrous tissue with a small space in between. The inner layer of the pericardium is called the **visceral pericardium** or **epicardium** (epi-KAR-de-um). It covers the heart the way an apple skin covers an apple. The outer layer of pericardium is called the **parietal pericardium.** It fits around the heart like a loose-fitting sack, allowing enough room for the heart to beat in it. It is easy to remember the difference between the *endocardium,* which lines the heart chambers, and the *epicardium,* which

Figure 11-2 Cutaway view of the front section of the heart showing the four chambers, the valves, the openings, and the major vessels. Arrows indicate the direction of blood flow: the blue areas represent unoxygenated blood and the red areas represent oxygenated blood.

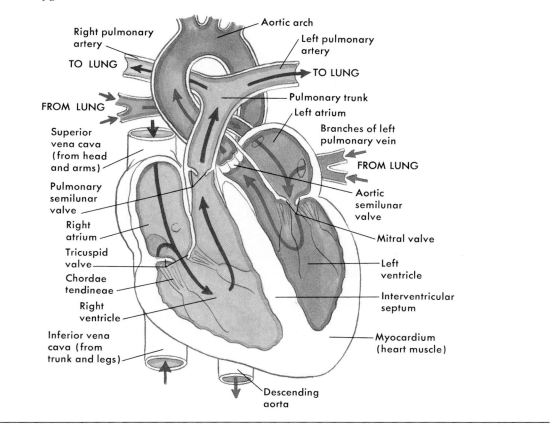

covers the surface of the heart, if you understand the meaning of the prefixes "endo-" and "epi-." "Endo-" comes from the Greek word meaning "inside" or "within," and "epi-" comes from the Greek word meaning "upon" or "on." The two pericardial layers slip against each other without friction when the heart beats because these membranes have moist, not dry, surfaces. A thin film of pericardial fluid furnishes the lubricating moistness between the heart and its enveloping pericardial sac. If the pericardium becomes inflamed, a condition called **pericarditis** (per-i-kar-di-tis) results.

Heart Valves

AV or atrioventricular valves Four valves keep blood flowing through the heart in only one direction (Figure 11-2). When the heart beats, the two atria contract first, forcing blood into the ventricles. Once filled, the two ventricles contract and force blood out of the heart. The two valves that separate the atrial chambers above from the ventricles below are called **AV** or **atrioventricular** (a-tre-o-ven-TRIK-u-lar) **valves.** The two AV valves are called the **bicuspid** or **mitral** (MI-tral) **valve,** located between the left atrium and ventricle, and the **tricuspid**

valve, located between the right atrium and ventricle. The AV valves prevent backflow of blood into the atria when the ventricles contract. Locate the AV valves in Figure 11-2. Note that a number of stringlike structures called **chordae tendineae** (KOR-de ten-DIN-e) attach the AV valves to the wall of the heart.

Semilunar valves The **semilunar** (sem-e-LU-nar) **valves** are located between the two ventricular chambers and the large arteries that carry blood away from the heart when contraction occurs. The ventricles, like the atria, contract together. Therefore the two semilunar valves both open and close at the same time. The **pulmonary semilunar valve** is located at the beginning of the pulmonary artery and allows blood to flow out of the right ventricle but prevents it from flowing back into the ventricle. The **aortic semilunar valve** is located at the beginning of the aorta and allows blood to flow out of the left ventricle up into the aorta but prevents backflow into this ventricle.

HEART SOUNDS

If a stethoscope is used to locate the apical beat of the heart, two distinct sounds can be heard. They are rhythmical and repetitive sounds that are often described as **lup-dup.**

The first or "lup" sound is caused by the vibration and abrupt closure of the AV valves as the ventricles contract. Closure of the AV valves prevents blood from rushing back up into the atria during contraction of the ventricles. This first sound is of longer duration and lower pitch than is the second. The pause between this first sound and the "dup" or second sound is shorter than that following the second sound when the "lup-dup" is repeated. The second or "dup" heart sound is caused by the closing of both the semilunar valves when the heart is in relaxation.

BLOOD FLOW THROUGH THE HEART

The heart acts as two separate pumps. The right atrium and the right ventricle function together to perform a task quite different from the left atrium and the left ventricle. When the heart "beats," first the two atria contract simulta-

neously. Then the two ventricles fill with blood and they, too, contract together. Although the upper and lower chambers of the heart contract together, the right and left sides of the heart act as separate pumps. As we study the blood flow through the heart, the separate functions of the two pumps will become clearer.

Note in Figure 11-2 that blood enters the right atrium through two large veins called the **superior vena** (VE-nah) **cava** (KA-va) and **inferior vena cava.** The right heart pump receives oxygen-poor blood from the veins. After entering the right atrium, it is pumped through the right AV or tricuspid valve and enters the right ventricle. When the ventricles contract, blood in the right ventricle is pumped through the pulmonary semilunar valve into the **pulmonary artery** and eventually to the lungs where oxygen is added and carbon dioxide is lost.

As you can see in Figure 11-2, blood rich in oxygen returns to the left atrium of the heart through four **pulmonary veins.** It then passes through the left AV or bicuspid valve into the left ventricle. When the left ventricle contracts, blood is forced through the aortic semilunar valve into the **aorta** (a-OR-ta) and is distributed to the body as a whole.

As you can tell from Figure 11-3, the two sides of the heart actually pump blood through two separate "circulations" and function as two separate pumps. The **pulmonary circulation** involves movement of blood from the right ventricle to the lungs, and the **systemic circulation** involves movement of blood from the left ventricle throughout the body as a whole.

CORONARY CIRCULATION

Blood flows into the heart muscle by way of two small vessels that are surely the most famous of all the blood vessels—the **right** and **left coronary arteries**—famous because *coronary disease* kills many thousands of people every year. The coronary arteries are the aorta's first branches. The openings into these small vessels lie behind the flaps of the aortic semilunar valves. In both *coronary thrombosis* and *coronary embolism* a blood clot occludes, or plugs up, some part of a coronary artery. Blood cannot pass

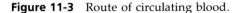

Figure 11-3 Route of circulating blood.

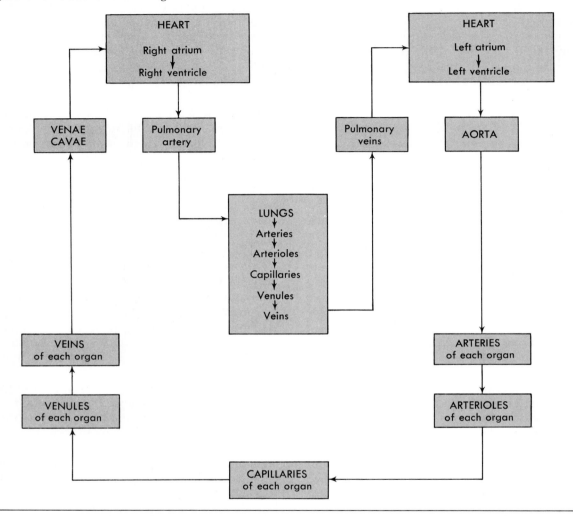

through the occluded vessel and so cannot reach the heart muscle cells it normally supplies. Deprived of oxygen, these cells soon die or are damaged (Figure 11-4). In medical terms, **myocardial** (mi-o-KAR-de-al) **infarction** (in-FARK-shun) occurs. Myocardial infarction is a common cause of death in middle-aged and elderly people. Recovery from a myocardial infarction is possible if the amount of heart tissue damaged was small enough so that nearby blood vessels can enlarge and resupply the damaged tissue.

The term **angina** (an-JI-nah) **pectoris** (PEK-tor-is) is used to describe the severe chest pain that occurs when the myocardium is deprived of adequate oxygen. It is often a warning that the coronary arteries are no longer able to supply enough blood and oxygen to certain areas of the heart. **Coronary bypass surgery** is a frequent treatment for angina. In this procedure veins are "harvested" or removed from other areas of the body and used to bypass partial blockages in coronary arteries.

F
i
c
a

portion of the coronary artery beyond the
point of blockage is damaged.

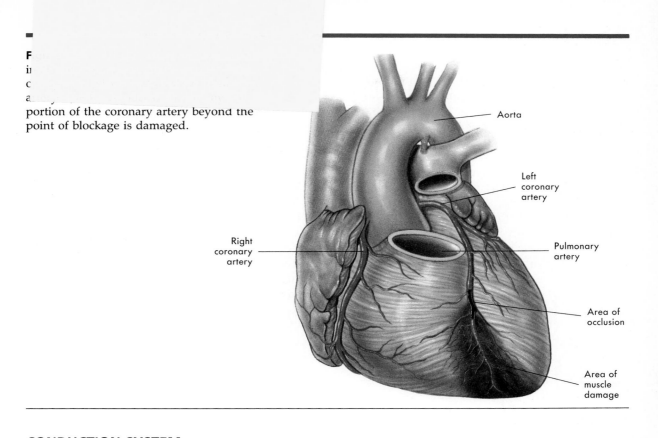

Aorta

Left
coronary
artery

Right
coronary
artery

Pulmonary
artery

Area of
occlusion

Area of
muscle
damage

CONDUCTION SYSTEM

Cardiac muscle is unique in that it is capable
of contracting even in the absence of an externally applied nervous stimulus. Even if completely removed from the body, in carefully controlled circumstances the heart is capable of beating for short periods of time. Four specialized
structures that are embedded in the wall of the
heart generate and then conduct impulses
through the heart muscle to cause first the atria
and then the ventricles to contract. The names
of the structures that make up this conduction
system of the heart are the **sinoatrial** (si-no-A-
tre-al) **node,** which is sometimes called the **SA
node** or the **pacemaker,** the **atrioventricular** (a-
tre-o-ven-TRIK-u-lar) or **AV node,** the **bundle
of His** or **AV bundle,** and the **Purkinje** (pur-
KIN-je) **fibers.** Impulse conduction normally
starts in the heart's pacemaker, namely, the SA
node. From here it spreads, as you can see in
Figure 11-5, in all directions through both atria.
This causes the atria to contract. When impulses

reach the AV node, it relays them by way of the
bundle of His and Purkinje fibers to the ventricles, causing them to contract. Normally, therefore, a ventricular beat follows each atrial beat.
Various diseases, however, can damage the
heart's conduction system and thereby disturb
the rhythmical beating of the heart. One such
disturbance is the condition commonly called
heart block. Impulses are blocked from getting
through to the ventricles, with the result that
the heart beats at a much slower rate than normal. A physician may treat heart block by im-
planting in the heart an **artificial pacemaker,** an
electrical device that causes ventricular contractions at a rate fast enough to maintain an adequate circulation of blood.

ELECTROCARDIOGRAM

The specialized structures of the heart's conduction system generate tiny electrical currents
that spread through surrounding tissues to the
surface of the body. This fact is of great clinical

Artificial heart

On December 3, 1982, a permanent artificial heart was implanted in a human being for the first time in medical history. Dr. Barney Clark, a retired dentist, was the recipient of the air-powered polyurethane device known as the Jarvik-7 artificial heart. Implantation of the permanent artificial pump became necessary when Dr. Clark's own diseased heart no longer responded to treatment. His contribution to science was unique. Although Dr. Clark died 112 days after the historic implant surgery as the result of multiorgan system failure, the artificial heart had beaten nearly 13 million times and was functioning normally until disconnected after Clark's death. The plastic organ consisted of two ventricular pumps or chambers that were sewn to the muscular walls of the right and left atria, which were left intact for that purpose when the diseased ventricles were excised. The two artificial ventricular chambers, as in the natural heart, were separate pumps in the Jarvik-7 device. They were held together by Velcro patches and connected to an exterior air compressor by tubes that emerged from the thoracic cavity. Rapid advances in the development of improved battery and pneumatically powered artificial hearts hold the promise of improving the quality of life for many critically ill patients.

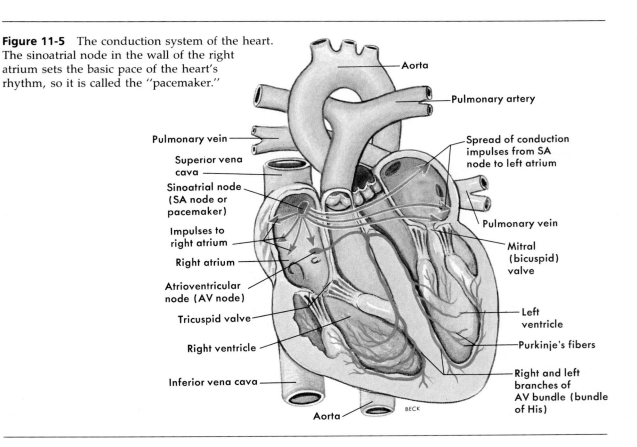

Figure 11-5 The conduction system of the heart. The sinoatrial node in the wall of the right atrium sets the basic pace of the heart's rhythm, so it is called the "pacemaker."

- Aorta
- Pulmonary artery
- Pulmonary vein
- Superior vena cava
- Sinoatrial node (SA node or pacemaker)
- Impulses to right atrium
- Right atrium
- Atrioventricular node (AV node)
- Tricuspid valve
- Right ventricle
- Inferior vena cava
- Aorta
- Spread of conduction impulses from SA node to left atrium
- Pulmonary vein
- Mitral (bicuspid) valve
- Left ventricle
- Purkinje's fibers
- Right and left branches of AV bundle (bundle of His)

BECK

Figure 11-6   Normal ECG deflections. **A,** P wave, QRS complex, and T wave. **B,** Relationship of ECG to cardiac muscle activity.

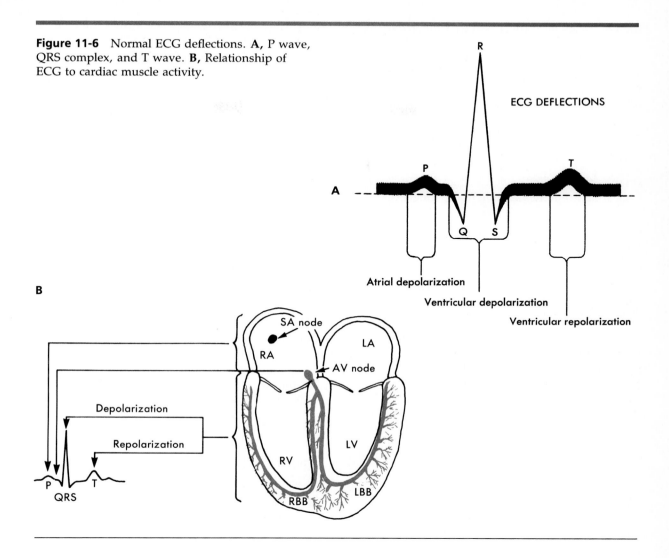

significance because these electrical signals can be picked up from the body surface and transformed into visible tracings by an instrument called an **electrocardiograph** (e-lek-tro-KAR-de-o-graf).

The **electrocardiogram** (e-lek-tro-KAR-de-o-gram) or **ECG** is the graphic record of the heart's electrical activity. Skilled interpretation of these ECG records may sometimes make the difference between life and death. A normal ECG tracing is shown in Figure 11-6.

A normal ECG tracing has three very char-

acteristic deflections or waves called the **P wave,** the **QRS complex,** and the **T wave.** These deflections represent the electrical activity that is associated with the contraction or relaxation of the atria or the ventricles. The term *depolarization* is used to describe the electrical activity associated with contraction of the heart muscle. *Repolarization* is associated with the relaxation phase of cardiac muscle activity. In the normal ECG shown in Figure 11-6 the small P wave occurs with depolarization of the atria. The QRS complex occurs as a result of depolarization of the ventricles, and the T wave results from electrical activity generated by repolarization of the ventricles. You may wonder why no visible record of atrial repolarization is noted in a normal ECG. Such a deflection does indeed occur, but it is masked or hidden by the large QRS complex that occurs at the same time.

Damage to cardiac muscle tissue caused by a myocardial infarction or disease that affects the conduction system of the heart will result in distinctive changes in the ECG. Therefore ECG tracings are extremely valuable in the diagnosis and treatment of heart disease.

BLOOD VESSELS

KINDS

Arteries, veins, and capillaries—these are the names of the three main kinds of blood vessels. **Arteries** carry blood away from the heart toward capillaries. **Veins** carry blood toward the heart away from capillaries. **Capillaries** carry blood from tiny arteries or **arterioles** into tiny veins or **venules.** The largest artery in the body is the aorta. The largest veins are the *superior vena cava* and the *inferior vena cava.* As noted above, the aorta carries blood out of the left ventricle of the heart, and the venae cavae return blood to the right atrium after the blood has circulated through the body.

STRUCTURE

Arteries, veins, and capillaries differ in structure. Examine Figures 11-7 and 11-8. Three coats or layers are found in both arteries and veins. The outermost layer or coat is called the **tunica externa.** Note that muscle tissue is found in the middle layer or **tunica media** of both arteries and veins. It is important to know, however, that the muscle layer is much thicker in arteries than in veins. Why is this important? Because the thicker muscle layer in the artery wall plays such a critical role in maintaining blood pressure and controlling blood distribution in the body. This is smooth muscle and so is controlled by the autonomic nervous system. A thin layer of elastic and white fibrous tissue covers an inner layer of endothelial cells called the **tunica interna** in both arteries and veins. The tunica interna is actually a single layer of endothelial cells that lines the inner surface of these vessels. When a surgeon cuts into the body, only arteries, arterioles, veins, and venules can be seen. Capillaries cannot be seen because they are microscopic vessels. The most important structural feature of capillaries is their extreme thinness—only one layer of flat, endothelium-like cells composes the capillary membrane. Instead of three layers or coats the capillary wall is composed of only one—the tunica interna. Substances such as glucose, oxygen, and wastes can quickly pass through it on their way to or from cells. Note in Figure 11-8 that smooth muscle cells called **precapillary sphincters** surround arterioles and help regulate the flow of blood into capillaries.

Median basilic vein

Intravenous injections are perhaps most often given into the median basilic vein at the bend of the elbow. Blood that is to be used for various laboratory tests is also usually removed from this vein.

FUNCTIONS

Arteries, veins, and capillaries serve different functions. Arteries and arterioles distribute blood from the heart to capillaries in all parts of the body. In addition, by constricting or dilating, arterioles help maintain arterial blood pressure at a normal level. Venules and veins collect blood from capillaries and return it to the heart. They also serve as blood reservoirs, since they can expand to hold a larger volume of blood or constrict to hold a much smaller amount. Capillaries function as exchange vessels. For example, glucose and oxygen move out of the blood in capillaries into interstitial fluid and on into cells. Carbon dioxide and various other substances move in the opposite direction, that is, into the capillary blood from the cells. Fluid also is exchanged between capillary blood and interstitial fluid. (See discussion in Chapter 19.)

Figure 11-7 **A,** Schematic drawings of an artery and a vein showing comparative thicknesses of the three coats: the outer coat or tunica externa, the muscle coat or tunica media, and the lining of the endothelium or the tunica interna. Note that the muscle and outer coats are much thinner in veins than in arteries and that veins have valves. **B,** Muscular artery and vein. Note the prominent elastic membrane and tunica media in the artery (left) and the more extensive tunica externa in the vein (right). (x140.)

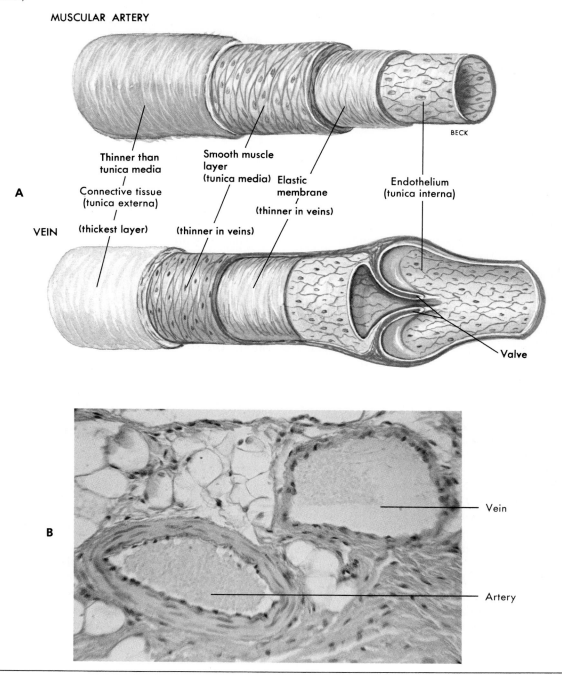

Figure 11-8 The walls of capillaries consist of only a single layer of endothelial cells. Note the thicker walls on the arteriole and the presence of smooth muscle cells called precapillary sphincters. Contraction of these muscle cells regulates blood flow into the capillary from the arteriole.

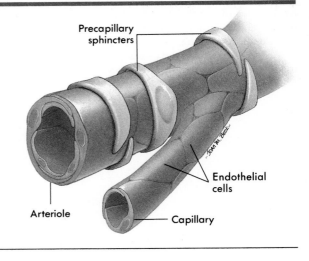

Precapillary sphincters

Endothelial cells

Arteriole

Capillary

Study Figure 11-9 to find out the names of the main arteries of the body and Figures 11-10 and 11-11 for the names of the main veins.

CIRCULATION

SYSTEMIC AND PULMONARY CIRCULATION

The term **circulation of blood** is self-explanatory, meaning that blood flows through vessels that are arranged to form a circuit or circular pattern. Blood flow from the left ventricle of the heart through blood vessels to all parts of the body and back to the right atrium of the heart has already been described as the **systemic circulation.** The left ventricle pumps blood into the aorta. From here it flows into arteries that carry it into the various tissues and organs of the body. As indicated in Figure 11-12, within each structure, blood moves from arteries to arterioles to capillaries. Here the vital two-way exchange of substances occurs between blood and cells. Next, blood flows out of each organ by way of its venules and then its veins to drain eventually into the inferior or superior vena cava. These two great veins return venous blood to the right atrium of the heart to complete the systemic circulation. But the blood has not quite come full circle back to its starting point in the left ven-

tricle. To do this and start on its way again, it must first flow through another circuit, referred to earlier as **pulmonary circulation.** Observe in Figure 11-12 that venous blood moves from the right atrium to the right ventricle to the pulmonary artery to lung arterioles and capillaries. Here, exchange of gases between blood and air takes place, converting the deep crimson that is typical of venous blood to the scarlet of arterial blood. This oxygenated blood then flows on through lung venules into four pulmonary veins and returns to the left atrium of the heart. From the left atrium it enters the left ventricle to be pumped again through the systemic circulation.

HEPATIC PORTAL CIRCULATION

The term **hepatic portal circulation** refers to the route of blood flow through the liver. Veins from the spleen, stomach, pancreas, gallbladder, and intestines do not pour their blood directly into the inferior vena cava as do the veins from other abdominal organs. Instead, they send their blood to the liver by means of the portal vein (Figure 11-13). Branches of the portal vein then empty into arterioles, which empty into the capillaries of the liver. Blood leaves the liver by way of the hepatic vein, which drains into the inferior vena cava. As noted in Figure 11-12, normally blood flows from arterioles to capillar-

Figure 11-9   Principal arteries of the body.

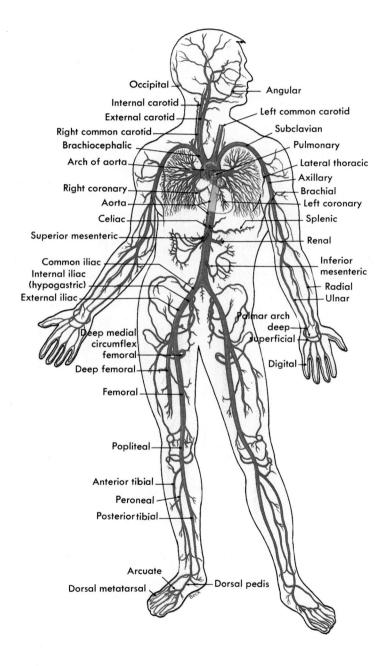

Figure 11-10   Principal veins of the body. Only the superficial veins are shown on the forearms and hands.

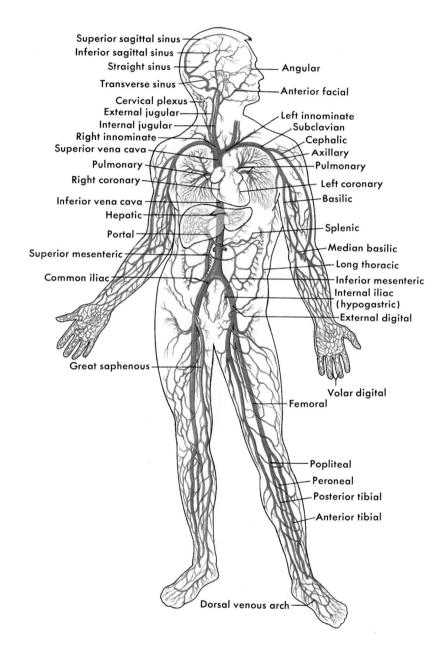

Superior sagittal sinus
Inferior sagittal sinus
Straight sinus
Transverse sinus
Cervical plexus
External jugular
Internal jugular
Right innominate
Superior vena cava
Pulmonary
Right coronary
Inferior vena cava
Hepatic
Portal
Superior mesenteric
Common iliac
Great saphenous

Angular
Anterior facial
Left innominate
Subclavian
Cephalic
Axillary
Pulmonary
Left coronary
Basilic
Splenic
Median basilic
Long thoracic
Inferior mesenteric
Internal iliac (hypogastric)
External digital
Volar digital
Femoral
Popliteal
Peroneal
Posterior tibial
Anterior tibial
Dorsal venous arch

Figure 11-11   Main superficial veins of the arm.

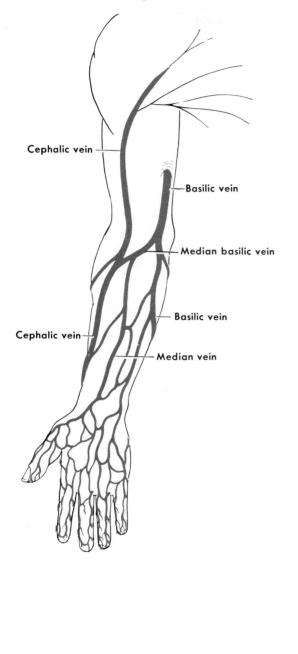

Cephalic vein

Basilic vein

Median basilic vein

Basilic vein

Cephalic vein

Median vein

Figure 11-12   Relationship of systemic and pulmonary circulation. As indicated by the numbers, blood circulates from the left ventricle of the heart to arteries, to arterioles, to capillaries, to venules, to veins, to the right atrium, to the right ventricle, to the lungs, and back to the left side of the heart, thereby completing a circuit.

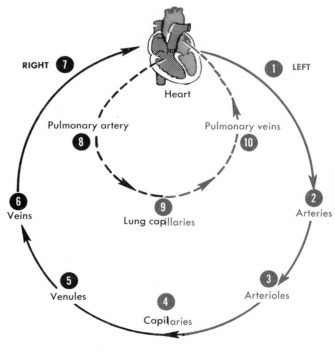

Figure 11-13 Hepatic portal circulation showing tributaries of the portal vein. In this very unusual circulation a vein is located between two capillary beds. The portal vein collects blood from capillaries in visceral structures located in the abdomen and empties into capillaries in the liver.

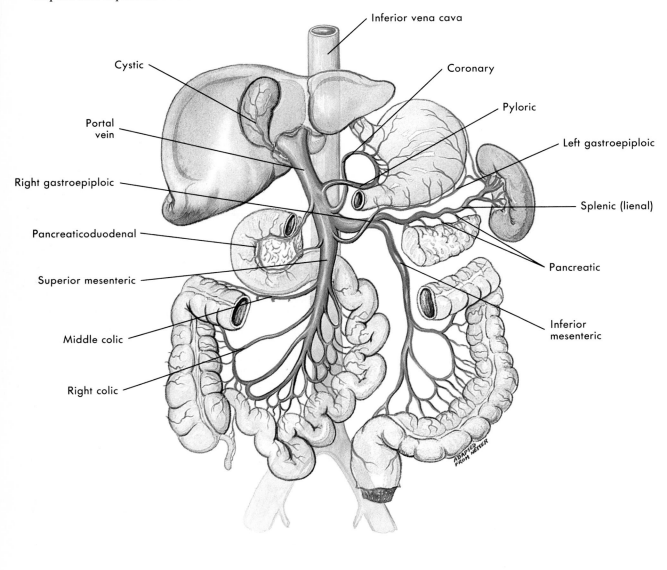

ies to veins, but blood flow is very unusual in the hepatic portal circulation. The portal vein is located between two capillary beds. The detour of venous blood through a second capillary bed in the liver before its return to the heart serves some valuable purposes. For example, when a meal is being absorbed, the blood in the portal vein contains a higher-than-normal concentration of glucose. Liver cells remove the excess glucose and store it as glycogen; therefore blood leaving the liver usually has a normal blood glucose concentration. Liver cells also remove and detoxify various poisonous substances that may be present in the blood returning to the intestines.

FETAL CIRCULATION

Circulation in the body before birth necessarily differs from circulation after birth. This is because the fetus must secure oxygen and food from maternal blood instead of from its own lungs and digestive organs.

In order for the exchange of nutrients and oxygen to occur between fetal and maternal blood, specialized blood vessels must carry the fetal blood to the **placenta** (plah-SEN-tah), where the exchange occurs, and then return it to the fetal body. Three vessels (shown in Figure 11-14 as part of the **umbilical cord**) accomplish this purpose. They are the two small **umbilical arteries** and a single much larger **umbilical vein.** Another structure unique to fetal circulation is called the **ductus venosus** (DUK-tus ve-NOS-us). As you can see in Figure 11-14, it is actually a continuation of the umbilical vein. It serves as a shunt, allowing most of the blood returning from the placenta to bypass the immature liver of the developing baby and empty directly into the inferior vena cava. Two other structures in the developing fetus allow blood to bypass the nonfunctional lungs, which remain collapsed until birth. The **foramen ovale** (fo-RA-men o-VAL-e) shunts blood from the right atrium directly into the left atrium, and the **ductus arteriosus** (DUK-tus ar-ter-i-O-sus) serves to connect the aorta and the pulmonary artery.

BLOOD PRESSURE

Perhaps a good way to understand blood pressure is to try to answer a few questions about it. What is blood pressure? Just what the words say—blood pressure is the pressure, or push, of blood.

Where does blood pressure exist? It exists in all blood vessels, but it is highest in the arteries and lowest in the veins. In fact, if we list blood vessels in order according to the amount of blood pressure in them and draw a graph of this, as in Figure 11-15, the graph looks like a hill, with aortic blood pressure at the top and vena caval pressure at the bottom. This blood pressure "hill" is spoken of as the *blood pressure gradient*. More precisely, the term blood pressure gradient means the difference between two blood pressures. The blood pressure gardient for the entire systemic circulation is the difference between the average, or mean, blood pressure in the aorta and the blood pressure at the termination of the venae cavae where they join the right atrium of the heart. The mean blood pressure in the aorta, given in Figure 11-15, is 100 millimeters of mercury (mm Hg) and the pressure at the termination of the venae cavae is 0. Therefore, with these typical normal figures, the systemic blood pressure gradient is 100 mm Hg (100 minus 0).

Why does blood pressure exist? What is its function? The blood pressure gradient is vitally involved in keeping the blood flowing. When a blood pressure gradient is present, blood circulates; conversely, when a blood pressure gradient is not present, blood does not circulate. For example, suppose that the blood pressure in the arteries were to decrease so that it became equal to the average pressure in arterioles. There would no longer be a blood pressure gradient between arteries and arterioles, and therefore there would no longer be a force to move blood out of arteries into arterioles. Circulation would stop, in other words, and very soon life itself would cease. This is why when arterial blood pressure is observed to be falling rapidly, whether in surgery or elsewhere in a hospital,

Figure 11-14 The fetal circulation.

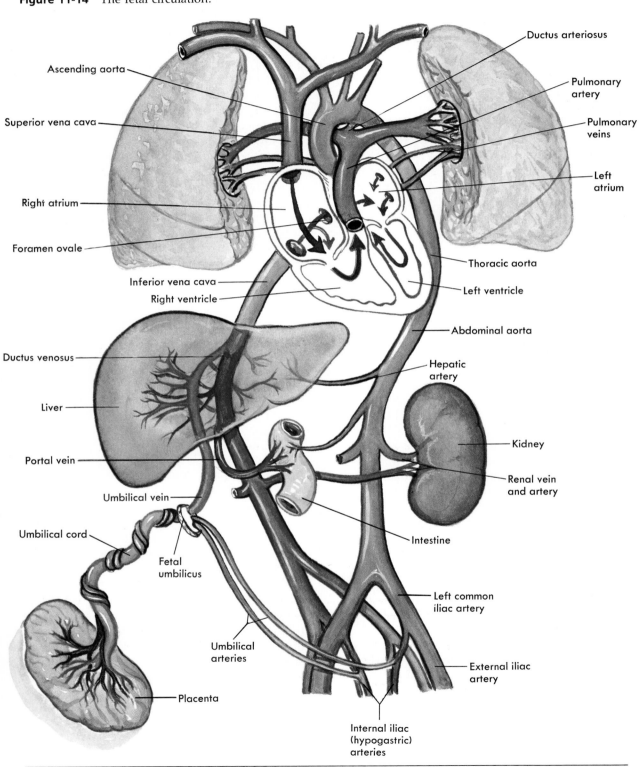

Ascending aorta

Superior vena cava

Right atrium

Foramen ovale

Inferior vena cava

Right ventricle

Ductus venosus

Liver

Portal vein

Umbilical vein

Umbilical cord

Fetal umbilicus

Umbilical arteries

Placenta

Ductus arteriosus

Pulmonary artery

Pulmonary veins

Left atrium

Thoracic aorta

Left ventricle

Abdominal aorta

Hepatic artery

Kidney

Renal vein and artery

Intestine

Left common iliac artery

External iliac artery

Internal iliac (hypogastric) arteries

Figure 11-15 Blood flows down a "blood pressure hill" from arteries, where blood pressure is highest, into arterioles, where it is somewhat lower, into capillaries, where it is still lower, and so on. All numbers on the graph indicate blood pressure measured in millimeters of mercury. The top figure, 100 mm, represents the average pressure in the aorta.

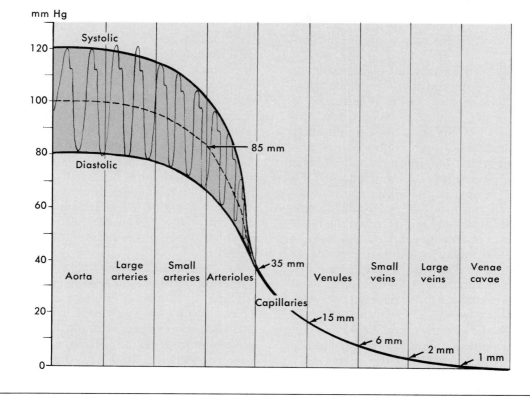

emergency measures are quickly started to try to reverse this fatal trend.

What we have just said in the preceding paragraph may start you wondering why high blood pressure (meaning, of course, high arterial blood pressure) and low blood pressure are bad for circulation. High blood pressure is considered bad for several reasons. For one thing, if it becomes too high, it may cause the rupture of one or more blood vessels (for example, in the brain, as happens in a stroke). But low blood pressure also can be dangerous. If arterial pressure falls low enough, circulation and life cease. Massive hemorrhage, for instance, kills in this way.

What causes blood pressure, and what makes blood pressure change from time to time? The direct cause of blood pressure is the volume of blood in the vessels. The larger the volume of blood in the arteries, for example, the more pressure the blood exerts on the walls of the arteries, or the higher the arterial blood pressure.

Conversely, the less blood in the arteries, the lower the blood pressure tends to be. Hemorrhage demonstrates well this relation between blood volume and blood pressure. In hemorrhage a pronounced loss of blood occurs, and this decrease in the volume of blood causes blood pressure to drop. In fact, the major sign of hemorrhage is a rapidly falling blood pressure.

The volume of blood in the arteries is determined by how much blood the heart pumps into the arteries and how much blood the arterioles drain out of them. (The volume of blood pumped into the arteries is called the *cardiac output*.) The diameter of the arterioles determines how much blood drains out of arteries into arterioles.

Both the strength and the rate of the heartbeat affect cardiac output and therefore blood pressure. Each time the left ventricle contracts, it squeezes a certain volume of blood (called the stroke volume) into the aorta and on into other arteries. The stronger each contraction is, the more blood it pumps into the aorta and arteries. Conversely, the weaker each contraction is, the less blood it pumps. Suppose that one contraction of the left ventricle pumps 70 milliliters of blood into the aorta, and suppose that the heart beats 70 times a minute. Seventy milliliters times 70 equals 4,900 milliliters. Almost 5 quarts of blood would enter the aorta and arteries every minute. Now suppose that the heartbeat were to become weaker and that each contraction of the left ventricle pumps only 50 instead of 70 milliliters of blood into the aorta. If the heart still contracts 70 times a minute, it will obviously pump much less blood into the aorta—only 3,500 milliliters instead of the more normal 4,900 milliliters per minute. This decrease in the heart's output of blood tends to decrease the volume of blood in the arteries, and the decreased arterial blood volume tends to decrease arterial blood pressure. In summary, the strength of the heartbeat affects blood pressure in this way—a stronger heartbeat tends to increase blood pressure and a weaker beat tends to decrease it.

The rate of the heartbeat may also affect arterial blood pressure. You might reason that when the heart beats faster, more blood would enter the aorta and that therefore the arterial blood volume and blood pressure would increase. This is true only if the stroke volume does not decrease sharply when the heart rate increases. Often, however, when the heart beats

faster, each contraction of the left ventricle takes place so rapidly that it squeezes out much less blood than usual into the aorta. For example, suppose that the heart rate speeded up from 70 to 100 times a minute and that at the same time its stroke volume decreased from 70 milliliters to 40 milliliters. Instead of a cardiac output of 70 × 70 or 4,900 milliliters per minute, the cardiac output would have changed to 100 × 40 or 4,000 milliliters per minute. Arterial blood volume would tend to decrease under these conditions and therefore blood pressure would also tend to decrease, even though the heart rate had increased. What generalization, then, can we make? We can only say that an increase in the rate of the heartbeat tends to increase blood pressure and a decrease in the rate tends to decrease blood pressure. But whether a change in the heart rate actually produces a similar change in blood pressure depends on whether the strength of the heart's beat also changes and in which direction (as indicated by the stroke volume).

Another factor that we ought to mention in connection with blood pressure is the viscosity of blood, or in plainer language, its stickiness. If blood becomes less viscous than normal, blood pressure decreases. For example, if a person suffers a hemorrhage, fluid will move into his blood from his interstitial fluid. This dilutes his blood and decreases its viscosity, and blood pressure then falls because of the decreased viscosity. As you may know, after hemorrhage either whole blood or plasma is preferred to saline solution for transfusions. The reason is that saline solution is not a viscous liquid and so cannot keep blood pressure at a normal level.

No one's blood pressure stays the same all the time. It fluctuates even in a perfectly healthy individual. For example, it goes up when a person exercises strenuously. Not only is this normal, but the increased blood pressure serves a good purpose. It tends to increase circulation, to bring more blood to muscles each minute, and thus to supply them with more oxygen and food for more energy.

A normal average arterial blood pressure is 120/80, or 120 millimeters of mercury systolic pressure (as the ventricles contract) and 80 millimeters of mercury diastolic pressure (as the ventricles relax).

PULSE

What you feel when you take a pulse is an artery expanding and then recoiling alternately. To feel a pulse, you must place your fingertips over an artery that lies near the surface of the body and over a bone or other firm background. Six such places are given below. Try to feel your pulse at each of these locations. (You can locate all of them on Figure 10-6, except the temporal and facial arteries.)

1 **Radial artery**—at the wrist
2 **Temporal artery**—in front of the ear or above and to the outer side of the eye
3 **Common carotid artery**—in the neck along the front edge of the sternocleidomastoid muscle at the level of the lower margin of the thyroid cartilage
4 **Facial artery**—at the lower margin of the lower jawbone on a line with the corners of the mouth
5 **Brachial artery**—at the bend of the elbow along the inner margin of the biceps muscle
6 **Dorsalis pedis artery**—on the front surface of the foot, just below the bend of the ankle joint

LYMPHATIC SYSTEM

The **lymphatic system** is not really a separate system of the body. It is part of the circulatory system, since it consists of **lymph,** a moving fluid that comes from the blood and returns to the blood by way of the lymphatic vessels. In addition to lymph and the lymphatic vessels, the system includes lymph nodes and specialized lymphatic organs such as the thymus, spleen, and tonsils.

Lymph forms in this way: blood plasma filters out of the capillaries into the microscopic spaces between tissue cells. Here the liquid is called **interstitial fluid** or tissue fluid. Most of the interstitial fluid goes back into the blood by the same route it came out, that is, back through

---

Lymph does not clot

If damage to the main lymphatic trunks in the thorax should occur as a result of penetrating injury, the flow of lymph must be stopped surgically or death will ensue. It is impossible to maintain adequate serum protein levels by dietary means if significant loss of lymph continues over time. As lymph is lost, rapid emaciation occurs, with a progressive and eventually fatal decrease in total blood fat and protein levels.

---

the capillary membrane. The remainder of the interstitial fluid enters tiny lymphatic capillaries to become lymph. It next moves on into larger lymphatics and finally enters the blood in veins in the neck region. The largest lymphatic vessel is the **thoracic duct.** Lymph from about three fourths of the body (Figure 11-16) eventually drains into the thoracic duct and from it goes back into the blood. The thoracic duct joins the left subclavian vein at the angle where the internal jugular vein also joins it (Figure 11-10). Lymph from the rest of the body drains into the right lymphatic ducts and on into the right subclavian vein.

Lymph serves a unique transport function by returning tissue fluid, proteins, fats, and other substances to the general circulation. It is important to realize, however, that the flow of lymph differs from the true "circulation" of blood seen in the cardiovascular system. Unlike vessels in the blood vascular system, the lymphatic vessels do not form a closed ring or circuit. Once lymph is formed from interstitial fluid in the soft tissues of the body, it flows only once through its system of lymphatic vessels before draining into the blood through the large neck veins. Lymph does not flow over and over again through vessels that form a circular route.

LYMPH NODES

Lymph nodes, or lymph glands as they are sometimes improperly called, are oval structures

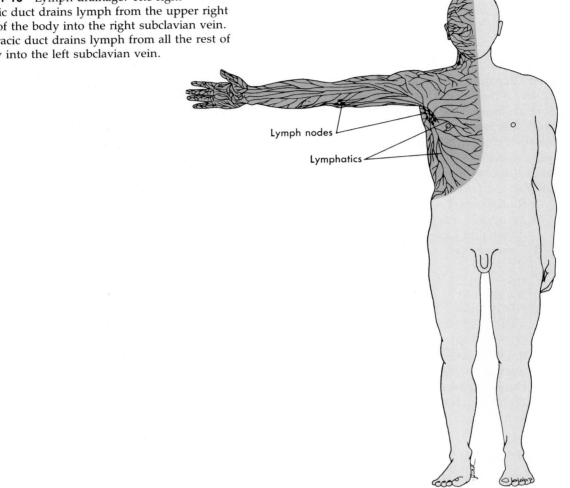

Figure 11-16 Lymph drainage. The right lymphatic duct drains lymph from the upper right quarter of the body into the right subclavian vein. The thoracic duct drains lymph from all the rest of the body into the left subclavian vein.

Lymph nodes

Lymphatics

located mainly in clusters along lymphatics (Figure 11-16). Some are as small as a pinhead and others as large as a lima bean. The structure of the lymph nodes makes it possible for them to perform two important functions: defense and white blood cell formation.

Defense Function: Filtration

Figure 11-17 shows the structure of a typical lymph node. Note that lymph enters the node through four **afferent** (from the Latin "to carry toward") **lymph vessels.** The afferent lymphatics deliver lymph to the node. In passing through the node, lymph is filtered so that injurious particles such as bacteria, soot, and cancer cells are removed and prevented from entering the blood and circulating all over the body. Lymph exits from the node through a single **efferent** (from the Latin "to carry away from") **lymph vessel.** Figure 11-18 shows how bacteria from an infected hair follicle are filtered from lymph passing through nearby nodes.

Figure 11-17   Structure of a lymph node. Several afferent lymphatics bring lymph to the node. Lymph is drained from the node by a single efferent vessel. Lymphatic tissue, densely packed with lymphocytes, composes the substance of the node.

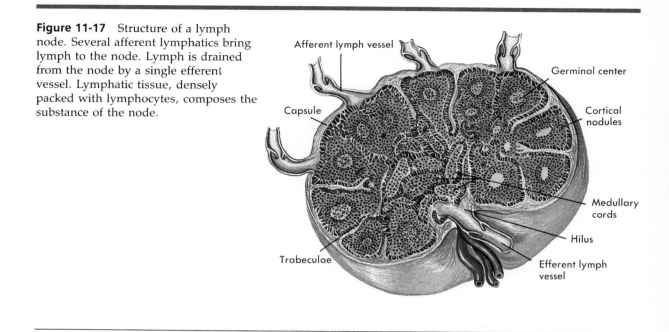

Figure 11-18   Diagrammatic representation of a skin section in which an infection surrounds a hair follicle. The yellow areas represent dead and dying cells (pus). Black dots around the yellow areas represent bacteria. Bacteria entering the node via the afferent lymphatics are filtered out.

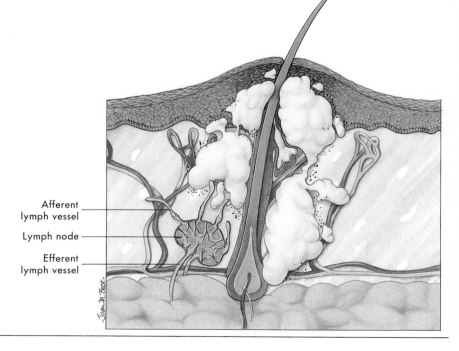

Figure 11-19 Lymphatic drainage of the breast. Note the extensive network of nodes that receive lymph from the breast.

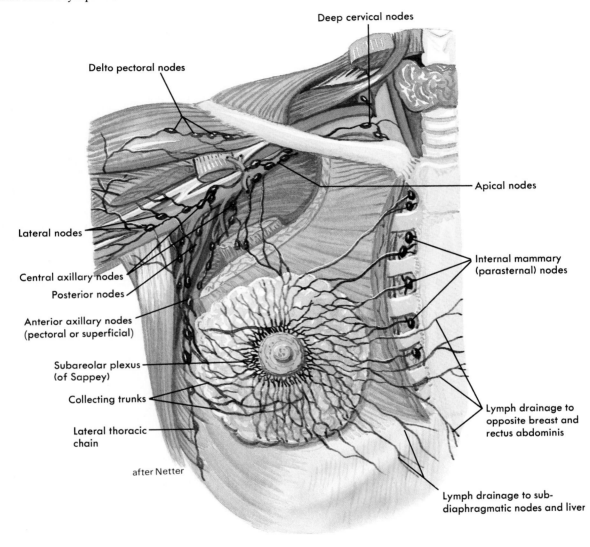

White Blood Cell Formation

The lymphatic tissue of lymph nodes forms lymphocytes and monocytes, the nongranular white blood cells described earlier in this chapter.

A nurse can use knowledge of lymph node location and function in many ways. For example, when taking care of a patient with an infected finger, the nurse should watch the elbow and axillary regions for swelling and tenderness of the lymph nodes. These nodes filter lymph returning from the hand and may become infected by the bacteria they trap. A surgeon uses knowledge of lymph node function when removing lymph nodes under the arms (axillary nodes) and in other areas during an operation for breast cancer. These nodes may contain cancer cells filtered out of the lymph drained from the breast. Cancer of the breast is one of the most common forms of this disease in women. Unfortunately, cancer cells from a single tumorous growth in the breast often spread to other areas of the body through the lymphatic system. Figure 11-19 shows how lymph from the breast drains into many different and widely placed nodes.

SPLEEN

The spleen remains one of the mystery organs of the body. Quite a bit of doubt still exists about its functions. The three lower ribs provide a protective shelter over the spleen, which is located in the upper left corner of the abdominal cavity just under the diaphragm. Except for blood vessels and nerves, the spleen connects with no other organs.

The spleen's main functions seem to be to form lymphocytes and monocytes (as do the lymph nodes) and to act as the body's blood bank. The spleen can store almost 1 pint of blood and quickly release it back into circulation when more blood is needed—during strenuous exercise and after hemorrhage, for instance.

OUTLINE SUMMARY

Heart

A Location, size, and position
1 Triangular organ located in mediastinum with two thirds of mass to left of body midline and one third to right; apex on diaphragm; shape and size of a closed fist (see Figure 11-1)
2 Cardiopulmonary resuscitation (CPR)—heart lies between sternum in front and bodies of thoracic vertebrae behind, rhythmic compresson of heart between sternum and vertebrae can maintain blood flow in cardiac arrest; if combined with artificial respiration procedure, can be life-saving

B Anatomy
1 Heart chambers (see Figure 11-2)
   a Two upper chambers called atria (receiving chambers)—right and left atria
   b Two lower chambers called ventricles (discharging chambers)—right and left ventricles
   c Wall of each heart chamber composed of cardiac muscle tissue called myocardium
   d Endocardium—smooth lining of heart chambers—inflammation of endocardium called endocarditis
2 Covering sac or pericardium
   a Pericardium is a two-layered fibrous sac with a lubricated space between the two layers

  b Inner layer called visceral pericardium or epicardium
  c Outer layer called parietal pericardium
3 Heart valves (see Figure 11-2)
  Four valves keep blood flowing in right direction through heart; prevent backflow (two AV or atrioventricular and two semilunar valves).
  a Tricuspid—at opening of right atrium into ventricle
  b Mitral (bicuspid)—at opening of left atrium into ventricle
  c Pulmonary semilunars—at beginning of pulmonary artery
  d Aortic semilunars—at beginning of aorta

C Heart Sounds
1 Two distinct heart sounds in every heartbeat or cycle—''lup-dup''
2 First (lup) sound is caused by vibration and closure of AV valves during contraction of heart
3 Second (dup) sound is caused by closure of the semilunar valves during relaxation of heart

D Blood Flow through the Heart
1 Heart acts as two separate pumps—the right atria and ventricle performing different functions from the left atria and ventricle
2 Sequence of blood flow: venous blood enters right atrium through superior and inferior vena cavae—passes from right atrium through tricuspid valve to right ventricle; from right ventricle blood passes through pulmonary semilunar valve to pulmonary artery to lungs—blood from lungs returns to left atrium and passes through bicuspid (mitral) valve to left ventricle. Blood in left ventricle is pumped through aortic semilunar valve into aorta and is distributed to body as a whole

E Coronary Circulation
1 Blood, which supplies oxygen and nutrients to myocardium of heart, flows through right and left coronary arteries
2 Blockage of blood flow through coronary arteries called myocardial infarction (heart attack) (Figure 11-4)
3 Angina pectoris—chest pain caused by inadequate oxygen to heart

F Conduction System (Figure 11-5)
1 SA (sinoatrial) node, the pacemaker—located in the wall of the right atrium near the opening of the superior vena cava
2 AV (atrioventricular) node—located in the right atrium along the lower part of the interatrial septum
3 AV (bundle of His) bundle—located in the septum of the heart
4 Purkinje fibers—located in the walls of the ventricles

G Electrocardiogram (Figure 11-6)
1 Specialized conduction system structures generate and transmit the electrical impulses that result in contraction of the heart
2 These tiny electrical impulses can be picked up on the surface of the body and transformed into visible tracings by a machine called an electrocardiograph
3 The visible tracing of these electrical signals is called an electrocardiogram or ECG
4 The normal ECG has three deflections or waves called the P wave, the QRS complex, and the T wave
  a P wave—associated with depolarization of atria
  b QRS complex—associated with depolarization of ventricles
  c T wave—associated with repolarization of ventricles

Blood Vessels

A Kinds
  1 Arteries—carry blood away from heart
  2 Veins—carry blood toward heart
  3 Capillaries—carry blood from arterioles to venules
B Structure—see Figures 11-7 and 11-8
C Functions
D Names of main arteries—see Figure 11-9
E Names of main veins—see Figure 11-10

Circulations

A Plan of circulation—refers to blood flow through vessels arranged to form a circuit or circular pattern
B Types of circulation
  1 Systemic circulation (Figure 11-12)
  2 Pulmonary circulation (Figure 11-12)
  3 Hepatic portal circulation (Figure 11-13)
  4 Fetal circulation (Figure 11-14)

Blood Pressure

A Blood pressure is push, or force of blood in blood vessels
B Highest in arteries, lowest in veins—see Figure 11-15
C Blood pressure gradient causes blood to circulate—liquids can flow only from area where pressure is higher to where it is lower
D Blood volume, heartbeat, and blood viscosity are main factors that produce blood pressure
E Blood pressure varies within normal range from time to time

Pulse

A Definition—alternate expansion and recoil of blood vessel wall
B Places where you can count the pulse easily—see p. 273

Lymphatic System (see Figures 11-16 to 11-19)

A Consists of lymphatic vessels, lymph nodes, lymph, and spleen
B Lymph—the fluid in the lymphatic vessels; lymph comes from blood by plasma-filtering fluid, some of which then enters the lymph capillaries to become lymph and be returned to blood by way of lymphatics; largest lymphatic is thoracic duct—drains lymph from all but upper right quarter of body into left subclavian vein
C Lymph nodes
  1 Located along certain lymphatics, usually in clusters; for example, at elbow, under arm, in groin, at knee
  2 Functions—filter out injurious particles such as microorganisms and cancer cells from lymph before it returns to blood; form some white blood cells (lymphocytes and monocytes)

Spleen

A Forms some white blood cells (lymphocytes and monocytes)
B Serves as blood bank for body—stores blood until needed and then releases it back into circulation

NEW WORDS

afferent
angina pectoris
arteriole
artery
artificial pacemaker
atrioventricular (AV)
  valve
atrium
bicuspid valve
capillary
cardiac output
cardiopulmonary
  resuscitation (CPR)
chordae tendineae
coronary circulation

diastolic pressure
ductus arteriosus
ductus venosus
efferent
electrocardiogram
  (ECG)
embolism
endocarditis
endocardium
epicardium
fetal circulation
foramen ovale
hepatic portal
  circulation
lymph

mitral valve
myocardial infarction
myocarditis
myocardium
P wave
pacemaker
pericarditis
pericardium
pulmonary circulation
pulse
Purkinje fibers
QRS complex
semilunar valve
sinoatrial node
systemic circulation

T wave
thoracic duct
thrombosis
tricuspid valve
umbilical
vein
ventricle
venule

CHAPTER TEST

1. The circulatory system is composed of two major subdivisions called the
   _____ subdivision and the _____ subdivision.

2. The blunt point of the lower edge of the heart is called the
   _____ .

3. Rhythmic compression of the heart combined with artificial respiration is
   called _____ _____ .

4. The two upper chambers of the heart are called _____ and
   the two lower chambers are called _____ .

5. Cardiac muscle tissue, which forms the wall of each heart chamber, is
   called the _____ .

6. Inflammation of the endocardial lining of the heart is called
   _____ .

7. The covering sac around the heart is called the _____ .

8. The two valves that separate the upper from the lower chambers of the
   heart are called _____ valves.

9. The valve located between the right atrium and ventricle is called the
   _____ valve.

10. Blood enters the right atrium through two large veins called the superior and inferior _____ _____ .

11. Blood passing through the pulmonary semilunar valve enters the _____ artery.

12. Severe chest pain that is caused by inadequate blood flow to the myocardium is called _____ _____ .

13. The sinoatrial or SA node is often called the _____ of the heart.

14. A graphic recording of the heart's electrical activity is called an _____ .

15. In a normal electrocardiogram the QRS complex occurs as a result of depolarization of the _____ .

16. Blood is carried toward the heart by large vessels called _____ .

17. The muscular middle layer of both arteries and veins is called the _____ _____ .

18. Blood flow from the left ventricle of the heart to all parts of the body and back to the right atrium of the heart is called the _____ circulation.

19. In portal circulation a portal vein is located between two _____ beds.

20. In fetal circulation blood is shunted from the right atrium directly into the left atrium by passing through the _____ _____ of the heart.

21. The volume of blood pumped into the arteries from the heart is called the _____ _____ .

22. The largest lymphatic vessel is called the _____ _____ .

23. Lymph enters a lymph node through several _____ vessels and exits from the node through a single _____ lymph vessel.

24. The lymphatic tissue of lymph nodes forms the nongranular _____ .

25. When a large lymphatic vessel is cut it must be surgically repaired because, unlike blood, lymph does not _____ .

Select the most correct answer from Column B for each statement in Column A. (Only one answer is correct.)

Column A	Column B
26. _____ blunt point of the heart	a. Pericarditis
27. _____ receiving chambers of heart	b. P wave
28. _____ inflamed heart covering	c. Coronary arteries
29. _____ attach valves to heart wall	d. Portal circulation
30. _____ supply blood to heart muscle	e. Heart attack
31. _____ myocardial infarction	f. Apex
32. _____ depolarization of atria	g. Ductus arteriosus
33. _____ vein between two capillary beds	h. Chordae tendineae
34. _____ fetal circulation	i. Thoracic duct
35. _____ largest lymph vessel	j. Atria

REVIEW QUESTIONS

1 Describe the position of the heart in the mediastinum.

2 What is cardiopulmonary resuscitation (CPR)?

3 Briefly explain what happens when blood clots, including what makes it start to clot.

4 You hear that a friend has a "coronary thrombosis." What does this mean to you?

5 Describe blood flow through the heart.

6 A patient has had an operation to repair the mitral valve. Where is the valve, and what is its function?

7 What are some differences between an artery, a vein, and a capillary?

8 Considering that the function of the circulatory system is to transport substances to and from the cells, do you think it is true that in one sense capillaries are our most important blood vessels? Give a reason for your answer.

9 The right ventricle of the heart pumps blood to and through only one organ. Which one?

10 What part of the heart pumps blood through the systemic circulation, that is, to and through all organs other than the lungs?

11 All blood returns from the systemic circulation to what part of the heart?

12 What part of the heart pumps blood through the pulmonary circulation, that is, to and through the lungs?

13 Blood returning from the pulmonary circulation (from the lungs, in other words) enters what part of the heart?

14 From which cavity of the heart does blood rich in oxygen leave the heart to be delivered to tissue capillaries all over the body?

15 How do arterial blood and venous blood differ with regard to their oxygen and carbon dioxide contents?

16 Does every artery carry arterial blood and every vein carry venous blood? If not, what exceptions are there?

17 Name the vein at the bend of the elbow into which substances are often injected and from which blood is sometimes withdrawn. (Check your answer with Figure 11-11.)

18 Nurses frequently have to take a patient's blood pressure. What vital function does blood pressure perform?

19 Sometimes a woman's arm becomes very swollen for a while after removal of a breast and the nearby lymph nodes and lymphat-ics, including some of those in the upper arm. Can you think of any reason why swelling occurs?

20 You could live without your spleen since it does not do anything vital for the body. What functions does it perform?

21 Discuss the two functions of lymph nodes.

22 How does the "circulation" of lymph differ from blood circulation?

OBJECTIVES

After you have completed this chapter, you should be able to:

1 Explain the generalized functions of the immune system and list the major cell types that are present.

2 Discuss the major types of immune system molecules and indicate how both antibodies and complement function.

3 Discuss and contrast the development and functions of B cells and T cells.

4 Compare and contrast humoral and cell-mediated immunity.

5 Define the following terms: *AIDS, interferon, opsonization, monoclonal antibodies, chemotactic factor, lymphotoxin.*

The immune system is the body's defense system. It defends us against three major kinds of enemies: microorganisms that invade our bodies, foreign tissue cells that have been transplanted into our bodies, and our own cells that have turned malignant (cancerous). The immune system functions to make us immune, that is, able to resist these enemies. Unlike other systems of the body, which are made up of groups of organs, the immune system is made up of billions of cells and trillions of molecules. This chapter presents information about these cells and molecules.

IMMUNE SYSTEM MOLECULES

The immune system functions because of the presence of adequate amounts of highly specialized protein molecules and unique cells.

The protein molecules most critical to immune system functioning are called **antibodies** (AN-ti-BOD-es) and **complement** (KOM-ple-ment). Antibody molecules outnumber the cells of the immune system by about 100 million to one.

ANTIBODIES
Definition

Antibodies, sometimes called **immunoglobulins** (im-u-no-GLOB-u-lins), are protein compounds that are normally present in the body. A defining characteristic of an antibody molecule is the presence of uniquely shaped concave regions called *combining sites* on its surface. Another defining characteristic is the ability of an antibody molecule to combine with a specific compound called an **antigen** (AN-ti-jen). All antigens are compounds whose molecules have small regions on their surfaces—called **epitopes** (EP-i-topes)—that are uniquely shaped to fit into the combining sites of a specific antibody molecule as precisely as a key fits into a specific lock. Antigens are usually foreign proteins. Most often antigens are molecules present in the surface membranes of invading cells, such as microorganisms, cancer cells, or transplanted tissue cells.

Functions

In general, antibodies function to produce what is called **humoral immunity** by making an-

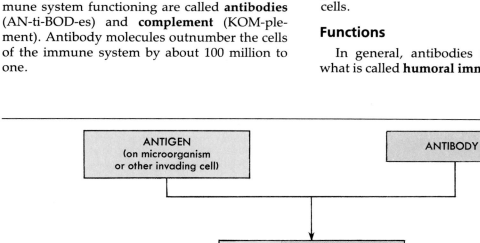

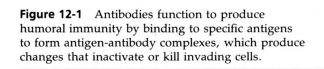

Figure 12-1 Antibodies function to produce humoral immunity by binding to specific antigens to form antigen-antibody complexes, which produce changes that inactivate or kill invading cells.

tigens unable to harm the body (Figure 12-1). To do this an antibody must first bind to its specific antigen, the one whose epitopes fit its combining sites. This forms an antigen-antibody complex. The antigen-antibody complex then acts in one or more ways to make the antigen, or the cell on which it is present, harmless. For example, if the antigen is a toxin, a substance poisonous to body cells, the toxin is neutralized, made nonpoisonous, by becoming part of an antigen-antibody complex. Or if antigens are molecules in the surface membranes of invading cells, when antibodies combine with them, the resulting antigen-antibody complexes may agglutinate the enemy cells, that is, make them stick together in clumps. Then macrophages or other phagocytes can rapidly destroy them by ingesting and digesting large numbers of them at one time.

Another important function of antibodies is called **opsonization** (op-so-ni-ZA-shun). This term is from a Greek word that means "to prepare for eating." It fits the process well. Certain **opsonins** (op-SO-nins), which are antibody fractions, help promote the attachment of phagocytic cells to the object they will engulf. As a result of opsonins, the contact between the phagocytic cell and its victim is enhanced and the object is more easily ingested. Opsonization of antigens by antibodies thus contributes to the efficiency of immune system phagocytic cells, which will be described on p. 287.

Monoclonal antibodies

Revolutionary new techniques that have permitted biologists to produce large quantities of pure and very specific antibodies have resulted in dramatic advances in medicine during the past 5 years. As a new medical technology, the development of **monoclonal antibodies** has been compared in importance with advances in recombinant DNA or genetic engineering.

Monoclonal antibodies are specific antibodies produced or derived from a population or culture of identical, or **monoclonal,** cells. In the past, antibodies produced by the immune system against a specific antigen had to be "harvested" from serum containing literally hundreds of other antibodies. The total amount of a specific antibody that could be recovered was very limited and the cost of recovery was high. Monoclonal antibody techniques are based on the ability of immune system cells to produce individual antibodies that bind to and react with very specific antigens. We know, for example, that if the body is exposed to the varicella virus of chicken pox, white blood cells will produce an antibody that will react very specifically with that virus and no other. In monoclonal antibody techniques, lymphocytes that are produced by the body after the injection of a specific antigen are "harvested" and then "fused" with other cells that have been transformed to grow and divide indefinitely in a tissue culture medium. The result is a rapidly growing population of identical or **monoclonal** cells that produce large quantities of a very specific antibody. Monoclonal antibodies have now been produced against a wide array of different antigens, including disease-producing organisms and various types of cancer cells.

The availability of very pure antibodies against specific disease-producing agents is the first step in the commercial preparation of diagnostic tests that can be used to identify viruses, bacteria, and even specific cancer cells in blood or other body fluids. The use of monoclonal antibodies may eventually serve as the basis for specific treatment of many human diseases.

Probably the most important way in which antibodies act is one we will consider last. It is called **complement fixation.** In many instances, when antigens that are molecules on a cell's surface combine with antibody molecules, they change the shape of the antibody molecule just slightly, but enough to expose two previously hidden regions. These are called **complement-binding sites.** Their exposure initiates a series of events that kill the cell on whose surface they take place. The next section describes these events.

COMPLEMENT
Definition

Complement is a group of inactive enzyme proteins normally present in blood. Formerly there were thought to be 9 complement enzymes. We now know that there are 14 complement proteins normally found in blood. They are identified by numbers that indicate their order of functioning.

Activation

Very briefly, complement is activated by antibodies. It begins with the binding of antibodies to antigens located on an invading cell's surface. This changes the shape of the antibody molecule, exposing its two complement-binding sites. This in turn activates complement enzyme 1 to bind to the exposed complement binding sites. This event activates complement enzyme 2 to bind to an adjacent antibody molecule. Then in rapid-fire succession the remaining complement enzymes are similarly activated and bound to antibody molecules. Together they form a doughnut-shaped assemblage—complete with a hole in the middle! Keep in mind that the antibody molecules are bound both to antigens in the invading cell's cytoplasmic membrane and to complement enzymes. The doughnut-shaped structure thus formed is fixed firmly on the invading cell's surface.

Function

Complement functions to kill invading cells of various types. How? In effect, by drilling a

Interferon

Interferon (in-ter-FER-on) is a small protein compound that plays a very significant role in producing immunity from viral infections. It is produced by T cells within hours after they have been infected by a virus. The interferon released from the T cells serves to protect other cells by *interfering* with the ability of the virus to reproduce as it moves from cell to cell. In the past, thousands of pints of blood had to be processed to harvest tiny quantities of leukocyte (T cell) interferon for study. Synthetic human interferon is now being "manufactured" in bacteria as a result of gene-splicing techniques and is available for the first time in quantities sufficient for clinical use. Synthetic interferon, when added to a nasal spray, can prevent the common cold and when injected will decrease the severity of many virus-related diseases including chickenpox and measles. Interferon is also being tested at many medical centers for its potential use as an anticancer drug. It has shown promise in treating several types of breast, skin, and blood cancer.

hole in their cytoplasmic membranes! Fluid then pours through the hole into the cell, swelling it, and soon bursting it wide open.

IMMUNE SYSTEM CELLS

The primary cells of the immune system include:

1 Lymphocytes
  a T lymphocytes
  b B lymphocytes
2 Phagocytes
  a Neutrophils
  b Monocytes
  c Macrophages

The most numerous cells of the immune system are the lymphocytes; they are ultimately re-

Figure 12-2 Human blood smear showing a neutrophil and a lymphocyte of the immune system. (x350.)

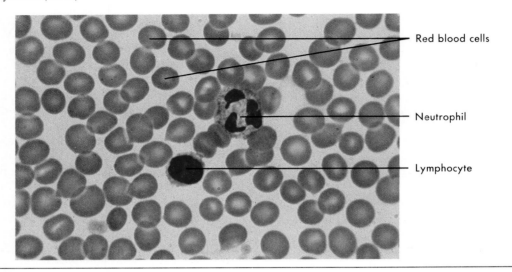

Red blood cells

Neutrophil

Lymphocyte

sponsible for antibody production. A million million strong, lymphocytes continually patrol the body, searching out any enemy cells that may have entered. Lymphocytes circulate in the body's fluids. Huge numbers of them wander vigilantly through most of its tissues. Lymphocytes densely populate the body's widely scattered lymph nodes and its other lymphatic tissues—especially the thymus gland in the chest and the spleen and liver in the abdomen.

In addition to the trillion or so lymphocytes, large numbers of phagocytes also function as part of the immune system. Recall from Chapter 10 that phagocytes are cells that can carry on phagocytosis, or ingestion and digestion of foreign cells or particles (see Figure 10-5). The most important phagocytes are neutrophils (see Figures 10-2 and 12-2) and monocytes. In addition, connective tissue cells named **macrophages** (MAK-ro-fa-ges) also function as phagocytes in the immune system (see Figure 10-3). The process of opsonization is especially important in macrophage activity. Highly specific opsonins or antibody fragments bind to and actually coat certain foreign particles. These opsonins serve as a

"flag" that alerts the macrophage to the presence of foreign material. They also help bind the phagocyte to the foreign material so that it can be engulfed more effectively.

LYMPHOCYTES
Types

There are two major types of lymphocytes (Figure 12-3), designated as B and T lymphocytes but usually called **B cells** and **T cells.**

Development of B Cells

All lymphocytes that circulate in the tissues arise from primitive cells in the bone marrow called stem cells and go through two stages of development. In chickens the first stage of B cell development occurs in a structure named the bursa of Fabricius—hence the name "B cells." Humans have no such bursa, and the structure in which B cells undergo their first stage of development is appropriately called the "bursa-equivalent structure." Its identity has not yet been positively established. Persuasive evidence, however, indicates the liver. The first stage of B cell development consists of stem cells

Figure 12-3 Lymphocytes and macrophages. In this electron micrograph the lymphocytes are small and spherical; the macrophages are larger and more irregular in form.

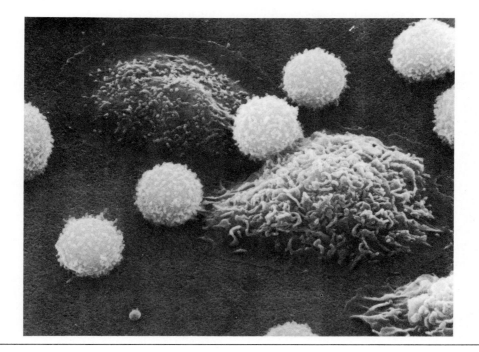

becoming transformed into immature B cells. The process begins shortly before birth and is completed by the time a human infant is a few months old. *Immature B cells* are small lymphocytes that have synthesized and inserted into their cytoplasmic membranes numerous molecules of one specific kind of antibody (Figure 12-4). These antibody-bearing immature B cells leave the bursa-equivalent structure where they were formed, enter the blood, and are transported to their new place of residence, chiefly the lymph nodes. There they act as seed cells. Each immature B cell undergoes repeated mitosis (cell division) and forms a clone of immature B cells. A **clone,** by definition, is a family of many identical cells all descended from one cell. Because all the cells in a clone of immature B cells have descended from one immature B cell, all of them bear the same surface antibody molecules as did their single ancestor cell.

The second stage of B cell development changes an immature B cell into an activated B cell. Not all immature B cells undergo this change. They do so only if an immature B cell comes in contact with certain protein molecules—antigens—whose shape fits the shape of the immature B cell's surface antibody molecules. If this happens, the antigens lock onto the antibodies and by so doing change the immature B cell into an activated B cell. Then the activated B cell, by dividing rapidly and repeatedly, develops into clones of two kinds of cells, namely, **plasma cells** and **memory cells.** *Plasma cells* secrete copious amounts of antibody into the blood—reportedly, 2,000 antibody molecules per second by each plasma cell for every second

Figure 12-4 B cell development takes place in two stages. First stage: Shortly before and after birth, stem cells develop into immature B cells. This occurs in bursa-equivalent structure, perhaps the liver. Second stage (occurs only if immature B cell contacts its specific antigen): Immature B cell develops into activated B cell, which divides rapidly and repeatedly to form a clone of plasma cells and a clone of memory cells. Plasma cells secrete antibodies capable of combining with specific antigen that caused immature B cell to develop into active B cell.

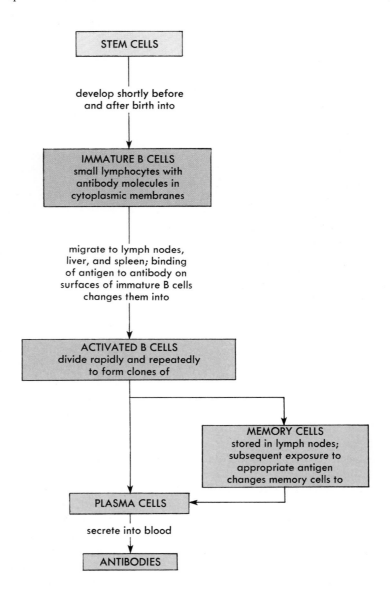

of the few days that it lives. Antibodies circulating in the blood constitute an enormous, mobile, ever-on-duty army.

Memory cells have the ability to secrete antibodies but do not immediately do so. They remain in reserve in the lymph nodes until such time as they may be contracted by the same antigen that led to their formation. Then, very quickly, the memory cells develop into plasma cells and secrete large amounts of antibody. Memory cells, in effect, seem to remember their ancestor-activated B cell's encounter with its appropriate antigen. They stand ready, at a moment's notice, to produce antibody that will combine with this antigen.

Function of B Cells

B cells function indirectly to produce *humoral immunity*. Recall that humoral immunity is resistance to disease organisms produced by the actions of antibodies circulating in body fluids. Activated B cells develop into plasma cells. Plasma cells secrete antibodies into the blood; they are the "antibody factories" of the body.

Development of T Cells

T cells, by definition, are lymphocytes that have undergone their first stage of development in the thymus gland. Stem cells from the bone marrow seed the thymus, and shortly before and after birth they develop into T cells. The newly formed T cells stream out of the thymus into the blood and migrate chiefly to the lymph nodes where they take up residence. Embedded in each T cell's cytoplasmic membrane are protein molecules so shaped that they can fit only one specific kind of antigen molecule. The second stage of T cell development takes place when and if a T cell comes into contact with its specific antigen. If this happens, the antigen binds to the protein on the T cell's surface, thereby changing the T cell into a sensitized T cell (Figure 12-5).

Functions of T Cells

Sensitized T cells function to produce cell-mediated immunity (Figure 12-6). As the name suggests, **cell-mediated immunity** is resistance to

Figure 12-5 T cell development. First stage takes place in thymus gland shortly before and after birth. Second stage occurs only if T cell contacts antigen, which combines with certain proteins present on T cell's surface.

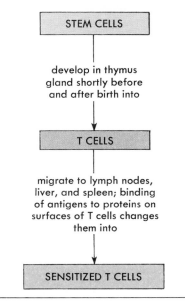

disease organisms resulting from the actions of cells—chiefly sensitized T cells. Some sensitized T cells kill invading cells directly (Figure 12-7). When bound to antigens present on an invading cell's surface, they release a substance called **lymphotoxin** (lim-fo-TOK-sin), which acts as a specific and lethal poison against the bound cell. Many sensitized cells produce their deadly effects indirectly by means of compounds known collectively as lymphokines, which they release into the area around enemy cells. Among these are a substance called chemotactic factor and another called macrophage activating factor. **Chemotactic factor** attracts macrophages into the neighborhood of the enemy cells. Then **macrophage activating factor** prods the assembled macrophages into destroying the cells by phagocytosing (ingesting and digesting) them.

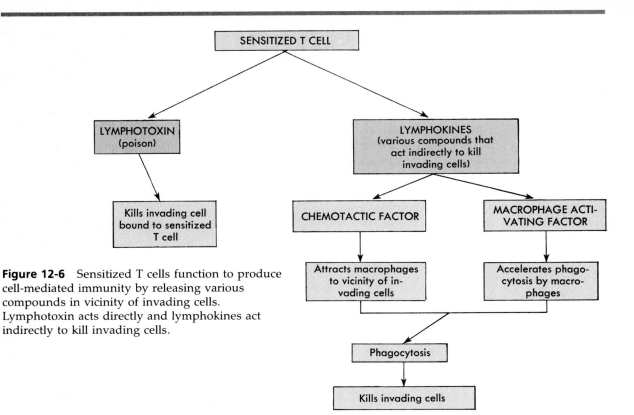

Figure 12-6 Sensitized T cells function to produce cell-mediated immunity by releasing various compounds in vicinity of invading cells. Lymphotoxin acts directly and lymphokines act indirectly to kill invading cells.

Figure 12-7 The white spheres seen in this scanning electron microscope view are T lymphocytes attacking a much larger cancer cell. T lymphocytes are a significant part of our defense against cancer and other types of foreign cells.

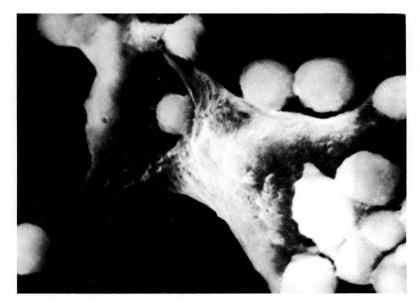

AIDS

AIDS or **acquired immune deficiency syndrome** was first described as a separate disease entity in 1981. The disease is caused by a retrovirus that enters the bloodstream and integrates its genes into the DNA of T cell lymphocytes. Viral reproduction kills the infected T cells and depletes the total number of T cells in the immune system pool. The virus was originally known as human T-lymphotropic virus III or *HTLV-III*. In 1987 the virus was renamed Human Immunodeficiency Virus (HIV) by an international panel on virus nomenclature. Because T cells play such an important role in cell-mediated immunity, any decrease in their numbers subjects victims of the disease to a whole array of so-called opportunistic infections. AIDS victims are unable to protect themselves from certain forms of cancer and many viral, protozoal, and fungal diseases.

The AIDS disease pattern in the United States shows that most victims are in three "high-risk" groups. Almost 70%-75% are homosexual or bisexual men, about 18% are intravenous drug users who share infected needles, and about 3% are hemophiliacs who have been infected with coagulation products obtained from diseased blood donors. However, this pattern is not typical in some countries, such as Africa, where the disease strikes larger numbers of the heterosexual population. There is considerable debate about the changing nature of the disease pattern worldwide. The number of reported cases in the United States reached 30,000 in 1986. It is difficult to predict the total number of potential AIDS cases because of the long latency period of 5 years or more between infection and the onset of symptoms in those infected persons who actually develop the disease. To complicate the picture even more, some people infected with the virus develop less severe AIDS-related conditions (ARC syndrome) such as chronic weight loss and swollen lymph nodes. They may remain in a sick but stable condition for years while others experience steady degeneration of their immune system and die of full-blown AIDS within months of the appearance of symptoms. Most investigators now believe the disease is much broader in scope than once thought.

OUTLINE SUMMARY

Immune System Cells

A Lymphocytes—type of white blood cells; most numerous of immune cells, estimated 1 trillion lymphocytes in body

B Phagocytes—cells that carry on phagocytosis, that is, ingestion and digestion of foreign cells or particles; include two types of white blood cells, neutrophils and monocytes, and connective tissue cells called macrophages

Immune System Molecules

Protein molecules, chiefly antibodies and complement; estimated 100 million times as many antibody molecules as lymphocytes

Lymphocytes

A Types—B cells and T cells

B Development of B cells—primitive stem cells migrate from bone marrow to "bursa-equivalent" structure, perhaps the liver

1 First stage—stem cells develop into immature B cells; takes place in bursa-equivalent structure during the few months before and after birth; immature B cells are small lymphocytes with antibody molecules (which they have synthesized) in their cytoplasmic membranes; migrate chiefly to lymph nodes

2 Second stage—immature B cell develops into activated B cell; initiated by immature B cell's contact with antigens, which bind to its surface antibodies; activated B cell, by dividing repeatedly, forms two clones of cells; plasma cells and memory cells; plasma cells secrete antibodies into blood; memory cells stored in lymph nodes; if subsequent exposure to antigen that activated B cell occurs, memory cells become plasma cells and secrete antibodies

C Function of B cells—indirectly, B cells produce humoral immunity; activated B cells develop into plasma cells; plasma cells secrete antibodies into blood; circulating antibodies produce humoral immunity

D Development of T cells—stem cells from bone marrow migrate to thymus gland

1 Stage 1—stem cells develop into T cells; occurs in thymus during few months before and after birth; T cells migrate chiefly to lymph nodes

2 Stage 2—T cells develop into sensitized T cells; occurs when, and if, antigen binds to T cell's surface proteins

E Functions of T cells—produce cell-mediated immunity; kill invading cells by releasing lymphotoxin that poisons cells and also by releasing chemicals that attract and activate macrophages to kill cells by phagocytosis

Antibodies (Immunoglobulins)

Antibodies bind to antigens to form antigen-antibody complexes that act in several ways to make antigens and cells with surface antigens harmless; for example, neutralization of toxins, agglutination of invading cells; binding of antigen to antibody changes shape of antibody molecule, exposing its complement-binding sites

Complement

A Definition—group of inactive enzymes normally present in blood

B Activation—complement activated by binding to antibodies that are already bound to antigens on surface of invading cell; activated complement enzymes assemble in form of doughnut-shaped structure on cell surface; hole in center of this structure

C Function—complement kills invading cells by "drilling" hole in their cytoplasmic membranes, which allows fluid to enter cell until it bursts

NEW WORDS

AIDS	clone	immunoglobulin	memory cells
antibodies	combining sites	interferon	monoclonal antibodies
antigen	complement	lymphotoxin	opsonins
B cells (lymphocytes)	complement fixation	macrophage	opsonization
cell-mediated immunity	epitopes	macrophage activating	plasma cells
chemotactic factor	humoral immunity	factor	T cells (lymphocytes)

CHAPTER TEST

1. The two protein molecules most critical to immune system functioning are called _____ and _____ .
2. Antibodies are sometimes called _____ .
3. Antibodies are characterized by their ability to combine with very specific compounds called _____ .
4. All antigens have small regions on their surfaces called _____ which "fit" into specific combining sites on antibody molecules.
5. Antibodies function to make antigens unable to harm the body—a process called _____ immunity.
6. Certain compounds help phagocytic cells to attach to and engulf foreign objects; these compounds are called _____ .
7. The most important way that antibodies act is called _____ _____ .
8. The small protein compound that plays an important role in immunity to viral infections is called _____ .
9. Specific antibodies artificially produced from cultures of identical cells are called _____ antibodies.
10. The most numerous cells of the immune system are the _____ .
11. The most important phagocytes of the immune system are the _____ and _____ .
12. B cells and T cells are the two major types of _____ in the immune system.
13. B cells function indirectly to produce _____ immunity in the body.
14. Sensitized T cells function to produce _____ immunity in the body.
15. The virus called HTLV-III or HIV is responsible for the disease called _____ .

Circle the T before each true statement and the F before each false statement.

T F   16. Antibodies are also called immunoglobulins.

T F   17. All antigens have epitopes on their surfaces.

T F   18. Cells of the immune system outnumber antibody molecules by about 100 million to one.

T F   19. Antibodies function to produce cell-mediated immunity.

T F   20. Opsonization is an important function of antibodies.

T F   21. Monoclonal antibodies cannot be used as the basis for specific treatment of human diseases.

T F   22. Complement fixation is an important way in which antibodies act.

T F   23. Interferon is a large lipid compound that produces immunity from bacterial infections.

T F   24. Lymphocytes are ultimately responsible for antibody production.

T F   25. Macrophages are capable of phagocytosis.

T F   26. Memory cells develop from sensitized T cells.

T F   27. Lymphokines act indirectly to kill invading cells.

T F   28. Sensitized T cells function to produce cell-mediated immunity.

T F   29. T cells begin as stem cells that develop in the thymus gland.

T F   30. Chemotactic factor plays no role in the process of phagocytosis.

REVIEW QUESTIONS

1 Explain what the terms *antigens* and *antibodies* mean.

2 What are combining sites? complement binding sites? epitopes?

3 Name four kinds of cells that constitute part of the immune system. Which are most numerous? How numerous are they estimated to be?

4 Name the two major immune system molecules.

5 How does the number of antibodies compare with the number of lymphocytes?

6 What are the two major types of lymphocytes called? What makes these names appropriate?

7 Describe briefly the two stages of development that B cells undergo.

8 Describe briefly the two stages of development that T cells undergo.

9 What cells secrete antibodies?

10 Explain the function of memory cells.

11 Describe the function of activated B cells.

12 Describe the function of sensitized T cells.

13 Immunoglobulins is a synonym for what word?

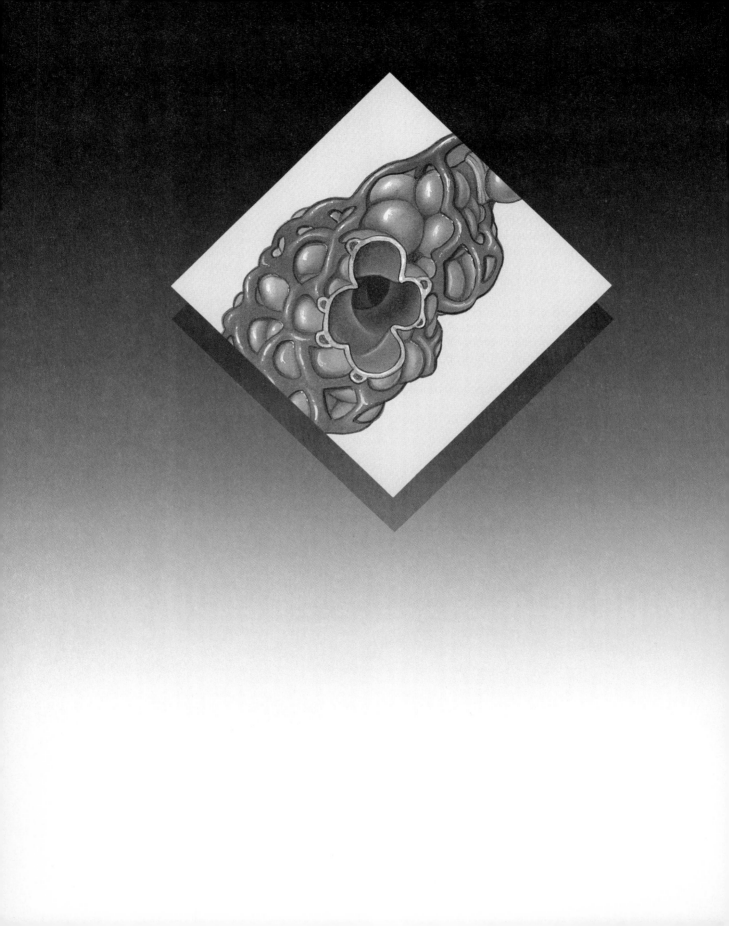

UNIT FIVE

Systems That Process and Distribute Foods and Eliminate Wastes

CHAPTER

13 The Digestive System

CHAPTER OUTLINE

Wall of the digestive tract

Mouth

Teeth
- Typical tooth

Salivary glands

Pharynx

Esophagus

Stomach

Small intestine

Liver and gallbladder

Pancreas

Large intestine

Appendix

Peritoneum
- Extensions

Digestion
- Carbohydrate digestion
- Protein digestion
- Fat digestion

Absorption

Metabolism
- Carbohydrate metabolism
- Fat metabolism
- Protein metabolism
- Metabolic rates

BOXED ESSAYS

Dental diseases
Heartburn
Ulcers
Pylorospasm
Oral rehydration therapy
Appendicitis

OBJECTIVES

After you have completed this chapter, you should be able to:

1. List in sequence each of the component parts or segments of the alimentary canal from the mouth to the anus and identify the accessory organs of digestion.
2. List and describe the four layers of the wall of the alimentary canal. Compare the lining layer in the esophagus, stomach, small intestine, and large intestine.
3. Discuss the basics of protein, fat, and carbohydrate digestion and give the end products of each of these processes.
4. Define and contrast mechanical and chemical digestion.
5. Define: *peristalsis, bolus, chyme, jaundice, pylorospasm, ulcer, diarrhea.*

The organs of the **digestive system** form an irregular-shaped tube, open at both ends, called the **alimentary** (al-e-MEN-tar-e) **canal** or the **gastrointestinal** (gas-tro-in-TES-ti-nal) **(GI) tract.** In the adult this hollow tube is about 29 feet long. Although this may seem strange, food or other material that enters the digestive tube is not really inside the body. Most parents of young children quickly learn that a button or pebble swallowed by their child will almost always pass unchanged and with little difficulty through the tract. Think of the tube as a passageway that extends through the body like a hallway through a building. Food must be broken down or **digested** and then absorbed through the walls of the digestive tube before it can actually enter the body and be used by cells. The breakdown or digestion of food material is both *mechanical* and *chemical* in nature. The teeth are used to physically break down food material before it is swallowed. The churning of food in the stomach then continues the mechanical breakdown process. Chemical breakdown results from the action of digestive enzymes and other chemicals acting on food as it passes through the GI tract. Ultimately, food is reduced to basic nutrients that can be absorbed through the lining of the intestinal wall and then distributed to body cells for use. This process of altering the chemical and physical composition of food so that it can be absorbed and used by body cells is known as digestion and is the function of the digestive system. Part of the digestive system, the large intestine, serves also as an organ of elimination, ridding the body of the waste material or **feces** resulting from the digestive process. Table 13-1 names both main and accessory digestive organs. Note that the *accessory organs* include the teeth, tongue, gallbladder, and appendix, as well as a number of glands that secrete their products into the digestive tube.

Foods undergo three kinds of processing in the body: **digestion, absorption,** and **metabolism.** Digestion and absorption are functions performed by the organs of the digestive system

Table 13-1 Organs of the digestive system

Main organs	Accessory organs
Mouth	Teeth and tongue
Pharynx (throat)	Salivary glands
Esophagus (foodpipe)	Parotid Submandibular Sublingual
Stomach	

Small intestine
  Duodenum
  Jejunum
  Ileum
	Liver
	
	Gallbladder
	
	Pancreas

Large intestine
  Cecum
  Colon
    Ascending colon
    Transverse colon
    Descending colon
    Sigmoid colon
  Rectum
  Anal canal
	Vermiform appendix

(Figure 13-1). Metabolism, on the other hand, is a function performed by all body cells. In this chapter we shall begin by describing digestive system organs and then discuss digestion, absorption, and metabolism.

The many terms used to describe the digestive system—or any other body system—make more sense and are easier to remember if you know the origins of some of the words. The word **aliment,** for example, is from a Latin term meaning "food." The Greek work *gastero* means "stomach" and *enteron* means "intestine." The table of common suffixes (word endings) at the end of this text lists *-logy* as meaning "study of." **Gastroenterology** (gas-tro-en-ter-OL-o-je) is the study of the stomach and intestines and their diseases. Scientific and medical terms may at first seem strange and confusing, but as your vocabulary increases, you will find that many

Figure 13-1   Location of digestive organs.

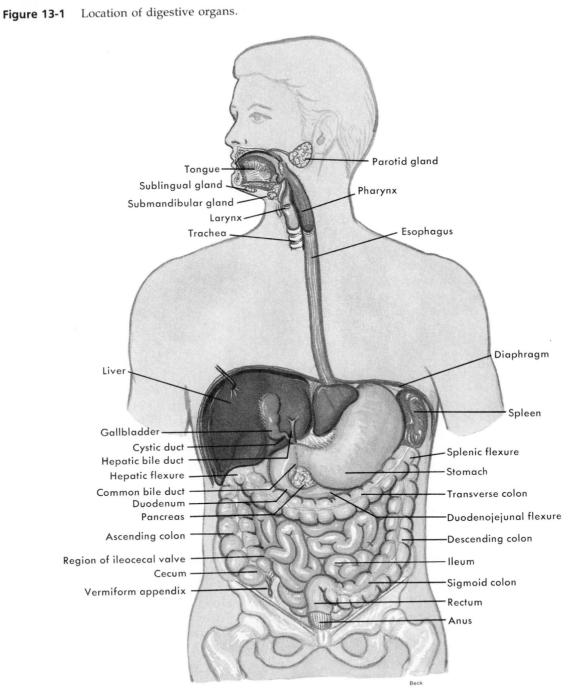

Tongue
Sublingual gland
Submandibular gland
Larynx
Trachea
Parotid gland
Pharynx
Esophagus
Liver
Diaphragm
Spleen
Gallbladder
Cystic duct
Hepatic bile duct
Hepatic flexure
Common bile duct
Duodenum
Pancreas
Ascending colon
Region of ileocecal valve
Cecum
Vermiform appendix
Splenic flexure
Stomach
Transverse colon
Duodenojejunal flexure
Descending colon
Ileum
Sigmoid colon
Rectum
Anus

Beck

Figure 13-2 Section of the small intestine showing the four layers typical of walls of the gastrointestinal tract. Circular folds of mucous membrane called plicae circularis increase the surface area of the lining coat.

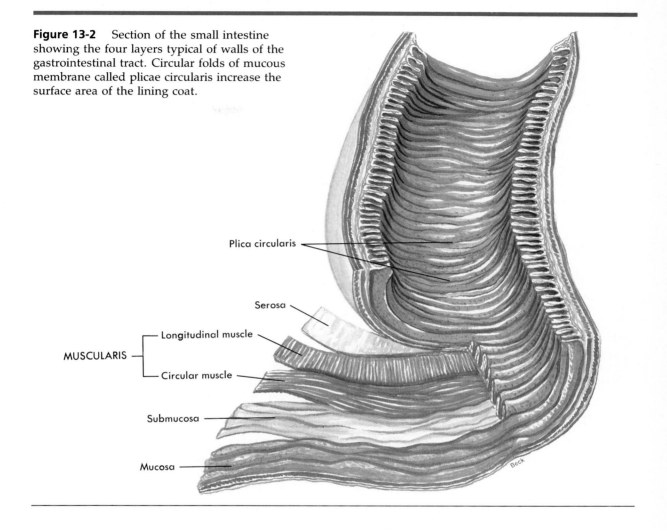

Plica circularis

Serosa

MUSCULARIS
— Longitudinal muscle
— Circular muscle

Submucosa

Mucosa

Beck

new terms are really combinations of inter-changeable word parts. For example, if you know that the word ending *-itis* attached to the name of a body part means inflammation of that part, it will be easy to remember that **gastritis** (gas-TRI-tis) is inflammation of the stomach and **enteritis** (en-ter-I-tis) is inflammation of the intestine.

WALL OF THE DIGESTIVE TRACT

The digestive tract has been described as a tube that extends from the mouth to the anus.

The wall of this digestive tube is fashioned of four layers of tissue (Figure 13-2). The inside or hollow space within the tube is called the **lumen.** The four layers, named from the inside coat to the outside of the tube, are as follows:

1 Mucosa or mucous membrane
2 Submucosa
3 Muscular coat or muscularis
4 Adventitia (ad-ven-TISH-ee-ah) or serosa.

Although the same four tissue coats form the various organs of the alimentary tract, their structures vary in different organs. The **mucosa** of the esophagus, for example, is composed of tough and stratified abrasion-resistant epithelium. The mucosa of the remainder of the tract is a delicate layer of simple columnar epithelium designed for absorption and secretion.

The **submucosa**, as the name implies, is a connective tissue layer that lies just below the mucosa. It contains many blood vessels and nerves. The two layers of muscle tissue called the **muscularis** serve an important function in the digestive process. By contraction of the muscular coat, food material is moved through the digestive tube. This movement of food through the tube is called **peristalsis** (per-i-STAL-sis). In addition, the contraction of the muscularis also assists in the mixing of food with digestive juice and in the further mechanical breakdown of larger food particles.

The **adventitia** or **serosa** is the outermost covering or coat of the digestive tube. In the abdominal cavity it is composed of the *parietal peritoneum*. The loops of the digestive tract are anchored to the posterior wall of the abdominal cavity by a large double fold of peritoneal tissue called the **mesentery** (MES-un-terr-ee), which is labeled in Figure 13-16.

MOUTH

The **mouth** or **oral cavity** is a hollow chamber having a roof, a floor, and side walls. Food enters or is ingested into the digestive tract through the mouth, and the process of digestion begins immediately. Like the remainder of the digestive tract, the mouth is lined with mucous membrane. It may be helpful if you review the structure and function of mucous membranes in Chapter 4. Typically, mucous membranes line hollow organs, such as the digestive tube, that open to the exterior of the body. Mucus produced by the lining of the GI tract helps keep it soft and pliable, thus aiding in the movement of food material through the tube.

The roof of the mouth is formed by the **hard** and **soft palates.** The hard palate is a bony structure in the anterior or front portion of the mouth, formed by parts of the palatine and maxillary bones. The soft palate is located above the pos-

terior or rear portion of the mouth. It is soft because it consists chiefly of muscle. Hanging down from the center of the soft palate is a cone-shaped process, the **uvula** (U-vu-lah). If you look in the mirror, open your mouth wide, and say "Ah," you can see your uvula. The uvula and soft palate prevent food and liquid from entering the nasal cavities above the mouth.

The floor of the mouth consists of the tongue and its muscles. The tongue is made of skeletal muscle covered with mucous membrane. It is anchored to bones in the skull and to the hyoid bone in the neck. A thin membrane called the **frenulum** (FREN-u-lum) attaches the tongue to the floor of the mouth. Occasionally the frenulum is too short to allow free movements of the tongue. Individuals with this condition cannot enunciate words normally and are said to be tongue-tied. Note in Figure 13-3 that the tongue can be divided into a blunt rear portion called the *root*, a pointed *tip*, and a central *body*.

Have you ever noticed the many small elevations on the surface of your tongue? They are named **papillae.** The largest of the papillae on the tongue are the **vallate** type which, as you can see in Figure 13-3, form an inverted V-shaped row of about ten to twelve mushroom-like elevations. Figure 13-4 is a photomicrograph of several vallate papillae. Students often confuse tongue papillae with taste buds. You can see the large vallate papillae by simply looking at the surface of your tongue with a mirror. The taste buds, however, are microscopic structures. Substances to be tasted are dissolved in the saliva and then enter a shallow depression called a **moat,** which surrounds the papilla. The taste buds (Figure 13-5) are specialized nerve endings located on the sides of the papillae.

TEETH

By the time a baby is 2 years old, the child probably has his full set of twenty baby teeth. When a young adult is somewhere between 17 and 24 years old, a full set of thirty-two permanent teeth is generally present. The average age for cutting the first tooth is about 6 months,

Figure 13-3   **A,** Surface of the tongue. **B,** Mouth cavity showing undersurface of tongue.

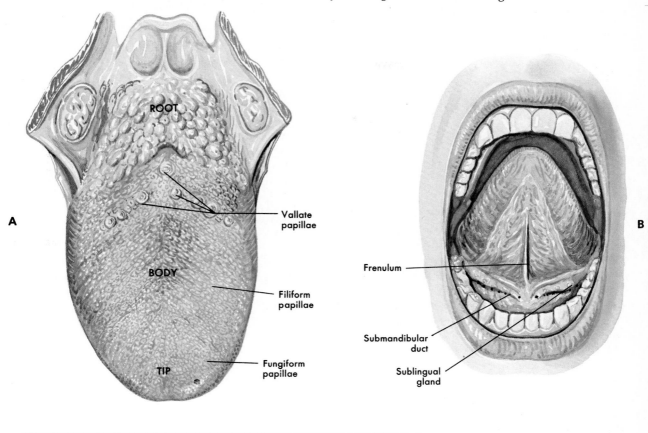

A

B

ROOT

BODY

TIP

Vallate
papillae

Filiform
papillae

Fungiform
papillae

Frenulum

Submandibular
duct

Sublingual
gland

Figure 13-4   Vallate papillae on surface of tongue. Taste buds are located on the sides of the papillae. Substances to be tasted must enter the open space around the papilla called a moat to contact the taste buds. (×35.)

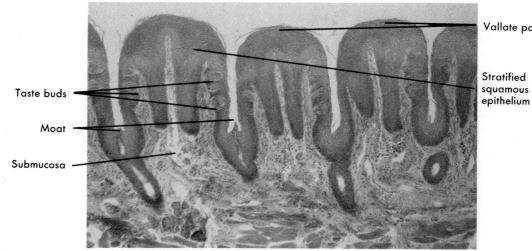

Taste buds

Moat

Submucosa

Vallate papilla

Stratified
squamous
epithelium

Figure 13-5   Taste buds. Enlargement of photomicrograph of taste buds in Figure 13-4. Arrow points to pore in outer surface of taste bud. Note that the pore opens into the moat or space surrounding the papilla. (×140.)

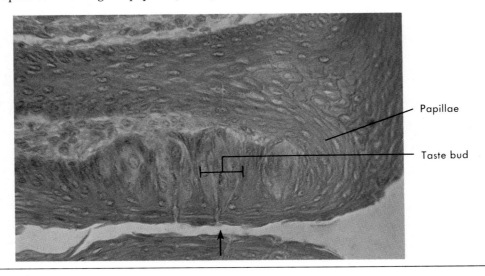

Papillae

Taste bud

and the average age for losing the first baby tooth and starting to cut the permanent teeth is about 6 years. Figures 13-6 and 13-7 give the names of the teeth and show which ones are lacking in the deciduous or baby set. Figure 13-8 illustrates tooth structure.

TYPICAL TOOTH

A typical tooth can be divided into three main parts: crown, neck, and root. The **crown** is that portion of the tooth that is exposed and visible in the mouth. It is covered by enamel—the hardest tissue in the body. Enamel is ideally suited to withstand the grinding that occurs during the chewing of hard and brittle foods. In addition to enamel, the outer shell of each tooth is covered by two other dental tissues—dentin and cementum (see Figure 13-8). Dentin makes up the greatest proportion of the tooth shell. It is covered by enamel in the crown and by cementum in the neck and root areas. Dentin contains

a pulp cavity consisting of connective tissue, blood and lymphatic vessels, and sensory nerves.

The **neck** of a tooth is the narrow portion, shown in Figure 13-8, that is surrounded by the pink gingiva or gum tissue. It joins the crown of the tooth to the root. It is the **root** that fits into the socket of either the upper or lower jaw.

The shape and placement of the teeth assist in their functions. The four major types of teeth are listed below.

1 Incisors
2 Canines
3 Premolars
4 Molars

Note in Figure 13-6 that the incisors have a sharp cutting edge. They serve a cutting function during **mastication** (mas-ti-KA-shun) or chewing of food. The canine teeth are sometimes called **cuspids.** They serve to pierce or tear the food that is being eaten. This tooth type is par-

Figure 13-6 The deciduous arch. Note that in the set of twenty temporary primary teeth there are no premolars (bicuspids) and there are only two pairs of molars in each jaw. Compare with permanent teeth shown in Figure 13-7.

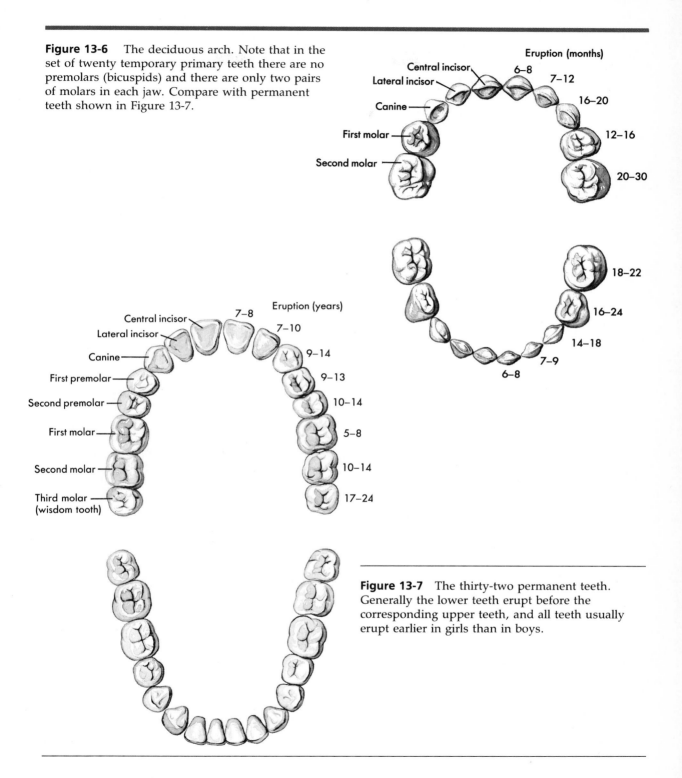

Eruption (months)

Central incisor — 6–8
Lateral incisor — 7–12
Canine — 16–20
First molar — 12–16
Second molar — 20–30

18–22
16–24
14–18
7–9
6–8

Eruption (years)

Central incisor — 7–8
Lateral incisor — 7–10
Canine — 9–14
First premolar — 9–13
Second premolar — 10–14
First molar — 5–8
Second molar — 10–14
Third molar (wisdom tooth) — 17–24

Figure 13-7 The thirty-two permanent teeth. Generally the lower teeth erupt before the corresponding upper teeth, and all teeth usually erupt earlier in girls than in boys.

Figure 13-8   A molar tooth sectioned to show its bony socket and details of its three main parts: crown, neck, and root. Enamel (over the crown) and cementum (over the neck and root) surround the dentin layer. The pulp contains nerves and blood vessels.

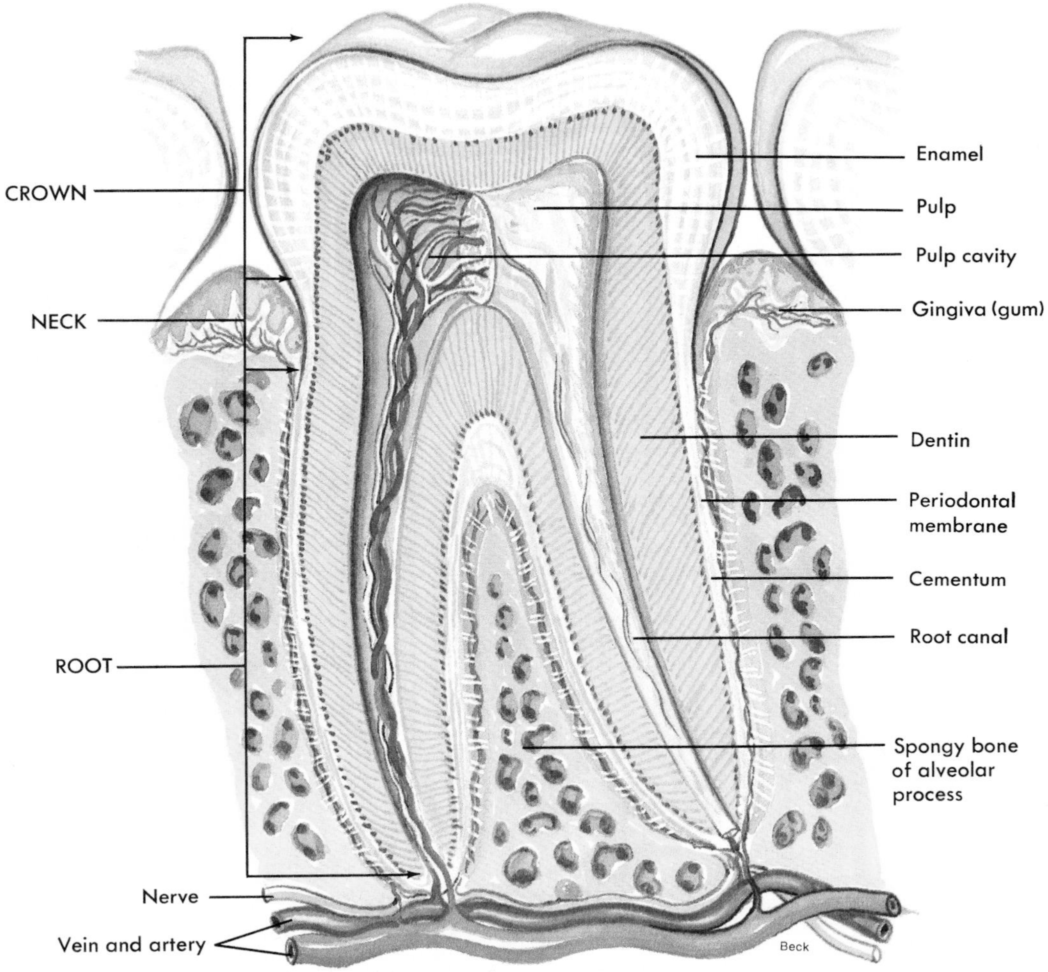

CROWN

NECK

ROOT

Nerve

Vein and artery

Enamel

Pulp

Pulp cavity

Gingiva (gum)

Dentin

Periodontal membrane

Cementum

Root canal

Spongy bone of alveolar process

Beck

ticularly apparent in meat-eating mammals such as dogs. Premolars or **bicuspids** and molars or **tricuspids** have rather large flat surfaces with either two or three grinding or crushing "cusps" on their surface. They provide for extensive breakdown of food material in the mouth. After food has been chewed, it is formed into a small rounded mass called a **bolus** (BO-lus) so that it can be swallowed.

SALIVARY GLANDS

Three pairs of salivary glands—the parotids, submandibulars, and sublinguals—secrete most (about 1 liter) of the saliva produced each day in the adult. The salivary glands (Fig. 13-9) are typical of the accessory glands that are associated with the digestive system. They are located outside of the digestive tube itself and must convey their secretions by way of ducts into the

Dental diseases

Tooth decay or "dental caries" is one of the most common diseases in the civilized world. It is a disease of the enamel, dentin, and cementum of teeth that results in the formation of a permanent defect called a **cavity.** Most people living in the United States, Canada, and Europe are significantly affected by the disease. Decay occurs on tooth surfaces where food debris, acid-secreting bacteria, and plaque accumulate. Regular and thorough brushing of the teeth is important in preventing decay. Reducing dietary intake of refined sugars, which promote growth of decay bacteria, will also reduce the rate of cavity formation. The introduction of fluoride to water supplies has proved to be the most effective and practical method of reducing the rate of tooth decay in children and is considered an important public health measure in many areas of the United States. If the disease is untreated, tooth decay results in infection, loss of teeth, and inflammation of the soft tissues in the mouth. Bacteria may also invade the paranasal sinuses or extend to the surface of the face and neck, causing serious complications.

Figure 13-9   The salivary glands.

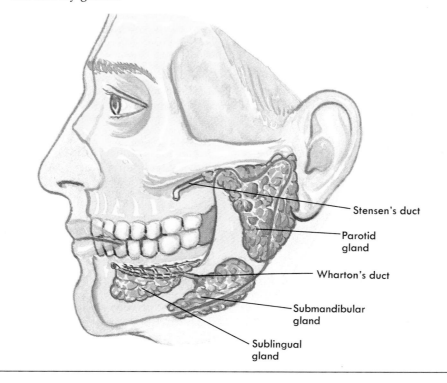

Stensen's duct

Parotid gland

Wharton's duct

Submandibular gland

Sublingual gland

Figure 13-10 Submandibular salivary gland. This gland produces mucus from the mucous acini and enzymatic secretion from the serous acini. Duct cross sections are also visible. ($\times 140$.)

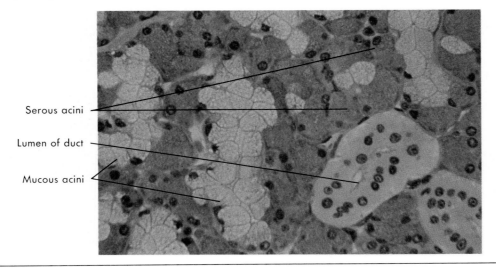

Serous acini

Lumen of duct

Mucous acini

tract. The parotid glands, largest of the salivary glands, lie just below and in front of each ear at the angle of the jaw—an interesting anatomical fact, because it explains why people who have mumps (an infection of the parotid gland) often complain that it hurts when they open their mouths or chew; these movements squeeze the tender inflamed gland. To see the openings of the ducts of the parotid glands or **Stensen's ducts,** look in a mirror at the insides of your cheeks opposite the second molar tooth on either side of the upper jaw. The ducts of the submandibular glands or **Wharton's ducts** open into the mouth on either side of the lingual frenulum (Figure 13-3). The ducts of the sublingual glands open into the floor of the mouth.

Saliva contains both mucus and a digestive enzyme called **salivary amylase** (AM-i-lase). Mucus moistens the food and allows it to pass with less friction through the esophagus and into the stomach. Salivary amylase begins the chemical digestion of carbohydrates. This dual function of saliva represented by production of both mucus and salivary amylase is apparent in the anatomy of the submandibular gland (Figure 13-10). Mucus is produced by **mucous acini,** and the digestive enzyme is produced in the **serous acini** of the gland.

PHARYNX

The **pharynx** is a tubelike structure made of muscle and lined with mucous membrane. Observe its location in Figure 13-1. Because of its location behind the nasal cavities and mouth, it functions as part of both the respiratory and digestive systems. Air must pass through the pharynx on its way to the lungs and food must pass through it on its way to the stomach. The pharynx as a whole is subdivided into three anatomical components that will be described in Chapter 14.

ESOPHAGUS

The **esophagus** (e-SOF-ah-gus) or foodpipe is the muscular, mucus-lined tube that connects the pharynx with the stomach. It is about 25 centimeters (10 inches) long. The esophagus serves as a simple passageway for movement of

Figure 13-11 Esophageal mucosa. Note the thick, abrasion-resistant nature of the lining. (×70.)

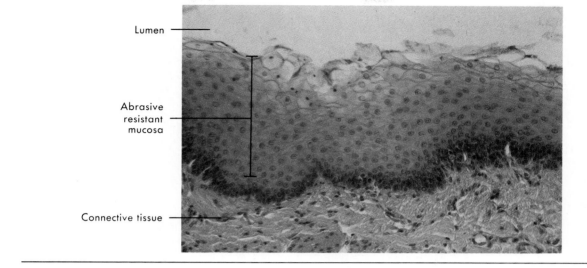

Lumen

Abrasive
resistant
mucosa

Connective tissue

food from the pharynx to the stomach and does not function in the digestive process. The mucosa of the esophagus (Figure 13-11) is modified to resist the abrasion of coarse food material that passes through it.

STOMACH

The **stomach** (Figure 13-12) lies in the upper part of the abdominal cavity just under the diaphragm. It serves as a pouch that food enters after it has been chewed, swallowed, and passed through the esophagus. The stomach looks small after it is emptied, not much bigger than a large sausage, but it expands considerably after a large meal. Have you ever felt so uncomfortably full after eating that you could not take a deep breath? If so, it probably meant that your stomach was so full of food that it occupied more space than usual and pushed up against the diaphragm. This made it hard for the diaphragm to contract and move downward as much as necessary for a deep breath.

After food has entered the stomach by passing through the muscular **cardiac sphincter**

(SFINGK-ter) at the end of the esophagus, the digestive process continues. Sphincters are rings of muscle tissue. The cardiac sphincter keeps food from reentering the esophagus when the stomach contracts.

Heartburn

Heartburn is often described as a burning sensation characterized by pain and a feeling of fullness beneath the sternum. It is a common problem caused by irritation of esophageal mucosa by acid stomach contents that reenter the esophagus. Even very small quantities of very acid stomach contents regurgitating back into the esophagus through the cardiac sphincter can cause discomfort and even inflammation. Total emptying of stomach contents back through the cardiac sphincter and ultimately up the esophagus and out of the mouth is called **vomiting** or **emesis** (EM-e-sis).

Figure 13-12   Stomach. Cutaway section shows muscle layers and interior mucosa thrown into folds called rugae.

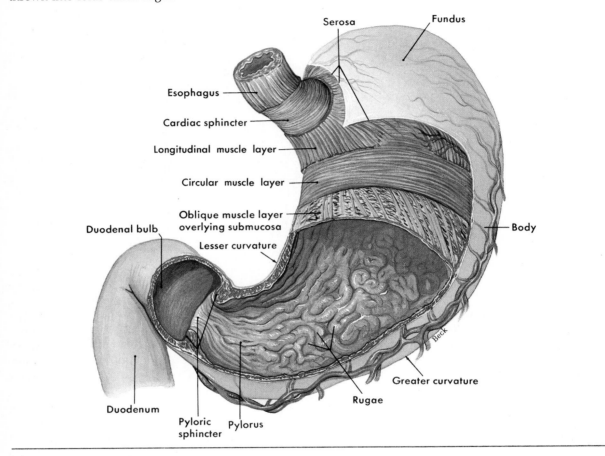

Contraction of the stomach's muscular walls mixes the food thoroughly with the gastric juice and breaks it down into a semisolid mixture called **chyme** (KIME). Gastric juice contains both hydrochloric acid and enzymes that function in the digestive process. Chyme formation is a continuation of the mechanical digestive process that begins in the mouth.

Note in Figure 13-12 that three layers of smooth muscle are present in the stomach wall. The muscle fibers that run lengthwise, around, and obliquely make the stomach one of the strongest internal organs—well able to break up food into tiny particles and to mix them thoroughly with the gastric juice to form chyme. Stomach muscle contractions result in **peristalsis,** which serves to propel food down the digestive tract. Mucous membrane lines the stomach; it contains thousands of microscopic **gastric glands** that secrete gastric juice and hydrochloric acid into the stomach. When the stomach is empty, its lining lies in folds called **rugae.**

The three divisions of the stomach shown in Figure 13-12 are the **fundus, body,** and **pylorus** (pi-LO-rus). The fundus is the enlarged portion to the left of and above the opening of the esoph-

Ulcers

An **ulcer** (Figure 13-13) is an open wound or sore in an area of the digestive system that is acted on by acid gastric juice. The two most common sites for ulcers are the stomach (gastric ulcers) and the duodenum (duodenal ulcers). Although most people think of ulcers as occurring in the stomach, most are duodenal. Ulcers cause disintegration, loss, and death of tissue as they erode the layers in the wall of the stomach or duodenum. Left untreated, ulcers cause persistent pain and may actually perforate the wall of the digestive tube, causing massive hemorrhage and widespread inflammation of the abdominal cavity and its contents. Usually perforation does not occur, but repeated small hemorrhages over long periods cause anemia. There is an old saying in medicine: "No acid, no ulcer." Most experts now agree that too much gastric acid secretion—that is, prolonged hyperacidity, is one of the most important factors in ulcer formation. If the protective layer of mucus and the dilution and buffering of acid gastric juices by swallowed food and by the alkaline juices of the small intestine are inadequate, then ulcers may form. Hyperacidity is also influenced by nervous system factors and by anxiety, other emotional states, and stress.

The drug cimetidine (Tagamet) and other medicines that reduce hydrochloric acid formation in the stomach are widely prescribed in the treatment of ulcers. In addition to excess acid, a bacterium called **Campylobacter pyloridis** (KAMP-pi-lo-bac-ter pi-LO-ri-dis), found in many ulcer patients, may also be a cause of both ulcers and chronic indigestion. The bacterium, which was first discovered in 1982 and linked to gastrointestinal inflammation in 1987, now joins acid as a potential cause of ulcers.

Pylorospasm

The pyloric sphincter is of clinical importance because **pylorospasm** (pi-LO-ro-spazm) is a fairly common condition in infants. The pyloric fibers do not relax normally in this condition. As a result, food is not allowed to leave the stomach and the infant vomits food instead of digesting and absorbing it. The condition is relieved by the administration of a drug that relaxes smooth muscles. Another abnormality of the pyloric sphincter is called **pyloric stenosis** (pi-LOR-ik ste-NO-sis), an obstructive narrowing of its opening.

agus into the stomach. The body is the central part of the stomach, and the pylorus is its lower narrow section, which joins the first part of the small intestine. Food is held in the stomach by the **pyloric** (pi-LOR-ik) **sphincter** muscle long enough for partial digestion to occur. The smooth muscle fibers of the sphincter stay contracted most of the time and thereby close off the opening of the pylorus into the small intestine. Note also in Figure 13-12 that the upper right border of the stomach is known as the **lesser curvature,** and the lower left border is called the **greater curvature.**

After food has been in the stomach for about 3 hours, the chyme will pass through the pyloric sphincter into the first part of the small intestine called the **duodenal** (du-o-DEE-nal) **bulb**—a site of frequent ulcer formation.

SMALL INTESTINE

The **small intestine** seems to be misnamed if you look at its length—it is roughly 20 feet long. It is noticeably smaller around, however, than the large intestine, so in this respect its name is appropriate. Different names identify different sections of the small intestine. In the order in which food passes through them, they are the **duodenum** (du-o-DEE-num), **jejunum,** (je-JOO-num), and **ileum** (IL-e-um).

Figure 13-13   Duodenal ulcer. The illustration depicts a deep ulceration in the duodenal wall extending as a crater through the entire mucosa and into the muscle layers.

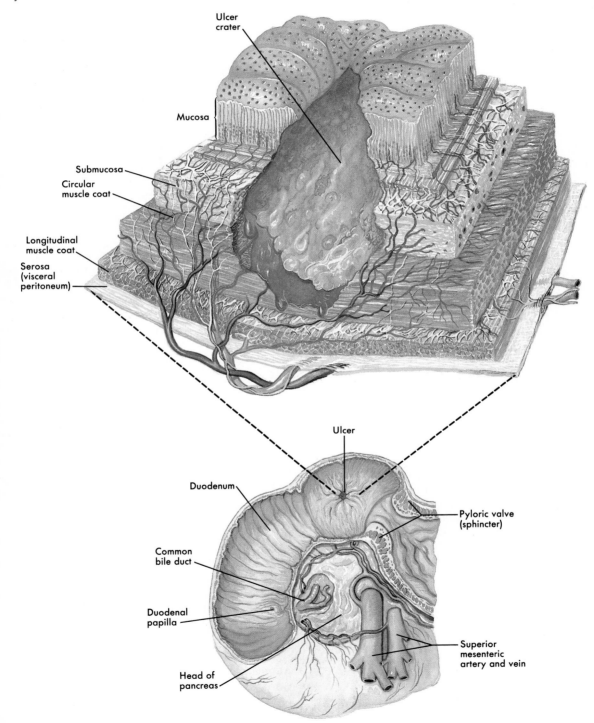

Figure 13-14 The small intestine. Note that the folds of mucosa are covered with villi and that each villus is covered with epithelium, which increases the surface area for absorption of food.

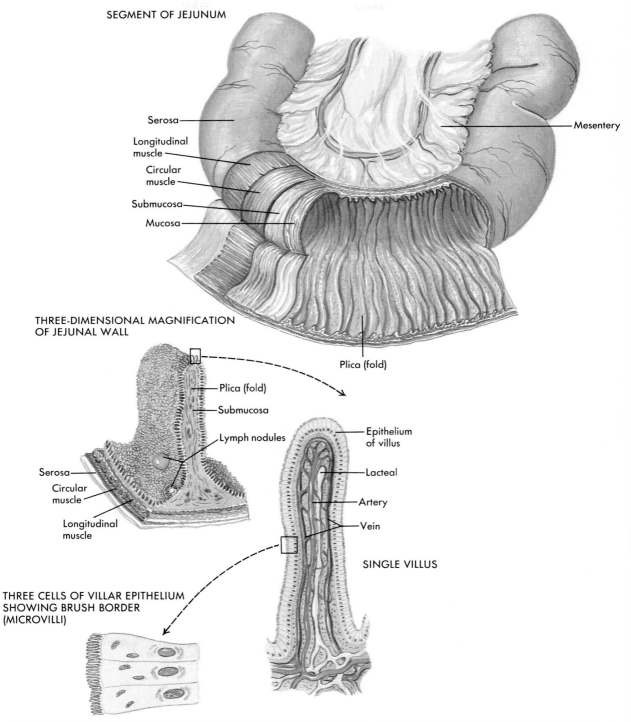

SEGMENT OF JEJUNUM

Serosa

Longitudinal muscle

Circular muscle

Submucosa

Mucosa

Mesentery

Plica (fold)

THREE-DIMENSIONAL MAGNIFICATION OF JEJUNAL WALL

Plica (fold)

Submucosa

Lymph nodules

Serosa

Circular muscle

Longitudinal muscle

Epithelium of villus

Lacteal

Artery

Vein

SINGLE VILLUS

THREE CELLS OF VILLAR EPITHELIUM SHOWING BRUSH BORDER (MICROVILLI)

The mucous lining of the small intestine, like that of the stomach, contains thousands of microscopic glands. They are called **intestinal glands,** and they secrete the intestinal digestive juice. Another structural feature of the lining of the small intestine makes it especially well suited to its function of food and water absorption; it is not perfectly smooth, as it appears to the naked eye. Instead, the intestinal lining is thrown into multiple circular folds called **plica** (PLI-kah) (Figures 13-2 and 13-14). These folds are themselves covered with thousands of tiny "fingers" called **villi** (VILL-i). Under the microscope they can be seen projecting into the hollow interior of the intestine. Inside each villus lies a rich network of blood capillaries that absorb the products of carbohydrate and protein digestion (sugars and amino acids). Millions and millions of villi jut inward from the mucous lining. Imagine the lining as perfectly smooth without any villi; think how much less surface area there would be for contact between capillaries and intestinal lining. Consider what an advantage a large contact area offers for faster absorption of food from

Figure 13-15 The gallbladder and bile ducts. Obstruction of either the hepatic or common bile duct by stone or spasm blocks the exit of the bile from the liver, where it is formed, and prevents bile from being ejected into the duodenum.

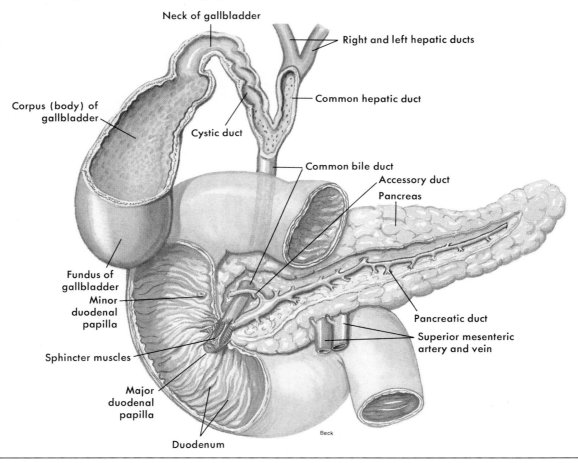

the intestine into the blood and lymph—one more illustration of the principle that structure determines function.

Note also in Figure 13-14 that each villus in the intestine contains a lymphatic vessel or **lacteal** that serves to absorb lipid or fat materials from the chyme passing through the small intestine. In addition to the thousands of villi that increase surface area in the small intestine, each villus is itself covered by epithelial cells, which have a brushlike border called **microvilli.** The microvilli further increase the surface area of each villus for absorption of nutrients.

The first subdivision of the small intestine or duodenum is where most of the chemical digestion occurs. The duodenum is C-shaped (Figure 13-15) and curves around the head of the pancreas. The acid chyme, which enters the duodenum from the stomach, first passes through a short, dilated area called the duodenal bulb. This area is the site of frequent ulceration (duodenal ulcers.) The middle third of the duodenum contains the openings of ducts that empty pancreatic digestive juice and bile from the liver into the small intestine. As you can see in Figure 13-15, the two openings are called the **minor** and **major duodenal papilla.** Occasionally a gallstone will block the major duodenal papilla, causing symptoms such as severe pain, jaundice, and digestive problems. Smooth muscle in the wall of the small intestine contracts to produce peristalsis, the wormlike movements that move food through the tract.

LIVER AND GALLBLADDER

The liver is such a large organ that it fills the entire upper right section of the abdominal cavity and even extends partway over onto the left side. Because its cells secrete a substance called **bile** into ducts, the liver is classified as an exocrine gland—in fact, it is the largest gland in the body.

Look now at Figure 13-15. First, identify the *hepatic ducts.* They drain bile out of the liver, a fact suggested by the name "hepatic," which comes from the Greek word for liver *(hepar).* Next, notice the duct that drains bile into the small intestine (duodenum), namely the *common bile duct.* It is formed by the union of the *common hepatic duct* with the *cystic duct.*

Because fats tend to form large globules, they must be broken down into smaller particles to increase the surface area for digestion. This is the function of bile. It acts to mechanically break up or **emulsify** (e-MUL-se-fi) fats. When chyme containing lipid or fat enters the duodenum, a mechanism that serves to contract the gallbladder and force bile into the small intestine is initiated. Fats in chyme stimulate or "trigger" the secretion of the hormone **cholecystokinin** (ko-le-sis-to-KIN-in) or **CCK** from the intestinal mucosa of the duodenum. This hormone then stimulates the contraction of the gallbladder and bile flows into the duodenum. Between meals, much of the bile moves up the cystic duct into the gallbladder, which is located on the undersurface of the liver. The gallbladder itself functions only to concentrate and store the bile that is produced in the liver.

Visualize a gallstone blocking the common bile duct shown in Figure 13-15. Bile could not then drain into the duodenum and leave the body in the feces. Therefore excessive amounts of bile would be absorbed into the blood. A yellowish skin discoloration called **jaundice** (JAWN-dis) would result. Obstruction of the common hepatic duct also leads to jaundice. Because bile cannot then drain out of the liver, excessive amounts of it are absorbed. Since bile is not resorbed from the gallbladder, no jaundice will occur if only the cystic duct is blocked.

In addition to secreting about a pint of bile a day, liver cells perform other functions necessary for healthy survival. They play a major role in the metabolism of all three kinds of foods. They help maintain a normal blood glucose concentration by carrying on complex and essential chemical reactions. Liver cells also carry on the first steps of protein and fat metabolism and synthesize several kinds of protein compounds. They release them into the blood, where they are called the blood proteins or plasma proteins. Prothrombin and fibrinogen, two of the plasma proteins formed by liver cells, play essential parts in blood clotting. (See p. 243.) Another protein made by liver cells, albumin, helps maintain normal blood volume. Liver cells detoxify various poisonous substances such as bacterial products and certain drugs. Liver cells store several substances, notably iron and vitamins A, B_{12}, and D.

PANCREAS

The pancreas lies behind the stomach in the concavity produced by the C shape of the duodenum. It is both an exocrine gland that secretes the pancreatic juice into ducts and an endocrine gland that secretes hormones into the blood. Pancreatic juice is the most important digestive juice. It contains enzymes that digest all three major kinds of foods. It also contains sodium bicarbonate, an alkaline substance that neutralizes the hydrochloric acid in the gastric juice that enters the intestines. Pancreatic juice enters the duodenum of the small intestine at the same place that bile enters. As you can see in Figure 13-15, both the common bile duct and the pancreatic duct open into the duodenum at the major duodenal papilla.

Between the cells that secrete pancreatic juice into ducts lie clusters of cells that have no contact with any ducts. These are the *islands of Langerhans*, which secrete the hormones of the pancreas described in Chapter 9.

LARGE INTESTINE

The **large intestine** is only about 1.5 meters or 5 feet in length. As the name implies, it has a much larger diameter than the small intestine. It forms the lower or terminal portion of the digestive tract. Undigested and unabsorbed food material enters the large intestine after passing through a sphincterlike structure (Figure 13-16) called the **ileocolic** (il-e-o-KOL-ik) **valve.** The word chyme is no longer appropriate in describing the contents of the large intestine. *Chyme,* which has the consistency of soup and is found in the small intestine, changes to the consistency of fecal matter as water and salts are resorbed during its passage to the exterior. During its movement through the large intestine,

Oral rehydration therapy

Simple diarrhea can result in life-threatening dehydration if significant water loss occurs during a short period of time. Diarrhea is one of the leading causes of infant mortality in developing countries. New attempts to educate mothers about the dangers of diarrhea and to provide them with a simple, easily prepared remedy have saved hundreds of thousands of lives in recent years. This treatment is called **oral rehydration therapy.** It involves liberal doses of an easily prepared solution containing sugar and salt. This salt-sugar solution replaces nutrients and electrolytes lost in diarrhea fluid. Because the replacement fluid can be prepared from readily available and inexpensive ingredients, it is particularly valuable in the treatment of infant diarrhea in third world countries.

material that escaped digestion in the small intestine is acted upon by bacteria. As a result of this bacterial action, additional nutrients may be released from cellulose and other fibers and absorbed. In addition to their digestive role, bacteria in the large intestine perform other important functions. They are responsible for the synthesis of vitamin K needed for normal blood clotting and for the production of some of the B complex vitamins. Once formed, these vitamins are absorbed from the large intestine and enter the blood.

Although important absorption of water, salts, and some vitamins does occur in the large intestine, this segment of the digestive tube is not as well suited for absorption as is the small intestine. Salts, especially sodium, are absorbed by active transport, and water is moved into the blood by osmosis. No villi are present in the mucosa of the large intestine. As a result, less surface area is available for absorption, and the efficiency and speed of movement of substances

Figure 13-16   Divisions of the large intestine.

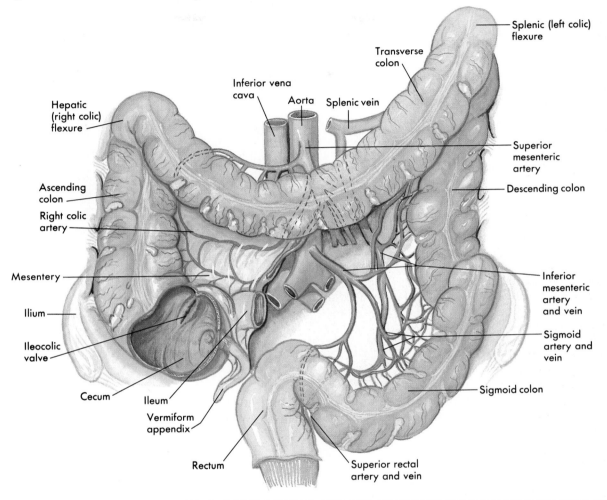

through the wall of the large intestine is much lower than in the small intestine. Normal passage of material through the large intestine takes about 3 to 5 days. If the rate of passage of material quickens, the consistency of the stools or fecal material will become more and more fluid, and **diarrhea** (di-ah-RE-ah) will result. If the time of passage through the large intestine is prolonged beyond 5 days, the condition is called **constipation.**

The subdivisions of the large intestine are listed below in the order in which food passes through them.

1 Cecum
2 Ascending colon
3 Transverse colon
4 Descending colon
5 Sigmoid colon
6 Rectum
7 Anal canal

These areas can be studied and identified by tracing the passage of material from its point of entry into the large intestine at the ileocolic valve to its elimination from the body through the external opening called the **anus.**

Figure 13-17 X-ray showing divisions of the large intestine. Note: *C*, cecum; *Hf*, hepatic flexure; *T*, transverse colon; *Sf*, splenic flexure; *D*, descending colon; *S*, sigmoid colon; and *R*, rectum.

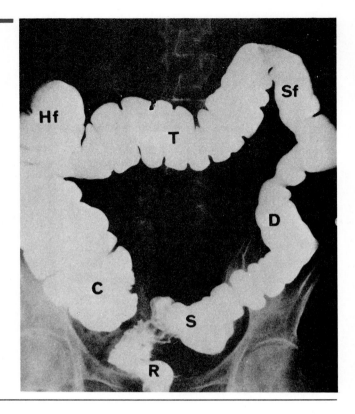

Note in Figures 13-16 and 13-17 that the ileocolic valve opens into a pouchlike area called the **cecum** (SEE-kum). The opening itself is about 5 or 6 cm above the beginning of the large intestine. Food residue in the cecum flows upward on the right side of the body in the **ascending colon.** The **hepatic** or **right colic flexure** is the bend between the ascending colon and the **transverse colon,** which extends across the front of the abdomen from right to left. The **splenic** or **left colic flexure** marks the point where the **descending colon** turns downward on the left side of the abdomen. The **sigmoid colon** is the S-shaped segment that terminates in the **rectum.** The terminal portion of the rectum is called the **anal canal,** which ends at the external opening or anus.

Two sphincter muscles stay contracted to keep the anus closed except during defecation. Smooth, or involuntary, muscle composes the **inner anal sphincter,** but striated, or voluntary, muscle composes the outer one. This anatomical fact sometimes becomes highly important from a practical standpoint. For example, often after a person has had a stroke, the voluntary anal sphincter at first becomes paralyzed. This means, of course, that the individual has no control at this time over bowel movements.

APPENDIX

The **vermiform appendix** (Latin *vermiformis* from *vermis*—"worm," *forma*—"shape") is, as the name implies, a wormlike tubular structure. Although it serves no known digestive function in humans, it is composed of lymphatic tissue and may play a minor role in the immunologic defense mechanisms of the body described in the last chapter. Note in Figure 13-16 that the appendix is directly attached to the cecum. The appendix contains a blind tubelike interior lumen that communicates with the lumen of the large intestine about 3 cm below the opening of

---

Appendicitis

The opening between the lumen of the appendix and the cecum is quite large in children and young adults—a fact of great clinical significance since trapped food or fecal material in the appendix will irritate and inflame its mucous lining causing **appendicitis.** The opening between the appendix and the cecum is often completely obliterated in the elderly, which explains the low incidence of appendicitis in the aged.

---

the ileocolic valve into the cecum. If the mucous lining of the appendix becomes inflamed, the resulting condition is the well-known affliction, **appendicitis.** As you can see in Figure 13-16, the appendix is located very close to the rectal wall. For patients with suspected appendicitis, a physician will often evaluate the appendix by a digital rectal examination.

PERITONEUM

The **peritoneum** is a large, moist, slippery sheet of serous membrane that lines the abdominal cavity and covers the organs located in it, including most of the digestive organs. The **parietal layer** of the peritoneum lines the abdominal cavity. The *visceral layer* of the peritoneum forms the outer or covering layer of each abdominal organ. The small space between the parietal and visceral layers is called the *peritoneal space.* It contains just enough peritoneal fluid to keep both layers of the peritoneum moist and able to slide freely against each other during breathing and digestive movements.

EXTENSIONS

The two most prominent extensions of the peritoneum are named the mesentery and the greater omentum. The **mesentery,** an extension of the parietal peritoneum, is shaped like a giant, plaited fan. Its smaller edge attaches to the lumbar region of the posterior abdominal wall and its long, loose outer edge encloses most of the

small intestine, anchoring it to the posterior abdominal wall. The **greater omentum** is an extension of the visceral peritoneum from the lower edge of the stomach, part of the duodenum, and the transverse colon. Shaped like a large apron, it hangs down over the intestines, and because spotty deposits of fat give it a lacy appearance, the greater omentum has been nicknamed the "lace apron." It usually envelops a badly inflamed appendix, walling it off from the rest of the abdominal organs.

DIGESTION

Digestion, a complex process that occurs in the alimentary canal, consists of various physical and chemical changes that prepare food for absorption. **Mechanical digestion** breaks food into tiny particles, mixes them with digestive juices, moves them along the alimentary canal, and finally eliminates the digestive wastes from the body. Chewing or mastication, swallowing or deglutition (deg-loo-TISH-un), peristalsis, and defecation are the main processes of mechanical digestion. **Chemical digestion** breaks down large, nonabsorbable food molecules into smaller, absorbable molecules—molecules that are able to pass through the intestinal mucosa into blood and lymph. Chemical digestion consists of numerous chemical reactions catalyzed by enzymes present in the following digestive juices: saliva, gastric juice, pancreatic juice, and intestinal juice.

CARBOHYDRATE DIGESTION

Very little digestion of carbohydrates (starches and sugars) occurs before food reaches the small intestine. Salivary amylase usually has little time to do its work because so many of us swallow our food so fast. Gastric juice contains no carbohydrate-digesting enzymes. But once the food reaches the small intestine, pancreatic and intestinal juice enzymes digest the starches and sugars. A pancreatic enzyme (amylase) starts the process by changing starches into a sugar, namely, maltose. Three intestinal enzymes—maltase, sucrase, and lactase—digest sugars by changing them into simple sugars, chiefly glucose (dextrose). Maltase digests maltose (malt sugar), sucrase digests sucrose (ordinary cane sugar), and lactase digests lactose (milk sugar). The end products of carbohydrate digestion are the so-called simple sugars, the most abundant of which is glucose.

PROTEIN DIGESTION

Protein digestion starts in the stomach. Two enzymes (rennin and pepsin) in the gastric juice cause the giant protein molecules to break up into somewhat simpler compounds. Then in the intestine other enzymes (trypsin in the pancreatic juice and peptidases in the intestinal juice) finish the job of protein digestion. Every protein molecule is made up of many amino acids joined together. When enzymes have split up the large protein molecule into its separate amino acids, protein digestion is completed. Hence the end product of protein digestion is amino acids. For obvious reasons, the amino acids are also referred to as "protein building blocks."

FAT DIGESTION

Not only very little carbohydrate digestion but also very little fat digestion occurs before food reaches the small intestine. An enzyme in gastric juice (gastric lipase) digests some fat in the stomach, but most fats go undigested until after bile in the duodenum emulsifies them— that is, breaks the fat droplets into very small droplets. After this takes place, the pancreatic enzyme (steapsin or pancreatic lipase) splits up the fat molecules into fatty acids and glycerol (glycerin). The end products of fat digestion, then, are fatty acids and glycerol.

Table 13-2 summarizes the main facts about chemical digestion. Enzyme names indicate the type of food digested by the enzyme. For example, the name *amylase* indicates that the enzyme digests carbohydrates (starches and sugars), *protease* indicates a protein-digesting enzyme, and *lipase* means a fat-digesting enzyme. When carbohydrate digestion has been completed, starches and complex sugars (polysaccharides) have been changed mainly to glucose, a simple sugar (monosaccharide). The end product of protein digestion, on the other hand, is

amino acids. Fatty acid and glycerol are the end products of fat digestion. Use information in Table 13-2 to answer questions 10 to 15, p. 331.

ABSORPTION

After food is digested, it is absorbed; that is, it moves through the mucous membrane lining of the small intestine into the blood and lymph. In other words, food absorption is the process by which molecules of amino acids, glucose, fatty acids, and glycerol go from the inside of the intestines into the circulating fluids of the body. Absorption of foods is just as essential a process as digestion of foods. The reason is fairly obvious. As long as food stays in the intestines, it cannot nourish the millions of cells that compose all other parts of the body. Their lives depend on the absorption of digested food and its transportation to them by the circulating blood.

Structural adaptations of the digestive tube, including folds in the lining mucosa, villi, and microvilli, increase the absorptive surface and the efficiency and speed of transfer of materials from intestinal lumen to body fluids. Many salts such as sodium are actively transported through the intestinal mucosa. Water follows by osmosis. Other nutrients are also actively transported into the blood of capillaries in the intestinal villi. Fats enter the lymphatic vessels or lacteals found in intestinal villi.

METABOLISM

A good phrase to remember in connection with the word metabolism is "use of foods" because basically this is what metabolism is—the use the body makes of foods once they have been digested, absorbed, and circulated to cells. It uses them in two ways: as an energy source and as building blocks for making complex chemical compounds. Before they can be used in these two ways, foods have to enter cells and undergo many chemical changes there. All chemical reactions that release energy from food molecules together make up the process of catabolism—a vital process, since it is the only way

Table 13-2 Chemical digestion

Digestive juices and enzymes	Enzyme digests (or hydrolyzes)	Resulting product*
Saliva		
Amylase (ptyalin)	Starch (polysaccharide)	Maltose (a disaccharide or double sugar)
Gastric juice		
Protease (pepsin) plus hydrochloric acid	Proteins, including casein	Proteoses and peptones (partially digested proteins)
Lipase (of little importance)	Emulsified fats (butter, cream, and so on)	**Fatty acids and glycerol**
Pancreatic juice		
Protease (trypsin)†	Proteins (either intact or partially digested)	Proteoses, peptides, and **amino acids**
Lipase (steapsin)	Bile—emulsified fats	**Fatty acids and glycerol**
Amylase (amylopsin)	Starch	Maltose
Intestinal juice (succus entericus)		
Peptidases	Peptides	**Amino acids**
Sucrase	Sucrose (cane sugar)	**Glucose and fructose‡** (Simple sugars of monosaccharides)
Lactase	Lactose (milk sugar)	**Glucose and galactose** (simple sugars)
Maltase	Maltose (malt sugar)	**Glucose** (grape sugar)

*Substances in boldface type are end products of digestion; that is, completely digested foods ready for absorption.
†Secreted in inactive form (trypsinogen); activated by enterokinase, an enzyme in the intestinal juice.
‡Glucose is also called dextrose; fructose is called levulose.

that the body has of supplying itself with energy for doing any of its many kinds of work. The many chemical reactions that build food molecules into more complex chemical compounds together constitute the process of anabolism. Catabolism and anabolism together make up the process of metabolism.

CARBOHYDRATE METABOLISM

Carbohydrates are the body's preferred energy food. Human cells catabolize glucose rather than other substances as long as enough glucose enters them to supply their energy needs. Two series of chemical reactions, occurring in a precise sequence, make up the process of glucose metabolism. **Glycolysis** is the name given the first series of reactions; **citric acid cycle** is the name of the second series. Glycolysis, as Figure 13-18 shows, changes glucose to pyruvic acid, and then the citric acid cycle changes the pyruvic acid to carbon dioxide. Glycolysis takes place in the cytoplasm of a cell, whereas the citric acid cycle goes on in the mitochondria, the cell's miniature power plants. Glycolysis uses no oxygen; it is an anaerobic process. The citric acid cycle, in contrast, is an oxygen-using or aerobic process.

While the chemical reactions of glycolysis and the citric acid cycle are going on, energy stored in the glucose molecule is being released. Almost instantaneously, however, more than half of it is put back into storage, not in glucose molecules but in the molecules of another compound, adenosine triphosphate (ATP). The rest

Figure 13-18   Scheme to show that catabolism breaks larger molecules down into smaller ones. Glycolysis splits one molecule of glucose (six carbon atoms) into two molecules of pyruvic acid (three carbon atoms each). The citric acid cycle converts each pyruvic acid molecule to three carbon dioxide molecules.

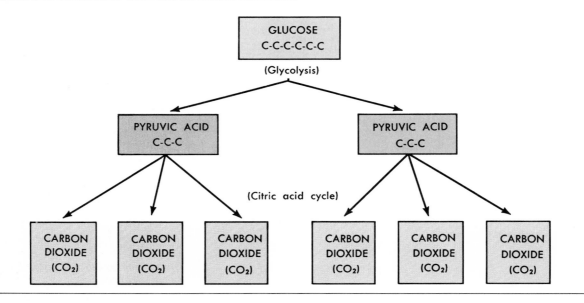

of the energy originally stored in the glucose molecule is released as heat. ATP serves as the direct source of energy for doing cellular work in all kinds of living organisms from one-cell plants to billion-cell animals, including man. Among biological compounds, therefore, ATP ranks as one of the most important. The energy stored in ATP molecules differs in two ways from the energy stored in food molecules: the energy in ATP molecules can be released almost instantaneously, and it can be used directly to do cellular work. Release of energy from food molecules occurs much more slowly because it accompanies the long series of chemical reactions that make up the process of catabolism. For some reason, energy released from food molecules cannot be used directly for doing cellular work. It must first be transferred to ATP molecules and be released explosively from them.

Although glucose is mainly catabolized, small amounts of it are anabolized. One process of glucose anabolism is called glycogenesis. Carried on chiefly by liver and muscle cells, **glycogenesis** consists of a series of reactions that join glucose molecules together, like so many beads in a necklace, to form glycogen, a compound sometimes called "animal starch."

Something worth noticing is that the amount of food in the blood normally does not change very much, not even when we go without food for many hours, when we exercise and use a lot of food for energy, or when we sleep and use little food for energy. The amount of glucose, for example, usually stays at about 80 to 120 milligrams in 100 milliliters of blood.

Several hormones help regulate carbohydrate metabolism to keep blood glucose normal. **Insulin** is one of the most important of these. It acts in some way not yet definitely known to make glucose leave the blood and enter the cells at a more rapid rate. As insulin secretion increases, more glucose leaves the blood and enters the cells. The amount of glucose in the blood therefore tends to decrease while the rate of glucose metabolism in cells tends to increase. Too little insulin secretion, such as occurs in diabetes mellitus, produces the opposite effects. Less glucose leaves the blood and enters the cells; more glucose therefore remains in the blood and less glucose is metabolized by cells. In other words, high blood glucose (hyperglycemia) and a low rate of glucose metabolism characterize insulin deficiency. Insulin is the only hormone that functions to lower the blood glucose level. Several other hormones, on the other hand, tend to increase it. Growth hormone secreted by the anterior pituitary gland, hydrocortisone secreted by the adrenal cortex, and epinephrine secreted by the adrenal medulla are three of the most important hormones that tend to increase blood glucose. More information about these hormones and others that help control metabolism appears on pp. 214 and 223.

FAT METABOLISM

Fats, like carbohydrates, are primarily energy foods. If cells have inadequate amounts of glucose to catabolize, they immediately shift to the catabolism of fats for their energy supply. This happens normally whenever a person goes without food for many hours. It happens abnormally in diabetic individuals. Because of an insulin deficiency, too little glucose enters the cells of a diabetic person to supply all of his energy needs. Result? The cells catabolize fats to make up the difference. In all persons, fats not needed for catabolism are anabolized and stored in adipose tissue.

PROTEIN METABOLISM

Protein catabolism occurs to some extent, but more important is protein anabolism, the process by which the body builds amino acids into complex protein compounds—for example, enzymes and proteins that form the structural part of the cell.

METABOLIC RATES

The **basal metabolic rate** (BMR) is the rate at which food is catabolized under basal condi-

tions—that is, when the individual is awake, is not digesting food, and is not adjusting to a cold external temperature. Or, stated differently, the basal metabolic rate is the number of calories of heat that must be produced per day by catabolism just to keep the body alive, awake, and comfortably warm. For the determination of the basal metabolic rate, the amount of oxygen the individual inhales in a specific length of time is measured, and this figure is used to calculate the basal metabolic rate, stated in calories.

The basal metabolic rate is an indirect measure of thyroid gland functioning. So, too, is the amount of protein-bound iodine (PBI) in venous blood. A higher than normal concentration of PBI indicates higher than normal secretion of thyroid hormone and higher than normal metabolic rate. PBI determination has replaced BMR determination in clinical use because PBI determination is a simpler procedure.

A person's BMR represents the amount of food that his body must catabolize each day for him simply to stay alive and awake in a comfortably warm environment. To provide energy for him to do muscular work and to digest and absorb any food, an additional amount of food must be catabolized. How much more food must be catabolized depends mainly on how much work the individual does. The more active he is, the more food his body must catabolize and the higher his total metabolic rate will be. The **total metabolic rate** (TMR) is the total amount of energy (expressed in calories) used by the body per day (Figure 13-19).

When the number of calories in your daily food intake equals your TMR, your weight remains constant (except for possible variations resulting from water retention or water loss). When your daily food contains more calories than your TMR, you gain weight; when your daily food contains fewer calories than your TMR, you lose weight. These weight control principles never fail to operate. Nature never forgets to count calories. Reducing diets make use of this knowledge. They contain fewer calories than the TMR of the individual eating the diet.

Figure 13-19 Factors that determine the basal and total metabolic rates.

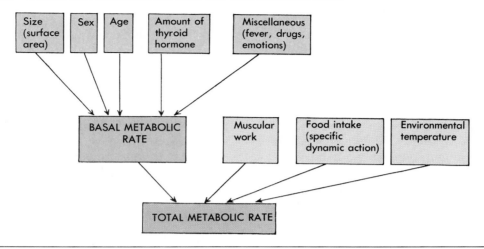

OUTLINE SUMMARY

Mouth
A Roof—formed by hard palate (parts of maxillary and palatine bones) and soft palate, an arch-shaped muscle separating mouth from nasopharynx; uvula, a downward projection of soft palate
B Floor—formed by tongue and its muscles; papillae, small elevations on mucosa of tongue; taste buds, found in many papillae; lingual frenulum, fold of mucous membrane that helps anchor tongue to floor of mouth

Teeth
A Twenty teeth in temporary set; average age for cutting first tooth about 6 months; set complete at about 2 years of age
B Thirty-two teeth in permanent set; 6 years about average age for starting to cut first permanent tooth; set complete usually between ages of 17 and 24 years
C Names of teeth—see Figure 13-7
D Structures of tooth—see Figure 13-8

Salivary Glands
A Parotid glands
B Submandibular glands
C Sublingual glands

Pharynx
Esophagus
Stomach
A Size—expands after large meal; about size of large sausage when empty
B Pylorus—lower part of stomach; pyloric sphincter muscle closes opening of pylorus into duodenum
C Wall—many smooth muscle fibers; contractions produce churning movements and peristalsis
D Lining—mucous membrane; many microscopic glands that secrete gastric juice and hydrochloric acid into stomach; mucous membrane lies in folds (rugae) when stomach is empty

Small Intestine

A Size —about 20 feet long but only an inch or so in diameter
B Divisions
  1 Duodenum
  2 Jejunum
  3 Ileum
C Wall—contains smooth muscle fibers that contract to produce peristalsis
D Lining—mucous membrane; many microscopic glands (intestinal glands) secrete intestinal juice; villi (microscopic finger-shaped projections from surface of mucosa into intestinal cavity) contain blood and lymph capillaries

Liver and Gallbladder

A Size and location—largest gland; fills upper right section of abdominal cavity and extends over into left side
B Functions—secretes bile; helps maintain normal blood glucose by carrying on process of glycogenesis; forms prothrombin, fibrinogen, and certain other blood proteins; also performs several other functions
C Ducts
  1 Hepatic—drains bile from liver
  2 Cystic—duct by which bile enters and leaves gallbladder
  3 Common bile—formed by union of hepatic and cystic ducts; drains bile from hepatic or cystic ducts into duodenum

Pancreas

A Location—behind stomach
B Functions
  1 Pancreatic cells secrete pancreatic juice into pancreatic ducts; main duct empties into duodenum
  2 Islands of Langerhans—cells not connected with pancreatic ducts; beta cells secrete insulin and alpha cells secrete glucagon into blood

Large Intestine

A Divisions
  1 Cecum
  2 Colon—ascending, transverse, descending, and sigmoid
  3 Rectum
B Opening to exterior—anus
C Wall—contains smooth muscle fibers that contract to produce churning, peristalsis, and defecation
D Lining—mucous membrane

Appendix

Blind tube off cecum; no known digestive functions in humans

Peritoneum

A Definitions—peritoneum, serous membrane lining abdominal cavity and covering abdominal organs; parietal layer of peritoneum lines abdominal cavity; visceral layer of peritoneum covers abdominal organs; peritoneal space lies between parietal and visceral layers
B Extensions—largest ones are the mesentery and greater omentum; mesentery is extension of parietal peritoneum, which attaches most of small intestine to posterior abdominal wall; greater omentum, or "lace apron," hangs down from lower edge of stomach and transverse colon over intestines

Digestion

Meaning—changing foods so that they can be absorbed and used by cells
A Mechanical digestion—chewing, swallowing, and peristalsis break food into tiny particles, mix them well with the digestive juices, and move them along the digestive tract
B Chemical digestion—breaks up large food molecules into compounds having smaller molecules; brought about by digestive enzymes

C Carbohydrate digestion—mainly in small intestine
  1 Pancreatic enzyme, amylopsin—changes starches to maltose
  2 Intestinal juice enzymes
    a Maltase changes maltose to glucose
    b Sucrase changes sucrose to glucose
    c Lactase changes lactose to glucose
D Protein digestion—starts in stomach; completed in small intestine
  1 Gastric juice enzymes, rennin and pepsin, partially digest proteins
  2 Pancreatic enzyme, trypsin, completes digestion of proteins to amino acids
  3 Intestinal enzyme, peptidases, complete digestion of partially digested proteins to amino acids
E Fat digestion
  1 Gastric lipase changes small amount of fat to fatty acids and glycerin in stomach
  2 Bile contains no enzymes but emulsifies fats (breaks fat droplets into very small droplets)
  3 Pancreatic lipase changes emulsified fats to fatty acids and glycerin in small intestine

Absorption

A Meaning—digested food moves from intestine into blood or lymph
B Where absorption occurs—foods and water from small intestine; water also absorbed from large intestine

Metabolism

A Meaning—use of foods by body cells for energy and building complex compounds
B Catabolism—breaks food molecules down into carbon dioxide and water, releasing their stored energy; oxygen used up in catabolism
C Anabolism—builds food molecules into complex substances
D Carbohydrates primarily catabolized for energy but small amounts are anabolized by glycogenesis (a series of chemical reactions that changes glucose to glycogen—occurs mainly in liver cells where glycogen is stored); glycogenolysis is process (series of chemical reactions) by which glycogen is changed back to glucose
E Blood glucose (imprecisely, blood sugar)—normally stays between about 80 and 120 milligrams per 100 milliliters of blood; *insulin* accelerates movement of glucose out of blood into cells, therefore tends to decrease blood glucose and increase glucose catabolism
F Fats both catabolized to yield energy and anabolized to form adipose tissue
G Proteins primarily anabolized and secondarily catabolized
H Metabolic rates
  1 Basal metabolic rate (BMR)—rate of metabolism when person is lying down, but awake, when about 12 hours have passed since last meal, and when environment is comfortably warm
  2 Total metabolic rate (TMR)—the total amount of energy, expressed in calories, used by the body per day
  3 Protein-bound iodine (PBI)—indirect measure of thyroid secretion and of metabolic rate

NEW WORDS

absorption
alimentary canal
appendicitis
ascites
bolus
cavity
chyme
diarrhea

digestion
emesis
enteritis
feces
frenulum
gastritis
gastroenterology

glycogenesis
glycogenolysis
jaundice
lumen
mastication
mesentery
papilla
peristalsis

peritoneum
plica
pylorospasm
rugae
ulcer
uvula
villus

CHAPTER TEST

1. In the digestive system the breakdown of food material is both _____ and _____ in nature.
2. Undigested waste material resulting from the digestive process is called _____ .
3. The study of the stomach, and the intestines, and of their diseases is called _____ .
4. Inflammation of the intestines is called _____ .
5. The liver and gallbladder are classified as _____ organs of digestion.
6. The hollow space within the digestive tube is called the _____ .
7. The inside or lining coat of the digestive tube is called the _____ .
8. The roof of the mouth is formed by the hard and soft _____ .
9. That portion of a tooth that is exposed and visible is called the _____ .
10. The teeth serve a cutting function during the chewing of food—a process called _____ .
11. Saliva contains the enzyme called salivary _____ .
12. Food moves from the pharynx to the stomach by passing through the _____ .
13. The semisolid mixture of food and gastric juice in the stomach is called _____ .
14. The movement of food through the digestive tract results from contractions called _____ .
15. The middle segment of the small intestine is called the _____ .

Select the most correct answer from Column B for each statement in Column A. (Only one answer is correct.)

Column A	Column B
16. _____ waste material	a. Liver
17. _____ inflammation of stomach	b. Emesis
18. _____ accessory organ of diges- tion	c. Enamel
	d. Fundus
19. _____ double fold of peritoneum	e. Feces
20. _____ hardest tissue in body	f. Jaundice
21. _____ vomiting	g. Mesentery
22. _____ semisolid mixture	h. Gastritis
23. _____ division of stomach	i. Lipase
24. _____ yellowish skin condition	j. Chyme
25. _____ digestive enzyme	

Circle the T before each true statement and the F before each false statement.

T F 26. The salivary glands are considered accessory organs of diges-tion.

T F 27. The left colic or splenic flexure is located between the trans-verse colon and ascending colon.

T F 28. The ileum is that portion of the small intestine found between the duodenum and jejunum.

T F 29. The large intestine is classified as an accessory organ of diges-tion.

T F 30. The adventitia or serosa is the outermost coat of the digestive tube.

T F 31. The uvula is attached to the soft palate.

T F 32. In humans there are 32 deciduous (baby) teeth and 20 perma-nent teeth.

T F 33. Dental caries is one of the most common diseases in the civi-lized world.

T F 34. No chemical digestion can occur in the mouth.

T F 35. The mucosa of the esophagus is thin, delicate, and easily in-jured.

T F 36. The pyloric sphincter is located between the esophagus and stomach.

T F 37. Most experts agree that excess stomach acid plays an important role in ulcer formation.

T F 38. Plica, villi, and microvilli all increase the surface area of the small intestine.

T F 39. The hormone cholecystokinin stimulates contraction of the gall-bladder.

T F 40. Numerous villi are present in the mucosa of the large intestine.

REVIEW QUESTIONS

1 What organs form the gastrointestinal tract?

2 Identify each of the following structures: jejunum, cecum, colon, duodenum, and ileum.

3 If you inserted 9 inches of an enema tube through the anus, the tip of the tube would probably be in what structure?

4 How many teeth should an adult have?

5 How many teeth should a child 2½ years old have? Would he have some of each of the following teeth: incisors, canines, premolars, and molars? If not, which ones would he not have?

6 Identify each of the following:

islands of Langerhans   rugae
parotid glands   villi
pylorus

7 What process do liver cells carry on that prevents the level of blood glucose from becoming dangerously high after a heavy meal?

8 What two processes do liver cells carry on that prevent a dangerously low blood glucose level from developing during fasting and between meals?

9 In what organ does the digestion of starches begin?

10 What digestive juice contains no enzymes?

11 Only one digestive juice contains enzymes for digesting all three kinds of food. Which juice is this? In what organ does it do its work?

12 What kind of food is not digested in the stomach?

13 Which digestive juice emulsifies fats?

14 What three digestive juices act on foods in the small intestine?

15 What juices digest carbohydrates? proteins? fats?

16 Explain as briefly and clearly as you can what each of the following terms means:

absorption   digestion
anabolism   metabolism
catabolism

17 Explain why you think the following statement is true or false: "If you do not want to gain or lose weight but just stay the same, you must eat just enough food to supply the calories of your BMR. If you eat more than this, you will gain; if you eat less than this, you will lose."

CHAPTER OUTLINE

OBJECTIVES

After you have completed this chapter, you should be able to:

1 Discuss the generalized functions of the respiratory system.

2 List the major organs of the respiratory system and describe the function of each.

3 Compare, contrast, and explain the mechanism responsible for the exchange of gases that occurs during internal and external respiration.

4 Identify and discuss the primary volume of air exchanged in pulmonary ventilation.

5 Define the following terms: *surfactant, respiratory membrane, epiglottis, pneumothorax, hypoxia, pulmonary ventilation, emphysema.*

No one needs to be told how important the **respiratory system** is. The respiratory system serves the body much as a lifeline to an oxygen tank serves a deep-sea diver. Think how panicky you would feel if suddenly your lifeline became blocked—if you could not breathe for a few seconds! Of all the substances that cells and therefore the body as a whole must have to survive, oxygen is by far the most crucial. A person can live a few weeks without food, a few days without water, but only a few minutes without oxygen. Constant removal of carbon dioxide from the body is just as important for survival as a constant supply of oxygen.

The organs of the respiratory system are designed to perform two basic functions—they serve as an: (1) **air distributor** and as a (2) **gas exchanger** for the body. The respiratory system ensures that oxygen is supplied to and carbon dioxide is removed from the body's cells. The process of respiration, therefore, is an important **homeostatic mechanism.** By constantly supplying adequate oxygen and by removing carbon dioxide as it forms, the respiratory system helps to maintain a constant environment that enables our body cells to function effectively.

In addition to air distribution and gas exchange, the respiratory system effectively (1) **filters,** (2) **warms,** and (3) **humidifies** the air we breathe. Respiratory organs or organs closely associated with the respiratory system, such as the **sinuses,** also influence speech or sound production and make the sense of smell or **olfaction** (ol-FAK-shun) possible. In this chapter the structural plan of the respiratory system will be considered first, then the respiratory organs will be discussed individually, and finally some facts about respiration that are especially important for an understanding of respiratory diseases will be given.

STRUCTURAL PLAN

Respiratory organs include the (1) **nose,** (2) **pharynx** (FAR-inks), (3) **larynx** (LAR-inks), (4) **trachea** (TRA-ke-ah), (5) **bronchi** (BRONG-ki), and the (6) **lungs.**

The basic structural design of this organ system is that of a tube with many branches ending in millions of extremely tiny, very thin-walled sacs called **alveoli** (al-VE-o-li). Figures 14-1 and 14-2 show the extensive branching of the "respiratory tree" in both lungs. Figure 14-2 shows the location on the chest or rib cage that lies over each area. Think of this air distribution system as an "upside-down tree." The trachea or windpipe then becomes the trunk and the bronchial tubes the branches. This idea will be developed when the types of bronchi and the alveoli are studied in more detail later in the chapter. A network of capillaries fits like a tight-fitting hairnet around each microscopic alveolus. Incidentally, this is a good place for us to think again about a principle already mentioned several times, namely, that structure determines function. The function of alveoli—in fact, the function of the entire respiratory system—is to distribute air close enough to blood for a gas exchange to take place between air and blood. The passive transport process of **diffusion,** which was described in Chapter 2, is responsible for the exchange of gases that occurs in the respiratory system. You may want to review the discussion of diffusion in Chapter 2 before you study the mechanism of gas exchange that occurs in the lungs and body tissues. Two facts about the structure of alveoli assist in diffusion and make them able to perform this function admirably. First, the wall of each alveolus is made up of a single layer of cells and so are the walls of the capillaries around it. This means that between the blood in the capillaries and the air in the alveolus there is a barrier probably less than $1/5,000$ of an inch thick! This extremely thin barrier is called the **respiratory membrane.** Note in Figure 14-3 that the surface of the respiratory membrane inside the alveolus is covered by a substance called **surfactant** (sur-FAK-tant). This important substance helps to reduce surface tension in the alveoli and keep them from collapsing as air moves in and out during respiration. Second, there are millions of alveoli. This means that together they make an enormous surface (approximately 1,100 square feet, an area many times larger than the surface of the entire body)

Infant respiratory distress syndrome (IRDS)

Infant respiratory distress syndrome or **IRDS** is a very serious, life-threatening condition that most often affects prematurely born infants of less than 37 weeks' gestation or those who weigh less than 2.2kg (5 lbs) at birth. IRDS is the leading cause of death among premature infants in the United States, claiming over 20,000 premature babies each year. The disease is characterized by a lack of **surfactant** (sur-FAK-tant) in the alveolar air sacs.

Surfactant is a phospholipid compound rich in **lecithin** (LES-i-thin), which is manufactured by specialized cells in the walls of the alveoli. Surfactant reduces surface tension between the alveolar walls and permits easy movement of air into and out of the lungs. The ability of the body to manufacture this important substance is not fully developed until shortly before birth—normally about 40 weeks following conception. Lack of surfactant is not a problem when the lungs are filled with amniotic fluid prior to birth. However,

at birth air replaces amniotic fluid, which is expelled from the lungs when breathing begins. As soon as a newborn infant takes its first breath, the alveoli must open and serve in gas exchange.

In newborn infants who are unable to manufacture surfactant, many air sacs collapse on expiration because of the increased surface tension. The effort required to reinflate these collapsed alveoli is much greater than that needed to reinflate normal alveoli with adequate surfactant. The baby soon develops labored breathing, and symptoms of respiratory distress appear shortly after birth.

Treatment of IRDS is aimed at keeping the alveoli open so that delivery and exchange of oxygen and carbon dioxide can occur. To accomplish this, a tube may be inserted into the respiratory tract and oxygen-rich air must be delivered under sufficient pressure to keep the alveoli from collapsing at the end of expiration. The procedure is called **PEEP** or **positive end expiratory pressure** treatment.

where larger amounts of oxygen and carbon dioxide can rapidly be exchanged.

RESPIRATORY TRACTS

The respiratory system is often divided into upper and lower tracts or divisions to assist in the description of symptoms associated with common respiratory problems such as a "cold." The organs of the upper respiratory tract are located outside of the thorax or chest cavity, whereas those in the lower tract or division are located almost entirely within it. The **upper respiratory tract** is composed of the nose, nasopharynx, oropharynx, laryngopharynx, and larynx. The **lower respiratory tract** or division consists of the trachea, all segments of the bronchial tree, and the lungs. The designation "upper respiratory infection" or URI is often used to describe a "head cold." Typically the symptoms of an upper respiratory infection involve the sinuses, nasal cavity, pharynx, and larynx, while the symptoms of a "chest cold" are similar to pneumonia and involve the organs of the lower respiratory tract.

RESPIRATORY MUCOSA

Before beginning the study of individual organs in the respiratory system, it is important to review the histology or microscopic anatomy of the **respiratory mucosa**—the membrane that

Figure 14-1 Structural plan of the respiratory organs showing the pharynx, trachea, bronchi, and lungs. The inset shows the grapelike alveolar sacs where the interchange of oxygen and carbon dioxide takes place through the thin walls of the alveoli. Capillaries surround the alveoli.

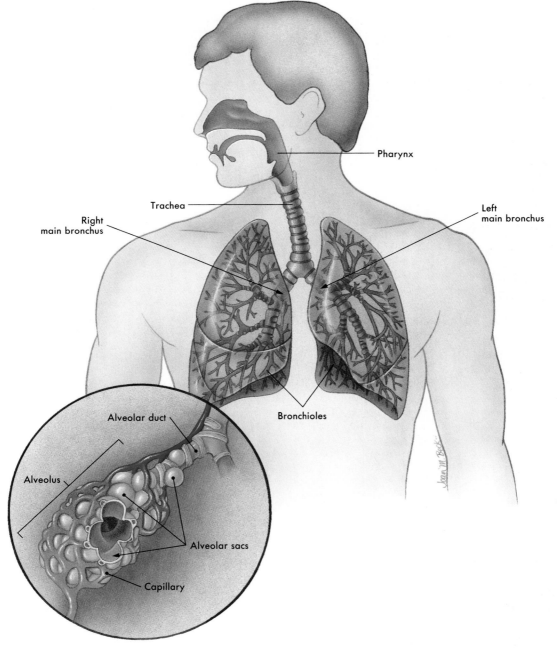

Figure 14-2   Location of lungs and trachea in relation to rib cage.

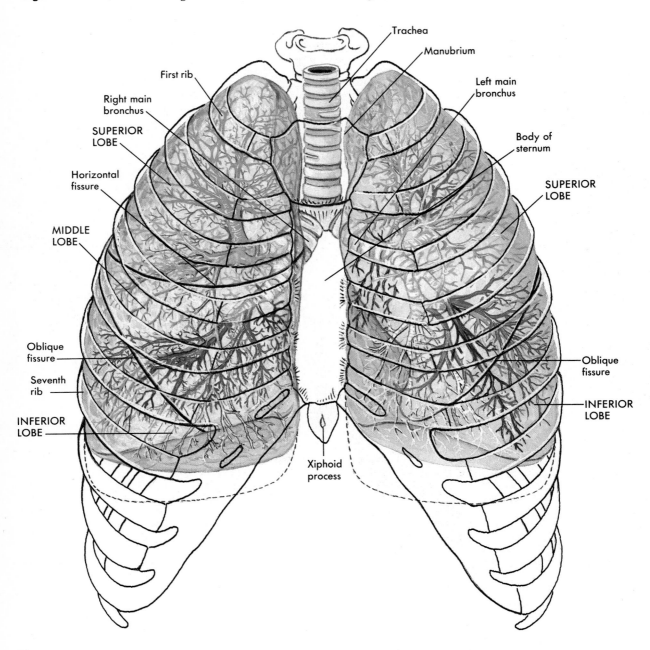

Figure 14-3   A small portion of the respiratory membrane greatly magnified. An extremely thin interstitial layer of tissue separates the endothelial cell and basement membrane on the capillary side from the epithelial cell and surfactant layer on the alveolar side of the respiratory membrane. The total thickness of the respiratory membrane is less than 1/5,000 of an inch!

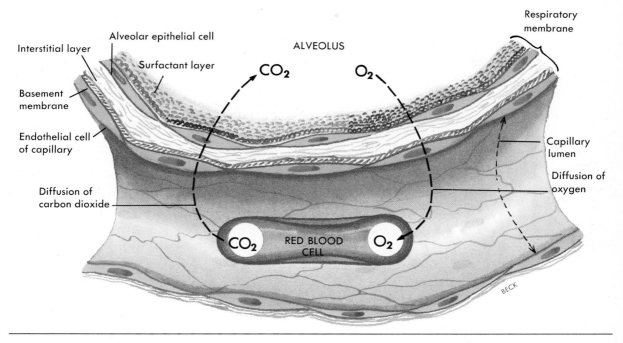

lines most of the air distribution tubes in the system. Recall that in addition to serving as air distribution passageways or gas exchange surfaces, the anatomical components of the respiratory tract and lungs function to *cleanse, warm,* and *humidify* inspired air. Air entering the nose is generally contaminated with one or more common irritants; examples include insects, dust, pollen, and bacterial organisms. A remarkably effective air purification mechanism removes almost every form of contaminant before inspired air reaches the alveoli or terminal air sacs in the lungs.

The layer of protective mucus that covers a large portion of the membrane that lines the respiratory tree serves as the most important air purification mechanism. Over 125 milliliters of respiratory mucus (sputum) is produced daily. It forms a continuous sheet called a "mucus blan-

ket" that covers the lining of the air distribution tubes in the respiratory tree. This layer of cleansing mucus moves upward to the pharynx from the lower portions of the bronchial tree on millions of hairlike cilia that cover the epithelial cells in the respiratory mucosa (Figure 14-4). The microscopic cilia that cover epithelial cells in the respiratory mucosa are unique in that they beat or move only in one direction. The result is movement of mucus toward the pharynx. Cigarette smoke paralyzes these cilia and results in accumulations of mucus and the typical smoker's cough, an effort to clear the secretions. Figure 14-10 shows a cross section through the windpipe or trachea. The details of the respiratory mucosa that lines the trachea are shown. Do not confuse the respiratory membrane with the respiratory mucosa! The **respiratory membrane** (Figure 14-3) separates the air in the al-

Figure 14-4 Respiratory mucosa lining the trachea. A layer of mucus can be seen covering the hairlike cilia. (×70.)

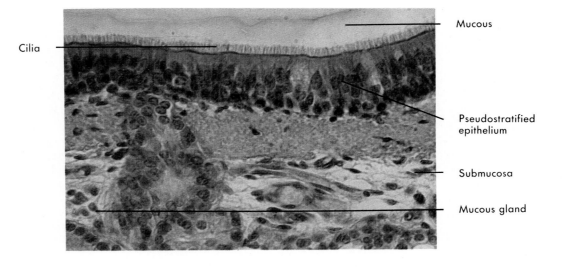

Cilia

Mucous

Pseudostratified epithelium

Submucosa

Mucous gland

Figure 14-5 Photomicrograph of lung showing parts of several alveoli. Note the proximity of the capillary to the alveolar wall. (×140.)

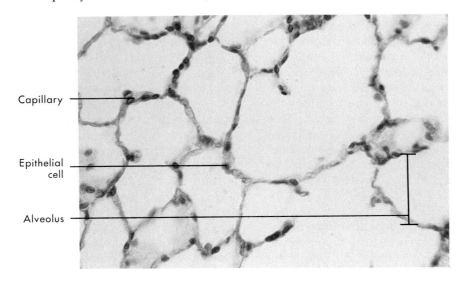

Capillary

Epithelial cell

Alveolus

veoli from the blood in surrounding capillaries. The **respiratory mucosa** (Figure 14-4) is covered with mucus and lines the tubes of the respiratory tree.

NOSE

Air enters the respiratory tract through the **external nares** (NA-rez) or nostrils. It then flows into the right and left **nasal cavities.** A partition called the *nasal septum* separates these two cavities, which are lined by a mucous membrane.

Nosebleeds

The nasal septum has a rich blood supply. **Nosebleeds** or **epistaxis** (ep-i-STAK-sis) often occur as a result of septal contusions caused by a direct blow to the nose.

The surface of the nasal cavities is moist from mucus and warm from blood flowing just under it. Nerve endings responsible for the sense of smell are located in the nasal mucosa. Four **paranasal sinuses**—frontal, maxillary, sphenoidal, and ethmoidal—drain into the nasal cavities (Figure 14-6). Because the mucosa that lines the sinuses is continuous with the mucosa that lines the nose, sinus infections, called **sinusitis** (si-nu-SI-tis), often develop from colds in which the nasal mucosa is inflamed. The paranasal sinuses are lined with a mucous membrane that assists in the production of mucus for the respiratory tract. In addition, these hollow spaces help lighten the skull bones and serve as resonant chambers for the production of sound.

Note in Figure 14-7 that three shelflike structures called **conchae** (KONG-ke) protrude into the nasal cavity on each side. Think of the mucosa-covered conchae as partitions that greatly increase the surface over which air must flow as it passes through the nasal cavity. As air moves over the conchae and through the nasal cavities, it is warmed and humidified. This anatomical information helps explain why breathing

through the nose is more effective in humidifying inspired air than is breathing through the mouth. If an individual who is sick requires supplemental oxygen, it is first bubbled through water in order to reduce the amount of moisture that would otherwise have to be removed from the lining of the respiratory tree to humidify it. Administration of "dry" oxygen pulls water from the mucosa and results in respiratory discomfort and irritation.

PHARYNX

The **pharynx** is the structure that many of us call the throat. Although only about 12.5 centimeters (5 inches) long, it can be divided into three portions. The uppermost part of the tube just behind the nasal cavities is called the **nasopharynx** (na-zo-FAR-inks). The portion behind the mouth is called the **oropharynx** (o-ro-FAR-inks). The last or lowest segment is called the **laryngopharynx** (lah-ring-go-FAR-inks). The pharynx as a whole serves the same purpose for the respiratory and digestive tracts as a hallway serves for a house. Air and food pass through the pharynx on their way to the lungs and the stomach respectively. Air enters the pharynx from the two nasal cavities and leaves it by way of the larynx; food enters it from the mouth and leaves it by way of the esophagus. The right and left **eustachian** (u-STA-ke-an) or auditory **tubes** open into the nasopharynx; they connect each middle ear with the nasopharynx (Figure 14-7). The lining of the eustachian tubes is continuous with the lining of both the nasopharynx and middle ear. Therefore, just as sinus infections can develop from colds in which the nasal mucosa is inflamed, middle ear infections can develop as a result of inflammation of the nasopharynx.

Two small masses of lymphatic tissue called **tonsils** can be found embedded in the mucous membrane of the pharynx. The **pharyngeal** (fa-RIN-je-al) **tonsils** or **adenoids** (AD-e-noids) are found in the nasopharynx. The **palatine tonsils** are located in the oropharynx. Both tonsils are generally removed in a **tonsillectomy** (ton-si-LEK-to-me). Although this surgical procedure is

Figure 14-6 Projection of paranasal sinuses and oral and nasal cavities on the skull and face. Note the direct connection between the sinuses and the nasal cavity.

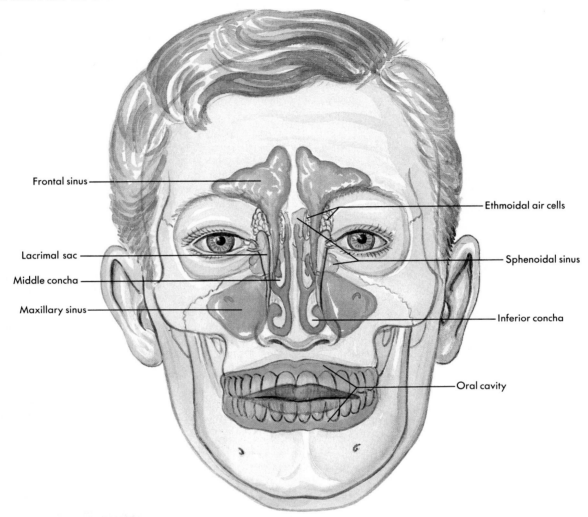

still rather common, the number reported each year continues to decrease as new and more effective antibiotics become available. Physicians now recognize the value of lymphatic tissue in the body defense mechanism and delay removal of the tonsils—even in cases of inflammation or **tonsillitis**—unless antibiotic treatment is ineffective. Swelling of the pharyngeal tonsils or adenoids caused by infection may make it difficult or impossible for air to travel from the nose into the throat. In these cases the individual is forced to breathe through the mouth.

LARYNX

The **larynx** or voice box, is located just below the pharynx. It is composed of several pieces of cartilage. You know the largest of these (the *thyroid cartilage*) as the "Adam's apple."

Two short fibrous bands called the **vocal**

Figure 14-7 Sagittal section through the face and neck. The nasal septum has been removed, exposing the lateral wall of the nasal cavity. Note the position of the conchae.

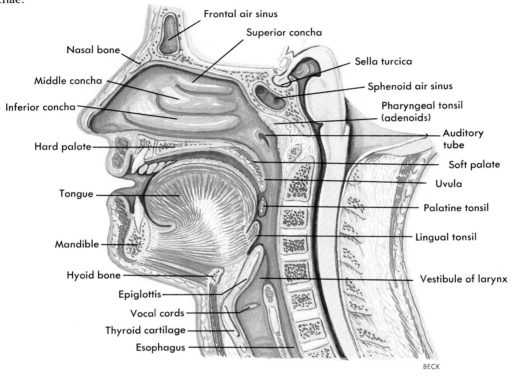

BECK

cords stretch across the interior of the larynx. Muscles that attach to the cartilages of the larynx can pull on these cords in such a way that they become either tense and short or relaxed and long. When they are tense and short, the voice sounds high pitched; when they are relaxed and long, it sounds low pitched.

To examine the larynx and vocal cords a physician will use a small round mirror that is attached at an angle to a long handle. It is called a **laryngeal mirror** (Figure 14-8). The mirror is inserted into the mouth and used to press the uvula and soft palate upward. In individuals who are very "gaggy," a deadening anesthetic liquid may be painted or sprayed on the throat tissues to eliminate the gag reflex. Figure 14-9 shows the larynx in sagittal section and as viewed from above with a laryngeal mirror. A

Figure 14-8 Position of the laryngeal mirror. The larynx and vocal cords as seen in the laryngeal mirror are illustrated in Fig. 14-9, *B*.

Figure 14-9 **A,** Sagittal section through the larynx. **B,** Larynx and vocal cords as viewed from above through a laryngeal mirror.

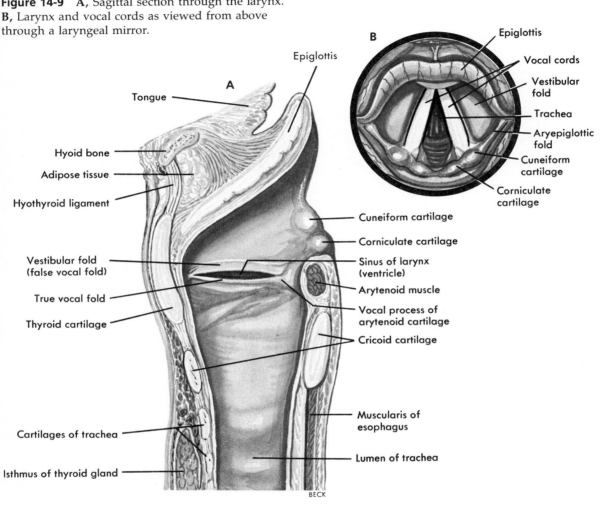

sagittal section was defined in Chapter 1 as a lengthwise cut or section that runs from front to back and divides an organ into right and left sections. In Figure 14-7 you are looking at the inside of the right half of the larynx. The left half has been removed. Compare Figures 14-7 and 14-9. The same view of the larynx is seen in both illustrations. However, in Figure 14-7 the larynx is surrounded by the other structures in the face and neck, and the details of its anatomy are less evident. Note in both illustrations that

another cartilage, the **epiglottis** (ep-i-GLOT-is) partially covers the opening into the larynx. It is the epiglottis that is sometimes referred to as the "trapdoor" cartilage. It closes off the larynx during swallowing and prevents food from entering the trachea.

TRACHEA

The **trachea** or windpipe is a tube about 11 centimeters (4½ inches) long that extends from

Figure 14-10 Cross section of the trachea. Inset shows the detail of the respiratory mucosa that lines the air passageway.

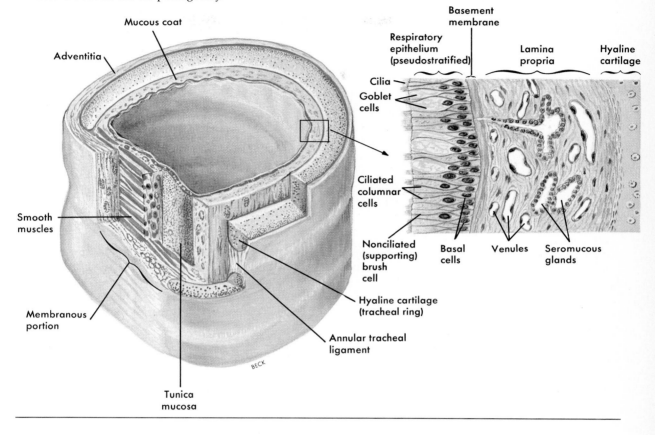

the larynx in the neck to the bronchi in the chest cavity (Figures 14-1 and 14-2). The trachea performs a simple but vital function—it furnishes part of the open passageway through which air can reach the lungs from the outside.

By pushing with your fingers against your throat about an inch above the sternum, you can feel the shape of the trachea or windpipe. Only if you use considerable force can you squeeze it closed. Nature has taken precautions to keep this lifeline open. Its framework is made of an almost noncollapsible material—15 or 20 C-shaped rings of cartilage placed one above the other with only a little soft tissue between them (Figure 14-10). Note in Figure 14-10 that the trachea is lined by respiratory mucosa. Specialized

glands below the ciliated epithelium help produce the blanket of mucus that is kept continually moving upward toward the pharynx.

Despite the structural safeguard of cartilage rings, closing of the trachea sometimes does occur. A tumor or an infection may enlarge the lymph nodes of the neck so much that they squeeze the trachea shut, or a person may aspirate (breathe in) a piece of food or something else that blocks the windpipe. Since air has no other way to get to the lungs, complete tracheal obstruction causes death in a matter of minutes. Choking on food and other substances caught in the trachea kills over 4,000 people each year and is the fifth major cause of accidental deaths in the United States. A lifesaving technique de-

Figure 14-11 **A,** Terminal bronchiole and air sacs (alveoli). **B,** Cross section of a normal bronchiole.

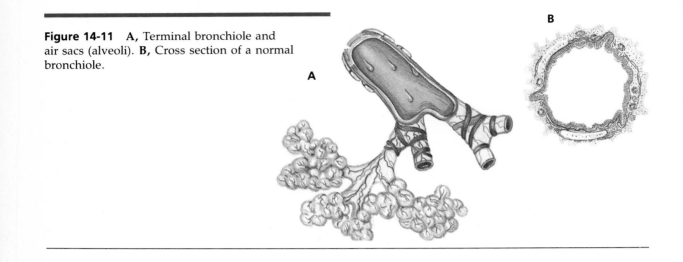

veloped by Dr. Henry Heimlich of Cincinnati, Ohio (see pp. 352 and 353), is now widely used to free the trachea of ingested food or other foreign objects that would otherwise block the airway and cause death in choking victims.

BRONCHI, BRONCHIOLES, AND ALVEOLI

Recall that one way to picture the thousands of air tubes that make up the lungs is to think of an upside-down tree. The trachea is the main trunk of this tree; the right bronchus (the tube leading into the right lung) and the left bronchus (the tube leading into the left lung) are the trachea's first branches or **primary bronchi.** In each lung they branch into smaller or **secondary bronchi,** which branch into **bronchioles.** The smallest bronchioles end in structures shaped like miniature bunches of grapes (Figures 14-1 and 14-11). The smallest bronchioles subdivide into microscopic-sized tubes called **alveolar ducts,** which resemble the main stem of a bunch of grapes. Each alveolar duct ends in several **alveolar sacs,** each of which resembles a cluster of grapes, and the wall of each alveolar sac is made up of numerous **alveoli,** each of which resembles a single grape. How well alveoli are suited to their function of exchanging gases be-

tween air and blood has been mentioned. They can perform this function effectively mainly because alveoli are extremely thin walled, each alveolus lies in contact with a blood capillary, and there are millions of alveoli in each lung.

Because the air tubes of the bronchial tree are embedded in connective tissue, you cannot see them by looking at the lungs from the outside. But if you cut into the lungs, the bronchi and bronchioles show up clearly.

LUNGS AND PLEURA

The **lungs** are fairly large organs. Figure 14-12 shows the relationship of the lungs to the rib cage at the end of a normal expiration. The two normal chest x-ray films in Figure 14-13 were taken at the end of a maximal inspiration. Note that the lungs in the chest x-ray films fill the entire chest cavity (all but the space in the center occupied mainly by the heart and large blood vessels). The narrow part of each lung, up under the collarbone, is its *apex;* the broad lower part, resting on the diaphragm, is its *base.* The *pleura* covers the outer surface of the lungs and lines the inner surface of the rib cage. The pleura resembles the peritoneum in structure and function but differs from it in location. Both are extensive, thin, moist, slippery membranes. Both line a large, closed cavity of the body and cover

Figure 14-12 Projection of the lungs and trachea in relation to rib cage and clavicles. Dotted line shows location of dome-shaped diaphragm at the end of expiration and before inspiration. Note that apex of each lung projects above the clavicle. Ribs 11 and 12 are not visible in this view.

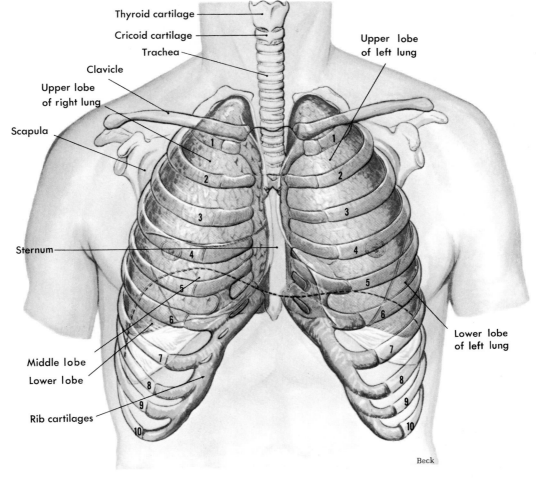

Beck

organs located in them. The *parietal pleura* lines the thoracic cavity, the *visceral pleura* covers the lungs, and the *pleural space* lies between the two pleural membranes.

Normally the pleural space contains just enough fluid to make both portions of the pleura moist and slippery and able to glide easily against each other as the lungs expand and deflate with each breath. However, the pleural space sometimes becomes distended with a large amount of fluid. The extra fluid presses on the lungs and makes it hard for the patient to breathe. In this case the physician may decide to remove the excess pleural fluid by means of a hollow, tubelike instrument that is pushed through the patient's chest wall into the pleural space. **Thoracentesis** (tho-rah-sen-TE-sis) is the medical term used to describe this procedure.

Pleurisy (PLOOR-i-se) and **pneumothorax** (nu-mo-THO-raks) are other terms having to do

Figure 14-13 Normal chest x-ray films. **A,** posterior view; **B,** anterior view. Both x-rays were taken at the end of a maximal inspiration.

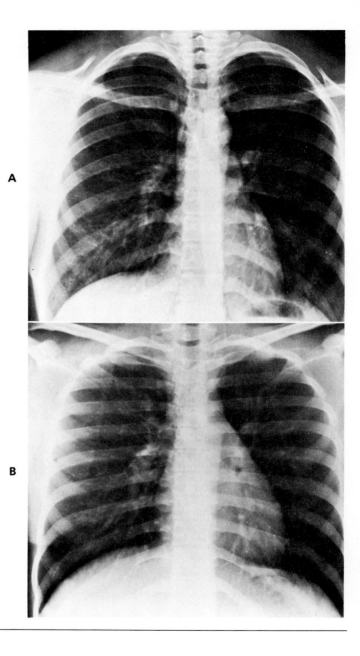

with pleura and pleural space. *Pleurisy* is an inflammation of the pleura. *Pneumothorax* is the presence of air in the pleural space on one side of the chest. The additional air increases the pressure on the lung on that side and causes it to collapse. While collapsed, the lung does not function in breathing.

RESPIRATION

Respiration means exchange of gases (oxygen and carbon dioxide) between a living organism and its environment. If the organism consists of only one cell, gases can move directly between it and the environment. If, however, the organism consists of billions of cells, as do our bodies then most of its cells are too far from the air for a direct exchange of gases. To overcome this difficulty, a pair of organs—the lungs—is provided where air and a circulating fluid (blood) can come close enough to each other for oxygen to move out of the air into blood while carbon dioxide moves out of the blood into air. Breathing, or **pulmonary ventilation,** is the process that moves air into and out of the lungs. It is the process that makes possible the exchange of gases between air in the lungs and in the blood. This exchange is often called **external respiration.** In addition, exchange of gases occurs between the blood and the cells of the body—a process called **internal respiration.**

MECHANICS OF BREATHING

Breathing involves not only the organs of the respiratory system but also the brain, spinal cord, nerves, certain skeletal muscles, and even some bones and joints. In breathing, nerve impulses stimulate the diaphragm to contract, and, as it contracts, its shape changes. The domelike shape of the diaphragm flattens out. Then, instead of protruding up into the chest cavity, it moves down toward the abdominal cavity. Thus the contraction or flattening of the diaphragm makes the chest cavity longer from top to bottom. Other muscle contractions raise the rib cage to make the chest cavity wider and greater in depth from front to back. As the chest cavity enlarges, the lungs expand along with it and air rushes into them and down into the alveoli. This part of respiration is called **inspiration.** For **expiration** to take place, the diaphragm and other respiratory muscles relax, making the chest cavity smaller and thereby squeezing air out of the lungs. Changes in the shape and size of the thoracic cavity result in changes in the air pressure within that cavity and in the lungs. It is the difference in air pressure that causes the actual movement of air into and out of the lungs.

EXCHANGE OF GASES IN LUNGS

Blood pumped from the right ventricle of the heart enters the pulmonary artery and eventually enters the lungs. It then flows through the thousands of tiny lung capillaries that are in close proximity to the air-filled alveoli (Figure 14-1). External respiration or the exchange of gases between the blood and alveolar air occurs by diffusion.

Diffusion is a passive process that results in movement *down a concentration gradient;* that is, substances move from an area of high concentration to an area of low concentration of the diffusing substance. Blood flowing through lung capillaries is low in oxygen. Oxygen is continually removed from the blood and used by the cells of the body. By the time it enters the lung capillaries it is low in oxygen content. Since alveolar air is rich in oxygen, diffusion will cause movement of oxygen from the area of high O_2 concentration (alveolar air) to the area of low O_2 concentration (capillary blood). Note in Figure 14-14 that most of the oxygen entering the blood combines with hemoglobin in the red blood cells to form **oxyhemoglobin** (ok-se-he-mo-GLO-bin) so that it can be carried to the tissues and used by the body cells.

Diffusion of carbon dioxide also occurs between blood in lung capillaries and alveolar air. Blood flowing through the lung capillaries is high in carbon dioxide. Most CO_2 is carried as bicarbonate ion in the blood. Some, as noted in Figure 14-14, combines with the hemoglobin in RBCs to form **carbaminohemoglobin** (kar-bam-i-no-he-mo-GLO-bin). As body cells remove oxygen from circulating blood, they add the waste product carbon dioxide to it. As a result, the blood in pulmonary capillaries eventually becomes low in oxygen and high in carbon dioxide. Diffusion of carbon dioxide results in its movement from an area of high concentration in the pulmonary capillaries to an area of low concentration in alveolar air. Then from the alveoli, carbon dioxide leaves the body in the expired air.

EXCHANGE OF GASES IN TISSUES

The exchange of gases that occurs between blood in tissue capillaries and the body cells is called *internal respiration.* As you would expect, the direction of movement of oxygen and carbon dioxide during internal respiration is just the opposite of that noted in the exchange that occurs during *external respiration* when gases are exchanged between the blood in the lung capillaries and the air in alveoli. As shown in Figure

Figure 14-14   Exchange of gases in lung capillaries. Drawing shows carbaminohemoglobin in a red blood cell dissociating to release CO_2, which diffuses out of blood into alveolar air. Simultaneously, O_2 is diffusing out of alveolar air into blood and associating with hemoglobin to form oxyhemoglobin.

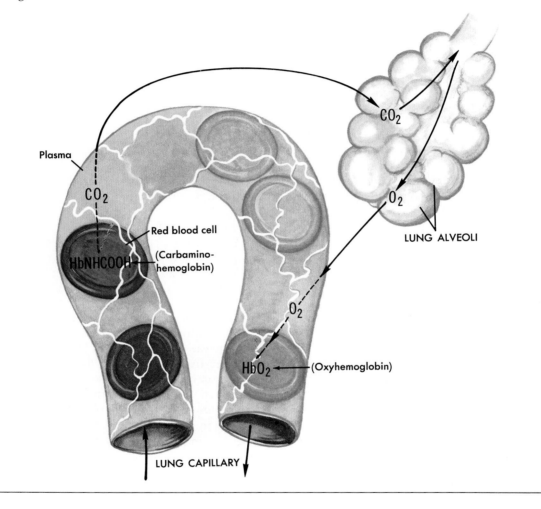

14-15, in the tissue capillaries oxyhemoglobin breaks down into oxygen and hemoglobin. Oxygen molecules move rapidly out of the blood through the tissue capillary membrane into the interstitial fluid and on into the cells that compose the tissues. The oxygen is used by the cells in their metabolic activities. Diffusion results in the movement of oxygen from an area of high concentration to an area of low concentration in the cells where it is needed. While this is happening, carbon dioxide molecules are leaving the cells, entering the tissue capillaries where bicarbonate (HCO_3^-) ions are formed and where hemoglobin molecules unite with the CO_2 to form carbaminohemoglobin. Once again, diffusion is responsible for the movement of carbon

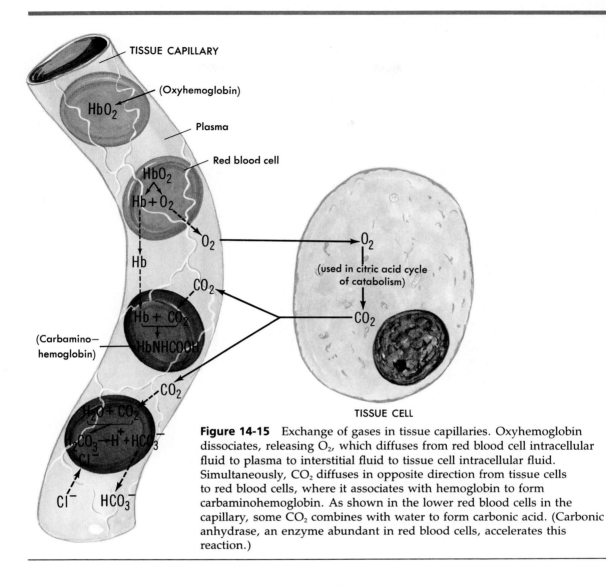

Figure 14-15 Exchange of gases in tissue capillaries. Oxyhemoglobin dissociates, releasing O_2, which diffuses from red blood cell intracellular fluid to plasma to interstitial fluid to tissue cell intracellular fluid. Simultaneously, CO_2 diffuses in opposite direction from tissue cells to red blood cells, where it associates with hemoglobin to form carbaminohemoglobin. As shown in the lower red blood cells in the capillary, some CO_2 combines with water to form carbonic acid. (Carbonic anhydrase, an enzyme abundant in red blood cells, accelerates this reaction.)

dioxide from an area of high concentration in the cells to an area of lower concentration in the capillary blood. In other words, oxygenated blood enters tissue capillaries and is changed into deoxygenated blood as it flows through them. In the process of losing oxygen, the waste product carbon dioxide is picked up and transported to the lungs for removal from the body.

VOLUMES OF AIR EXCHANGED IN PULMONARY VENTILATION

A special device called a **spirometer** is used to measure the amount of air exchanged in breathing. Figure 14-16 illustrates the various pulmonary volumes, which can be measured as a subject breathes into a spirometer. Ordinarily we take 500 milliliters (about a pint) of air into our lungs. Because this amount comes and goes

Figure 14-16   During normal, quiet respirations, the atmosphere and lungs exchange about 500 ml of air (tidal volume). With a forcible inspiration, about 3,300 ml more air can be inhaled (inspiratory reserve volume). After a normal inspiration and normal expiration, approximately 1,000 ml more air can be forcibly expired (expiratory reserve volume). Vital capacity is the largest amount of air that can enter and leave the lungs during respiration. Residual volume is the air that remains trapped in the alveoli.

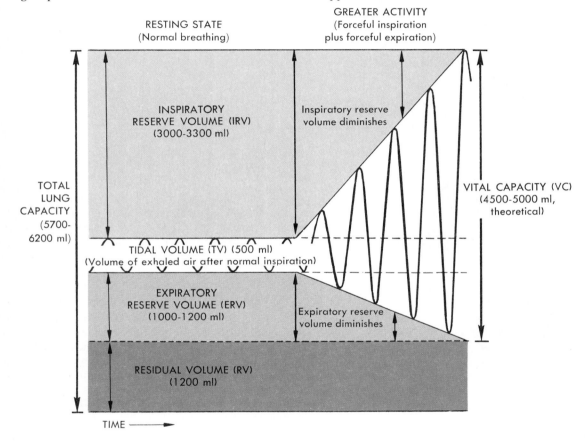

regularly like the tides of the sea, it is referred to as the **tidal volume (TV).** The largest amount of air that we can breathe in and out in one inspiration and expiration is known as the **vital capacity (VC).** In normal young men this is about 4,800 milliliters. Tidal volume and vital capacity are frequently measured in patients with lung or heart disease, conditions that often lead to abnormal volumes of air being moved in and out of the lungs.

Observe the area in Figure 14-16 that repre-

sents the **expiratory reserve volume (ERV).** This is the amount of air that can be forcibly exhaled after expiring the tidal volume. Compare this with the area in Figure 14-16 that represents the **inspiratory reserve volume (IRV).** The IRV is the amount of air that can be forcibly inspired over and above a normal inspiration. As the tidal volume increases, both the ERV and IRV will decrease. Note in Figure 14-16 that vital capacity is the total of tidal volume, inspiratory reserve volume, and expiratory reserve volume—or expressed in another way: VC = TV + IRV + ERV. **Residual volume (RV)** is simply the air that remains in the lungs after the most forceful expiration.

By now you know that the body uses oxygen to obtain energy for the work it has to do. Briefly, energy is stored in foods and the oxidation reactions of catabolism make this energy available for all kinds of cellular work (pp. 322 to 325). Therefore, the more work the body does, the more oxygen must be delivered to its millions of cells. One way this is accomplished is by increasing the rate and depth of respirations. Although we may take only sixteen breaths a minute when we are not moving about, when we are exercising we take considerably more than this. Not only do we take more breaths, we also breathe in more air with each breath. Instead of about a pint of air, we may breathe deeply enough to take in several pints—sometimes even up to the limit of our vital capacity.

The way respirations are made to increase during exercise is an interesting example of the body's automatic regulation of its vital functions. When we are exercising, cells carry on catabolism at a faster rate than usual. This increased metabolic rate means that more carbon dioxide is formed in the cells and that more carbon dioxide enters the blood in the tissue capillaries and increases venous blood's carbon dioxide concentration. This higher carbon dioxide concentration has a stimulating effect on the neurons of the respiratory center of the medulla. They send out more impulses to the respiratory muscles, and as a result respiration increases, becoming both faster and deeper. Therefore, as more air moves in and out of the lungs per minute, more carbon dioxide leaves the blood, more oxygen enters it, and more energy becomes available for doing the extra work of exercise.

To help supply cells with more oxygen when they are doing more work, automatic adjustments occur not only in respirations, but also in circulation. Most notably, the heart beats faster and harder and therefore pumps more blood through the body each minute. This means that the millions of red blood cells make more round trips between the lungs and tissues each minute and so deliver more oxygen per minute to tissue cells.

DISORDERS OF RESPIRATORY SYSTEM

Many different kinds of diseases and injuries may be responsible for respiratory disorders. For example, tuberculosis and lung cancer may destroy part of the lungs and pneumonia may plug up alveoli. A brain hemorrhage may depress the respiratory center, causing respirations to become slow and labored or even to stop completely. In recent years a disease rarely heard of in our grandfathers' generation—**emphysema** (em-fi-SEE-mah)—has become increasingly common. In this disease the walls of many of the alveoli become greatly overstretched. As the alveoli enlarge, their walls rupture and then fuse into large irregular spaces. Air becomes trapped in the lungs. As a result less air than normal is exhaled and inhaled, and less oxygen and carbon dioxide are exchanged between alveolar air and blood. Emphysema victims therefore develop **hypoxia** (hi-POK-se-ah). A synonym for hypoxia is oxygen deficiency. If a person is hypoxic, the cells receive less oxygen than what is needed for normal functioning. Lung disease is not the only abnormality that can produce hypoxia. Anemia, for example, can also cause hypoxia. In many types of anemia an individual's red blood cells contain less hemoglobin than normal and so they transport less oxygen than normal. Although in certain types of anemia the hemoglobin present in individual red blood cells may be normal, the decrease in total numbers of red cells will still result in generalized hemoglobin deficiency and hypoxia.

In addition to emphysema, **bronchial asthma** and **chronic bronchitis** are also common respiratory diseases that affect the terminal bronchioles and alveoli. In both chronic bronchitis and asthma the air tubes are narrowed and the accumulation of pus or mucus will slow the action of cilia in the respiratory mucosa. Bronchial asthma is characterized by difficult expiration

Heimlich maneuver

The Heimlich maneuver is an effective and often lifesaving technique that can be used to open a windpipe that is suddenly obstructed. The maneuver (see figures, opposite) uses air already present in the lungs to expel the object obstructing the trachea. Most accidental airway obstructions result from pieces of food aspirated during a meal; the condition is sometimes referred to as a "café coronary." Other objects such as chewing gum or balloons are frequently the cause of obstructions in children. Individuals trained in emergency procedures must be able to tell the difference between airway obstruction and other conditions such as heart attacks that produce similar symptoms. The key question they must ask the person who appears to be choking is, "Are you choking?" A person with an obstructed airway will not be able to speak, even while conscious. The Heimlich maneuver, if the victim is standing, consists of the rescuer's grasping the victim with both arms around the victim's waist just below the rib cage and above the navel. The rescuer makes a fist with one hand, grasps it with the other, and then delivers an upward thrust against the diaphragm just below the xiphoid process of the sternum. Air trapped in the lungs is compressed, forcing the object that is choking the victim out of the airway.

Technique if victim can be lifted (see A)

1 Rescuer stands behind the victim and wraps both arms around the victim's chest slightly below the rib cage and above the navel. Victim is allowed to fall forward with head, arms, and chest over the rescuer's arms.
2 Rescuer makes a fist with one hand and grasps it with the other hand, pressing thumb side of fist against victim's abdomen just below the end of the xiphoid process and above the navel.
3 The hands only are used to deliver the upward subdiaphragmatic thrusts. It is performed with sharp flexion of the elbows, in an upward rather than inward direction, and is usually repeated four times. It is very important *not* to compress the rib cage or actually press on the sternum during the Heimlich maneuver.

Technique if victim has collapsed or cannot be lifted (see B)

1 Rescuer places victim on floor face up.
2 Facing victim, rescuer straddles the victim's hips.
3 Rescuer places one hand on top of the other, with the bottom hand on the victim's abdomen slightly above the navel and below the rib cage.
4 Rescuer performs a forceful upward thrust with the heel of the bottom hand, repeating several times if necessary.

and labored breathing. In asthmatic patients muscles surrounding the bronchioles become spastic. As the muscles contract they narrow the air tubes, which often fill with thick mucus. As a result it is common to hear a whistling sound, or "wheeze," during an asthmatic attack. Compare the appearance of the normal bronchiole and terminal alveoli seen in Figure 14-11 with the appearance of these structures in emphysema, chronic bronchitis, and bronchial asthma (Figure 14-17).

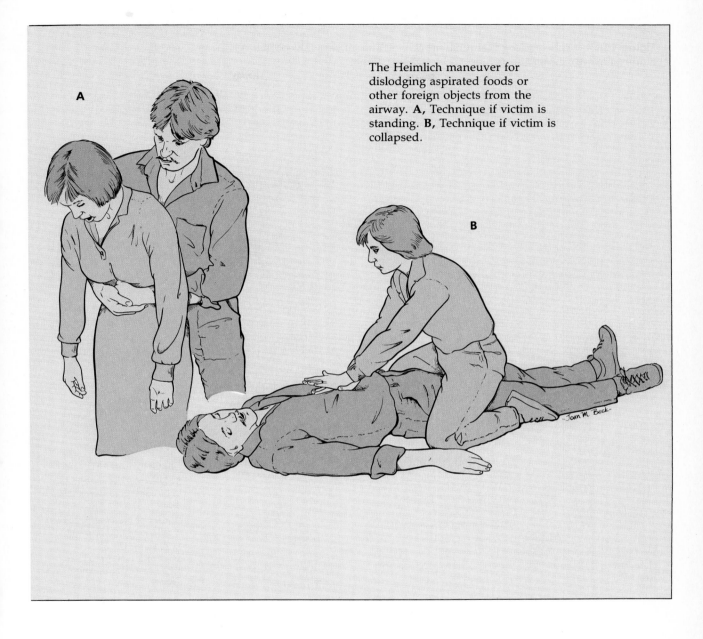

The Heimlich maneuver for dislodging aspirated foods or other foreign objects from the airway. **A,** Technique if victim is standing. **B,** Technique if victim is collapsed.

Figure 14-17   Appearance of terminal bronchiole and air sacs (alveoli) in: **A,** emphysema; **B,** chronic bronchitis; **C,** bronchial asthma.

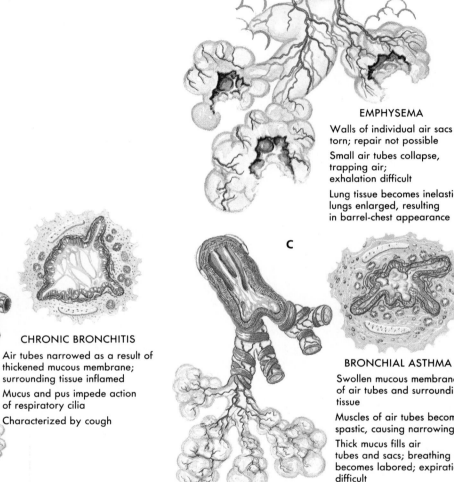

A

EMPHYSEMA

Walls of individual air sacs torn; repair not possible

Small air tubes collapse, trapping air; exhalation difficult

Lung tissue becomes inelastic; lungs enlarged, resulting in barrel-chest appearance

B

CHRONIC BRONCHITIS

Air tubes narrowed as a result of thickened mucous membrane; surrounding tissue inflamed

Mucus and pus impede action of respiratory cilia

Characterized by cough

C

BRONCHIAL ASTHMA

Swollen mucous membranes of air tubes and surrounding tissue

Muscles of air tubes become spastic, causing narrowing

Thick mucus fills air tubes and sacs; breathing becomes labored; expiration difficult

OUTLINE SUMMARY

Structural Plan

Basic plan of respiratory system similar to an inverted tree if it were hollow; leaves of tree would be comparable to alveoli, with the microscopic sacs enclosed by networks of capillaries

Respiratory Tracts
Respiratory Mucosa (Figure 14-4)

A Specialized membrane that lines the air distribution tubes in the respiratory tree

B Over 125 ml of mucus (sputum) produced each day forms a "mucus blanket" over much of the respiratory mucosa

C Mucus serves as an air purification mechanism by trapping inspired irritants such as dust and pollen

D Cilia covering mucosa beat in only one direction, moving mucus upward to pharynx for removal

Nose

A Structure
  1 Nasal septum separates interior of nose into two cavities
  2 Mucous membrane lines nose
  3 Frontal, maxillary, sphenoidal, and ethmoidal sinuses drain into nose

B Functions
  1 Warms and moistens air inhaled
  2 Contains sense organs of smell

Pharynx

A Structure
  1 Pharynx (throat) about 12.5 centimeters (5 inches) long
  2 Divided into nasopharynx, oropharynx, and laryngopharynx
  3 Two nasal cavities, mouth, esophagus, larynx, and eustachian tubes all have openings into pharynx
  4 Adenoids and openings of eustachian tubes open into nasopharynx; tonsils found in oropharynx
  5 Mucous membrane lines pharynx

B Functions
  1 Passageway for food and liquids
  2 Air distribution; passageway for air

Larynx (Voice Box)

A Structure
  1 Several pieces of cartilage form framework
    a Thyroid cartilage (Adam's apple) is largest
    b Epiglottis partially covers opening into larynx
  2 Mucous lining
  3 Vocal cords stretch across interior of larynx

B Functions
  1 Air distribution; passageway for air to move to and from lungs
  2 Voice production

Trachea (Windpipe)

A Structure
  1 Tube about 11 centimeters (4½ inches) long that extends from larynx into the thoracic cavity
  2 Mucous lining
  3 C-shaped rings of cartilage hold trachea open

B Function—passageway for air to move to and from lungs

C Obstruction
  1 Blockage of trachea occludes airway and if complete will cause death in minutes
  2 Tracheal obstruction causes over 4000 deaths annually in the U.S.
  3 Heimlich maneuver (pp. 352 and 353) is lifesaving technique used to free trachea of obstructions

Bronchi, Bronchioles, and Alveoli

A Structure
  1 Trachea branches into right and left bronchi.

2 Each bronchus branches into smaller and smaller tubes called bronchioles

3 Bronchioles end in clusters of microscopic alveolar sacs, the walls of which are made up of alveoli

B Function

1 Bronchi and bronchioles—air distribution; passageway for air to move to and from alveoli

2 Alveoli—exchange of gases between air and blood

Lungs and Pleura

A Structure

1 Size—large enough to fill chest cavity except for middle space occupied by heart and large blood vessels

2 Apex—narrow upper part of each lung, under collarbone

3 Base—broad lower part of each lung; rests on diaphragm

4 Pleura—moist, smooth, slippery membrane that lines chest cavity and covers outer surface of lungs; prevents friction between lungs and chest wall during breathing

5 Respiratory membrane—see Figure 14-3

B Function—breathing (pulmonary ventilation and pulmonary respiration)

Respiration

A Mechanics of breathing

1 Contraction of diaphragm and of chest-elevating muscles enlarges chest cavity, expands lungs, and causes air to move down into lungs

2 Relaxation of diaphragm and of chest elevators decreases size of chest cavity, deflates lungs, and causes air to move out of lungs

B Exchange of gases in lungs

1 Carbaminohemoglobin breaks down into carbon dioxide and hemoglobin

2 Carbon dioxide moves out of lung capillary blood into alveolar air and out of body in expired air

3 Oxygen moves from alveoli into lung capillaries

4 Hemoglobin combines with oxygen, producing oxyhemoglobin

C Exchange of gases in tissues

1 Oxyhemoglobin breaks down into oxygen and hemoglobin

2 Oxygen moves out of tissue capillary blood into tissue cells

3 Carbon dioxide moves from tissue cells into tissue capillary blood

4 Hemoglobin combines with carbon dioxide, forming carbaminohemoglobin

D Volumes of air exchanged in pulmonary ventilation—see Figure 14-16

1 Volumes of air exchanged in breathing can be measured with a spirometer

2 Tidal volume (TV) amount normally breathed in and out with each breath

3 Vital capacity (VC)—largest amount of air that one can breathe in and out in one inspiration and expiration

4 Expiratory reserve volume (ERV)—amount of air that can be forcibly inhaled after expiring the tidal volume

5 Inspiratory reserve volume (IRV)—amount of air that can be forcibly inhaled after a normal inspiration

6 Residual volume (RV)—air that remains in the lungs after the most forceful expiration

7 Rate—usually about 16 to 20 breaths a minute; must faster during exercise

Disorders of Respiratory System

A Tuberculosis

B Lung cancer

C Pneumonia

D Brain hemorrhage

E Emphysema

F Chronic bronchitis

G Bronchial asthma

NEW WORDS

adenoids	expiratory reserve	pharynx	surfactant
bronchi	volume (ERV)	pleurisy	tidal volume
bronchial asthma	Heimlich maneuver	pneumothorax	(TV)
carbaminohemoglobin	hyperventilation	residual volume	thoracentesis
conchae	hypoxia	(RV)	tonsillectomy
chronic bronchitis	inspiratory reserve	respiration	tonsillitis
emphysema	volume (IRV)	respiratory	trachea
epistaxis	larynx	membrane	vital capacity
	oxyhemoglobin	spirometer	(VC)

CHAPTER TEST

1. Organs of the respiratory system serve to function as both an air _____ and as a gas _____ .

2. The respiratory system effectively _____ , _____ , and _____ the air we breathe.

3. The exchange of gases in the respiratory system is dependent on the passive transport process of _____ .

4. Surface tension in the alveoli is reduced by a substance called _____ .

5. The trachea is located in the _____ respiratory tract.

6. Nosebleeds are described by the medical term _____ .

7. The frontal, maxillary, sphenoidal, and ethmoidal sinuses are classified as _____ sinuses.

8. That portion of the pharynx located behind the mouth is called the _____ .

9. The adenoids or pharyngeal tonsils are located in the _____ .

10. The voice box is also known as the _____ .

11. During swallowing, food is prevented from entering the trachea by the _____ .

12. The presence of air in the pleural space is called _____ .

13. The exchange of gases between air in the lungs and the blood is called _____ respiration.

14. Most oxygen entering the blood combines with hemoglobin in the red blood cells to form _____ .

15. The largest amount of air we can breathe in and out in one inspiration and expiration is known as the _____ .

Select the most correct answer from Column B for each statement in Column A. (Only one answer is correct.)

Column A	Column B
16. _____ lack of surfactant	a. Diffusion
17. _____ thin walled sacs	b. Thyroid cartilage
18. _____ passive transport process	c. Adenoids
19. _____ nosebleeds	d. Breathing
20. _____ pharyngeal tonsils	e. Hypoxia
21. _____ "Adam's apple"	f. IRDS
22. _____ lines thoracic cavity	g. Parietal pleura
23. _____ pulmonary ventilation	h. Epistaxis
24. _____ oxygen deficiency	i. Emphysema
25. _____ alveoli torn	j. Alveoli

REVIEW QUESTIONS

1 Discuss the location, microscopic structure, and functions of the respiratory mucosa.
2 Identify (list) the paranasal air sinuses.
3 Discuss the functions of the nose in respiration.
4 What and where are the pharynx and larynx?
5 What structures open into the pharynx?
6 What are the anatomical subdivisions of the pharynx?
7 Where are the tonsils and adenoids located?
8 Why does sinusitis or middle ear infection occur so frequently after a common cold?
9 What function do the C-shaped rings of cartilage serve in the trachea?
10 What and where is the "Adam's apple"?
11 Define the following terms:

    parietal pleura   pneumothorax
    pleural space   visceral pleura
    pleurisy

12 Do breathing and respiration mean the same thing? Define each term.
13 Briefly explain how O_2 and CO_2 can move between alveolar air, blood, and tissue cells.
14 Explain the following equation:

$$VC = TV + IRV + ERV$$

15 What does the term "residual volume" mean?
16 How can a brain hemorrhage affect the respiratory system?
17 Discuss the anatomy of the "respiratory membrane."
18 Compare the structure, location, and functions of the respiratory membrane and the respiratory mucosa.
19 Discuss the Heimlich maneuver.

CHAPTER
15 The Urinary System

OBJECTIVES

After you have completed this chapter, you should be able to:

1 Identify the major organs of the urinary system and give the generalized function of each.

2 Name the parts of a nephron and describe the role each component plays in the formation of urine.

3 Explain the importance of filtration, tubular resorption, and tubular secretion in urine formation.

4 Discuss the mechanisms that control urine volume.

5 Define the following terms: *uremia, micturition, urinary retention, urinary suppression, incontinence, anuria, renal calculi.*

As you might guess from its name, the urinary system performs the functions of producing urine and eliminating it from the body. What you might not guess so easily is how essential these functions are for the maintenance of homeostasis and healthy survival. The constancy of body fluid volumes and the levels of many important chemicals are dependent on normal urinary system function. Unless the urinary system operates normally, the normal composition of blood cannot long be maintained, and serious consequences soon follow. The kidneys function to "clear" or clean the blood of the many waste products that are continually produced as a result of metabolism of foodstuffs in the body cells. As nutrients are burned for energy, the waste products that are produced must be removed from the blood or they quickly accumulate to toxic levels—a condition called **uremia** (u-RE-me-ah) or **uremic poisoning.** The kidneys also play an important role in maintaining the electrolyte, water, and acid-base balance in the body. In this chapter we shall discuss the structure and function of each of the urinary's system's organs. There are two kidneys, two ureters, one bladder, and one urethra (Figure 15-1). We shall also mention briefly some disease conditions produced by abnormal functioning of the urinary system.

KIDNEYS

LOCATION

To locate the kidneys on your own body, stand erect and put your hands on your hips with your thumbs meeting over your backbone. In this position your kidneys lie just above your thumb—in short, the kidneys are located just above the waistline. Usually the right kidney is a little lower than the left. They are located under the muscles of the back and behind the parietal peritoneum (the membrane that lines the abdominal cavity). This is a convenient anatomical fact. Because of it a surgeon can operate on a kidney without cutting through the peritoneum. A heavy cushion of fat normally encases each kidney and helps hold it in place. In an extremely thin person who lacks this cushion,

one or both kidneys may drop down, a condition called **renal ptosis** (TO-sis), enough to put a kink in the tubes that drain urine out of the kidney. This, of course, obstructs urine flow.

Note the relatively large size of the renal arteries in Figure 15-1. Normally a little over 10% of the total blood pumped by the heart each minute will enter the kidneys. The rate of blood flow through this organ is among the highest in the body. This is understandable, since one of the main functions of the kidney is to remove waste products from the blood. Maintaining a high rate of blood flow and normal blood pressure in the kidney is essential for the formation of urine.

INTERNAL STRUCTURE

If you were to slice through a kidney from side to side and open it like the pages of a book, you would see the structures shown in Figure 15-2. Identify each of the following:

1 **Cortex** (KOR-teks)—the outer part of the kidney; (the word *cortex* comes from the Latin word for bark or rind, so the cortex of an organ is its outer layer; kidneys, brain, and adrenal glands all have a cortex); only the cortex of the kidney contains the renal corpuscles of the nephron units (see microscopic structure, (Figure 15-3)

2 **Medulla** (me-DUL-ah)—the inner portion of the kidney

3 **Pyramids** (PIR-ah-mids)—the triangular-shaped divisions of the medulla of the kidney

4 **Papilla** (pah-PIL-ah) (pl. *papillae*)—narrow, innermost end of a pyramid

5 **Pelvis**—the kidney or renal pelvis is an expansion of the upper end of a ureter (the tube that drains urine into the bladder)

6 **Calyx** (KA-liks) (pl. *calices*)—each calyx is a division of the renal pelvis; opening into each calyx is the papilla of a pyramid

MICROSCOPIC STRUCTURE

More than a million microscopic-sized units named **nephrons** (NEF-rons) make up each kidney's interior. The shape of a nephron is unique, unmistakable, and admirably suited to its func-

Figure 15-1 Locations of urinary system organs.

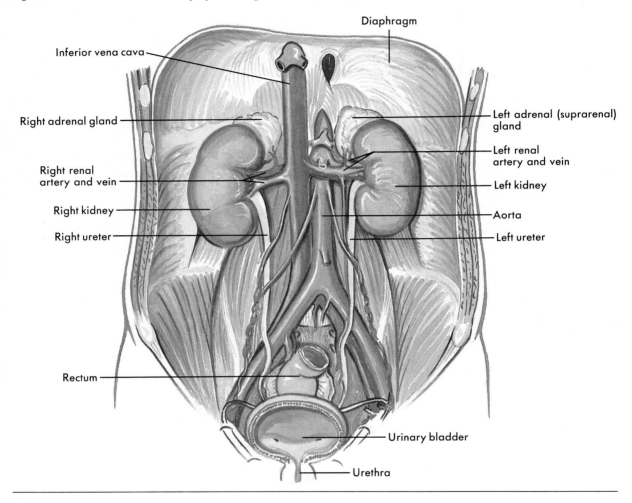

tion of producing urine. It looks a little like a tiny funnel with a very long stem, but an unusual stem in that it is highly convoluted, that is, has many bends in it. Locate each of the following parts of a nephron in Figures 15-3 to 15-5:

1 RENAL CORPUSCLE
  a Bowman's capsule—the cup-shaped top of a nephron. It is the sack-like Bowman's capsule that surrounds the glomerulus
  b Glomerulus (glo-MER-u-lus)—a network of blood capillaries tucked into

Bowman's capsule. Note in Figures 15-4 and 15-5 that the small artery (arteriole) that delivers blood to the glomerulus, **(afferent arteriole)** is larger in size than the blood vessel that drains blood from it **(efferent arteriole).** This explains the high blood pressure that exists in the glomerular capillaries. This high pressure is required to filter wastes from the blood.

2 RENAL TUBULE
  a Proximal convoluted tubule—the first segment of a renal tubule, called proximal because it lies nearest the tubule's

Figure 15-2   Coronal section through right kidney.

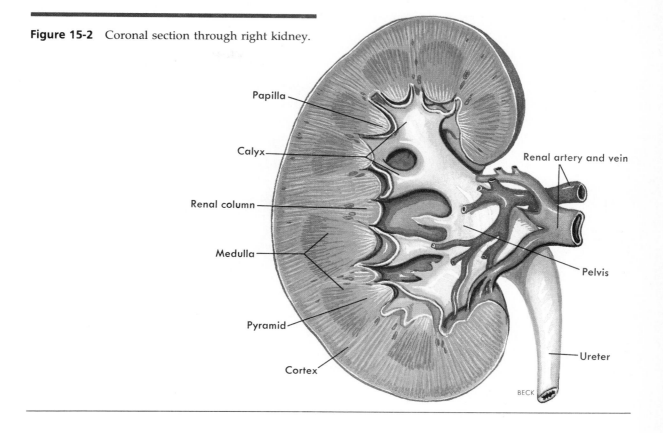

Papilla

Calyx

Renal column

Medulla

Pyramid

Cortex

Renal artery and vein

Pelvis

Ureter

BECK

origin from Bowman's capsule and convoluted because it has several bends in it

b Loop of Henle (HEN-le)—the extension of the proximal tubule; observe that the loop of Henle consists of a straight descending limb, a loop, and a straight ascending limb

c Distal convoluted tubule—the part of the tubule distal to the ascending limb of Henle, the extension of the ascending limb

d Collecting tubule—a straight, that is, not convoluted, part of a renal tubule; distal tubules of several nephrons join to form a single collecting tubule

In summary, nephrons, the microscopic units of a kidney, have two main parts, a *renal corpuscle* (Bowman's capsule with glomerulus) and a *renal tubule*.

FUNCTION

The kidneys are vital organs. The function they perform, that of forming urine, is essential for maintaining life. Early in the proces of urine formation, fluid, elctrolytes, and wastes from metabolism are *filtered* from the blood and enter the nephron unit. Additional wastes may be actively *secreted* into the tubules of the nephron while substances that are useful to the body are *resorbed* into the blood. Normally the kidneys vary the amount of many substances leaving the blood so that they equal the amounts of them entering the blood from various sources. In short, they adjust their output to equal their intake. By so doing, the kidneys play an essential part in maintaining homeostasis. Homeostasis cannot be maintained—nor can life itself—if the kidneys fail and the condition is not soon corrected. If kidney function ceases because of injury or disease, life can be maintained by using

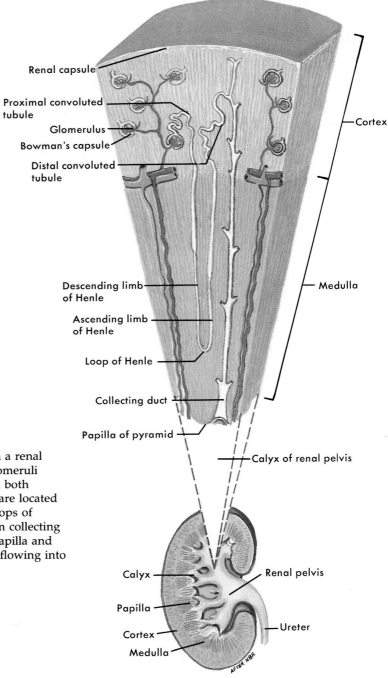

Renal capsule

Proximal convoluted tubule

Glomerulus

Bowman's capsule

Distal convoluted tubule

Descending limb of Henle

Ascending limb of Henle

Loop of Henle

Collecting duct

Papilla of pyramid

Cortex

Medulla

Calyx of renal pelvis

Calyx

Papilla

Cortex

Medulla

Renal pelvis

Ureter

Figure 15-3 Magnified wedge cut from a renal pyramid. Note that renal corpuscles (glomeruli surrounded by Bowman's capsules) and both proximal and distal convoluted tubules are located in cortex of kidney. Medulla contains loops of Henle and collecting tubules. Urine from collecting ducts exits from pyramid through the papilla and enters the calyx and renal pelvis before flowing into the ureter.

Figure 15-4 Scanning electron photomicrograph of a glomerulus. Note that the blood vessel that brings blood to the glomerulus (afferent arteriole) has a larger diameter than the blood vessel that drains blood from it (efferent arteriole).

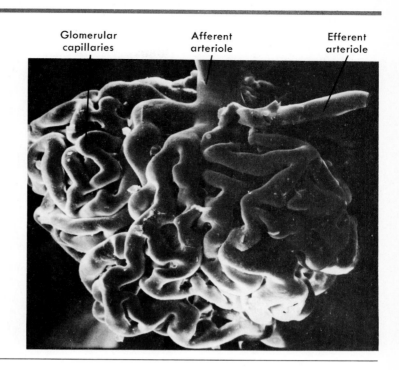

Glomerular capillaries   Afferent arteriole   Efferent arteriole

an **artificial kidney** to cleanse the blood of wastes (see below).

HOW KIDNEYS FORM URINE

The kidneys' 2 million or more nephrons form urine by three processes: filtration, resorption, and secretion. Urine formation begins with the process of **filtration,** which goes on continually in the renal corpuscles (glomeruli plus Bowman's capsules encasing them). Blood flowing through the glomeruli exerts pressure, and this glomerular blood pressure is sufficiently high to push water and dissolved substances out of the glomeruli into the Bowman's capsules. Briefly, glomerular blood pressure causes filtration through the glomerular-capsular membrane. If glomerular blood pressure drops below a certain level, filtration and urine formation cease. Hemorrhage, for example, may cause a precipitous drop in blood pressure followed by kidney failure.

Glomerular filtration normally occurs at the rate of 125 milliliters (ml) per minute. The following simple calculations may help you visualize how enormous this volume is:

$$125 \times 60 = 7{,}500 \text{ ml} = \text{Glomerular filtration rate per hour}$$

$$7{,}500 \times 24 = 180{,}000 \text{ ml or } 180 \text{ liters}$$
(about 190 quarts) = Volume of fluid filtered out of glomerular blood per day

Obviously no one ever excretes anywhere near 180 liters of urine per day. Why? Because most of the fluid that leaves the blood by glomerular filtration, the first process in urine formation, returns to the blood by the second process—resorption.

Resorption, by definition, is the movement of substances out of the renal tubules into the blood capillaries located around the tubules (peritubular capillaries). Substances resorbed are water, glucose and other nutrients, and sodium and other ions. Resorption begins in the proximal convoluted tubules and continues in

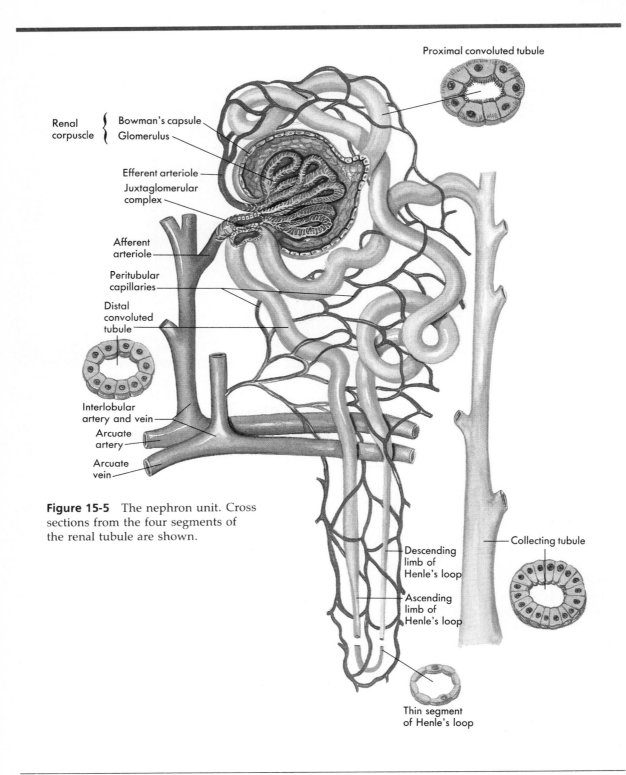

Proximal convoluted tubule

Renal corpuscle {
Bowman's capsule
Glomerulus

Efferent arteriole
Juxtaglomerular complex

Afferent arteriole

Peritubular capillaries

Distal convoluted tubule

Interlobular artery and vein

Arcuate artery

Arcuate vein

Figure 15-5 The nephron unit. Cross sections from the four segments of the renal tubule are shown.

Descending limb of Henle's loop

Ascending limb of Henle's loop

Collecting tubule

Thin segment of Henle's loop

The artificial kidney

The artificial kidney is a mechanical device that uses the principle of dialysis to remove or separate waste products from the blood. In the event of kidney failure, the process, appropriately called **hemodialysis** (Greek *haima*—"blood"; *alysis*—"separate") is literally a reprieve from death for the patient. During a hemodialysis treatment, a semipermeable membrane is used to separate large (nondiffusible) particles such as blood cells from small (diffusible) ones such as urea and other wastes. The figure shows blood from the radial artery passing through a porous (semipermeable) cellophane tube that is housed in a tanklike container. The tube is surrounded by a bath or dialyzing solution containing varying concentrations of electrolytes and other chemicals. The pores in the membrane are small and allow only very small molecules, such as urea, to escape into the surrounding fluid. Larger molecules and blood cells cannot escape and are returned through the tube to reenter the patient via a wrist vein. By constantly replacing the bath solution in the dialysis tank with freshly mixed solution, levels of waste materials can be kept at low levels. As a result, wastes such as urea in the blood will rapidly pass into the surrounding wash solution. For a patient with complete kidney failure, two or three hemodialysis treatments a week are required. New dialysis methods are now being developed, and dramatic advances in treatment are expected in the next few years.

One relatively new technique used in the treatment of renal failure is called **continuous ambulatory peritoneal dialysis** *(CAPD)*. In this procedure, 1 to 3 L of sterile dialysis fluid is introduced directly into the peritoneal cavity through an opening in the abdominal wall. Peritoneal membranes in the abdominal cavity serve to transfer waste products from the blood into the dialysis fluid, which is then drained back into a plastic container after about 2 hours. This technique is less expensive than hemodialysis and does not require the use of complex equipment.

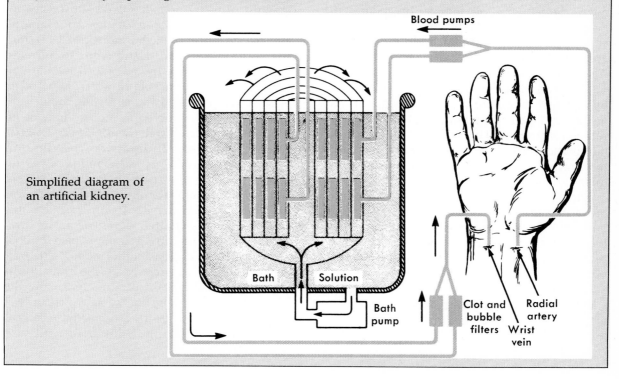

Simplified diagram of an artificial kidney.

the loop of Henle, distal convoluted tubules, and collecting tubules.

Large amounts of water—approximately 178 liters per day—are resorbed by osmosis from the proximal tubules. In other words, nearly 99% of the 180 liters of water that leave the blood each day by glomerular filtration returns to the blood by proximal tubule resorption.

The nutrient glucose is entirely resorbed from the proximal tubules. It is actively transported out of them into peritubular capillary blood. None of this valuable nutrient is wasted by being lost in the urine. However, exceptions to this normal rule do occur. For example, in *diabetes mellitus*, if blood glucose concentration increases above a certain level, the tubular filtrate then contains more glucose than kidney tubule cells can resorb. Some of the glucose therefore remains behind in the urine. Glucose in the urine **(glycosuria)** (gli-ko-SU-re-ah), is a well-known sign of diabetes mellitus.

Sodium ions and other ions are only partially resorbed from renal tubules. For the most part sodium ions are actively transported back into blood from the tubular urine. The amount of sodium resorbed varies from time to time; it depends largely on salt intake. In general the greater the amount of salt intake, the less the amount of salt resorption and therefore the greater the amount of salt excreted in the urine. Also, the less the salt intake, the greater the salt resorption and the less salt excreted in the urine. By varying the amount of salt resorbed, the body usually can maintain homeostasis of the blood's salt concentration. This is an extremely important matter because cells are damaged by either too much or too little salt in the fluid around them.

Secretion is the process by which substances move into urine in the distal and collecting tubules from blood in the capillaries around these tubules. In this respect secretion is resorption in reverse. Whereas resorption moves substances out of the urine into the blood, secretion moves substances out of the blood into the urine. Substances secreted are hydrogen ions, potassium ions, ammonia, and certain drugs. Hydrogen ions, potassium ions, and drugs are secreted by

being actively transported out of the blood into tubular urine. Ammonia is secreted by diffusion. Kidney tubule secretion plays a crucial role in maintaining the body's acid-base balance (see Chapter 20).

In summary, three processes occurring in successive portions of the nephron accomplish the function of urine formation (Figure 15-6 and Table 15-1):

1 **Filtration**—of water and dissolved substances out of the blood in the glomeruli into Bowman's capsules
2 **Resorption**—of water and dissolved substances out of the kidney tubules back into the blood (Note that this process prevents substances needed by the body from being lost in the urine. Usually 97% to 99% of the water filtered out of the glomerular blood is retrieved from the tubules.)
3 **Secretion**—of hydrogen ions (H^+), potassium ions (K^+), and certain drugs

CONTROL OF URINE VOLUME

The body has ways to control both the amount and the composition of the urine that it secretes. It does this mainly by controlling the amount of water and dissolved substances resorbed by the convoluted tubules. For example, a hormone (antidiuretic hormone or ADH) from the posterior pituitary gland tends to decrease the amount of urine by making distal and collecting tubules permeable to water. If no ADH is present, both distal and collecting tubules are practically impermeable to water, so little or no water is resorbed from them. When ADH is present in the blood, distal and collecting tubules are permeable to water and water is resorbed from them. As a result, water is lost from the body or more water is retained—whichever way you wish to say it. At any rate, for this reason ADH is accurately described as the "water-retaining hormone." You might also think of it as the "urine-decreasing hormone." The hormone aldosterone, secreted by the adrenal cortex, plays an important part in controlling the kidney tubules' resorption of salt. Primarily it stimulates the tubules to resorb sodium salts at a faster rate. Secondarily, aldosterone tends also

Figure 15-6   Diagram showing the steps in urine formation in successive parts of a nephron: filtration, resorption, and secretion.

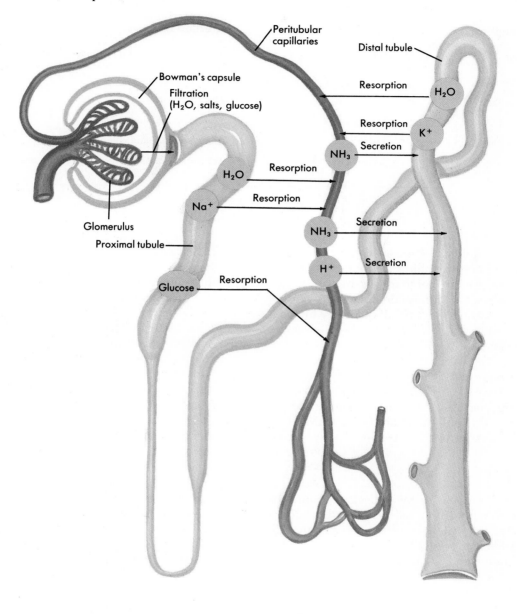

Table 15-1 Functions of parts of nephron in urine formation

Part of nephron	Process in urine formation	Substances moved and direction of movement
Glomerulus	Filtration	Water and solutes (for example, sodium and other ions, glucose and other nutrients) filter out of glomeruli into Bowman's capsules
Proximal tubule	Resorption	Water and solutes
Loop of Henle	Resorption	Sodium and chloride ions
Distal and collecting tubules	Resorption	Water, sodium, and chloride ions
	Secretion	Ammonia, potassium ions, hydrogen ions, and some drugs

to increase tubular water resorption. The term "salt- and water-retaining hormone" therefore is a descriptive nickname for aldosterone.

Sometimes the kidneys do not excrete normal amounts of urine—as a result of kidney disease, cardiovascular disease, or stress. Here are some terms associated with abnormal amounts of urine:

1 **Anuria** (ah-NU-re-ah)—literally, absence of urine
2 **Oliguria** (ol-i-GU-re-ah)—scanty urine
3 **Polyuria** (pol-e-U-re-ah)—unusually large amounts of urine

URETERS

Urine drains out of the collecting tubules of each kidney into the renal pelvis and on down the ureter into the urinary bladder (Figure 15-1). The **renal pelvis** is the basinlike upper end of the ureter and is located inside the kidney. Ureters are narrow tubes less than ¼ inch wide but 10 to 12 inches long. Mucous membrane lines both ureters and renal pelves. Note in Figure 15-7 that the ureter has a thick muscular wall. Contraction of the muscular coat produces peristaltic-type movements that assist in moving urine down the ureters into the bladder. The lining membrane of the ureters is richly supplied with sensory nerve endings.

URINARY BLADDER

The empty urinary bladder lies in the pelvis just behind the pubic symphysis. When full of urine it projects upward into the lower portion of the abdominal cavity.

Elastic fibers and involuntary muscle fibers in the wall of the urinary bladder make it well suited for its functions of expanding to hold variable amounts of urine and then contracting to empty itself. Mucous membrane lines the urinary bladder. The lining is loosely attached to the deeper muscular layer so that when the bladder is empty it is very wrinkled and lies in folds called *rugae*. When the bladder is filled the inner surface is smooth. Note in Figure 15-8 that one triangular shaped area on the back or posterior surface of the bladder is free of rugae. This area, called the *trigone*, is always smooth. Here the lining membrane is tightly fixed to the deeper muscle coat. The trigone extends between the openings of the two ureters above and the point of exit of the urethra below.

Attacks of *renal colic*—pain caused by the passage of a kidney stone—have been described in medical writings since antiquity. Kidney stones will cause intense pain if they have sharp edges or are large enough to distend the walls or cut the lining of the ureters or urethra as they pass from the kidneys to the exterior.

Figure 15-7 Cross section of ureter. Note the thick layer of muscle surrounding the tube. (×160.)

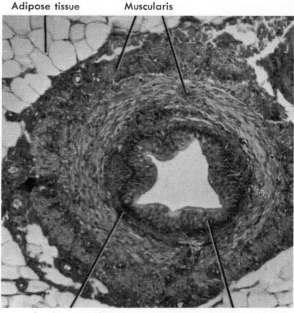

Adipose tissue   Muscularis

Lamina propria   Transitional epithelium

Removal of kidney stones using ultrasound

Current statistics suggest that approximately 1 in every 1000 adults in the United States will suffer from kidney stones or **renal calculi** (KAL-ku-li) at some point in their life.

Although symptoms of excruciating pain are common, many kidney stones are small enough to pass out of the urinary system spontaneously. If this is possible, no specific therapy is required other than treatment for pain and antibiotics if the calculi are associated with infection. Larger stones, however, may obstruct the flow of urine and are much more serious and difficult to treat.

Until recently only traditional surgical procedures were effective in removing relatively large stones that formed in the calyces and renal pelvis of the kidney. In addition to the risks that always accompany major medical procedures, surgical removal of stones from the kidneys frequently requires rather extensive hospital and home recovery periods, lasting 6 weeks or more. A new technique that uses ultrasound to pulverize the stones so that they can be flushed out of the urinary tract without surgery is currently in limited use in several hospitals across the United States. The large and highly specialized ultrasound generator required for the procedure is called a **lithotriptor** (li-tho-TRIP-ter). Using a lithotripter, doctors can break up the stones with ultrasound waves without making an incision. Recovery time is minimal and patient costs are reduced.

Figure 15-8 The male urinary bladder, cut to show the interior. Note how the prostate gland surrounds the urethra as it exits from the bladder.

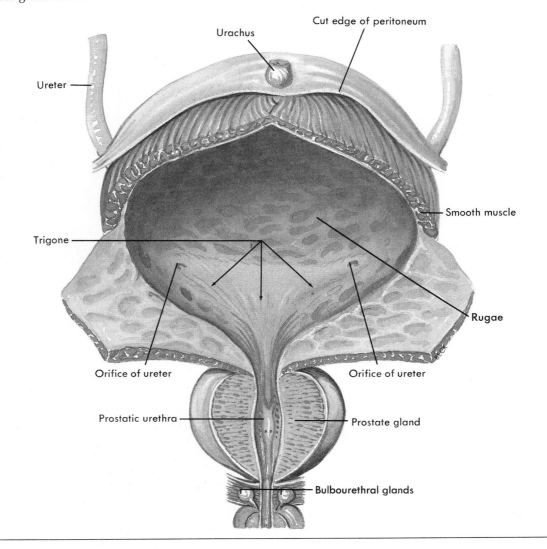

URETHRA

To leave the body, urine passes from the bladder, down the urethra, and out of its external opening, the **urinary meatus.** In other words, the urethra is the lowermost part of the urinary tract. The same sheet of mucous membrane that lines the renal pelves, ureters, and bladder ex-

tends down into the urethra too—a structural feature worth noting because it accounts for the fact that an infection of the urethra may spread upward throughout the urinary tract. The urethra is a narrow tube; it is only about 1½ inches long in a woman but about 8 inches long in a man. In the male the urethra serves a dual func-

Urinary catheterization

Urinary catheterization is the passage or insertion of a hollow tube or catheter through the urethra into the bladder for the withdrawal of urine. It is a medical procedure commonly performed on many patients who experience problems with urinary retention. Correct catheterization procedures require aseptic techniques to prevent the introduction of infectious bacteria into the urinary system. Recent clinical studies have now proved what many health professionals have suspected for many years; improper catheterization techniques cause bladder infections (cystitis) in large numbers of hospitalized patients. One landmark study confirmed the high percentage of catheterized patients who develop cystitis (almost 8%) and found that of those who developed such infections, a significant number died. The results of the study are indeed sobering and point out the need for extensive training of health professionals in this area.

tion: (1) as the terminal portion of the urinary tract and (2) as the passageway for movement of the reproductive fluid (semen) from the body. The female urethra serves only the urinary tract.

MICTURITION

The terms **micturition** (mik-tu-RISH-un), **urination** (u-ri-NA-shun), and **voiding** all refer to the passage of urine from the body or the emptying of the bladder. This is a completely reflex action in infants or very young children. Although there is considerable variation between individuals, most children between 2 and 3 years of age learn to urinate voluntarily and also to inhibit voiding if the urge comes at an inconvenient time.

Two **sphincters** (SFINGK-ters) or rings of muscle tissue guard the pathway leading from the bladder. The **internal urethral sphincter** is located at the bladder exit, and the **external urethral sphincter,** which is also called the **compressor urethrae,** circles the urethra just below the neck of the bladder. When contracted, both sphincters seal off the bladder and allow urine to accumulate without leakage to the exterior. The internal urethral sphincter is involuntary, and the external urethral sphincter is composed of striated muscle and is under voluntary control.

The muscular wall of the bladder permits this organ to accommodate a considerable volume of urine with very little increase in pressure until a volume of 300 to 400 ml is reached. As the volume of urine increases, the need to void may be noticed at volumes of 150 ml, but micturition in adults does not normally occur much below volumes of 350 ml. As the bladder wall stretches, nervous impulses are transmitted to the second, third, and fourth sacral segments of the spinal cord and an **emptying reflex** is initiated. The reflex causes contraction of the muscle of the bladder wall and relaxation of the internal sphincter. Urine then enters the urethra. If the external sphincter, which is under voluntary control, is relaxed, micturition will occur. Voluntary contraction of the external sphincter can suppress the emptying reflex until the bladder is filled to capacity with urine and loss of control occurs. Contraction of this powerful sphincter is also capable of voluntary abrupt termination of urination.

Higher centers in the brain also function in micturition by integrating bladder contraction and internal and external sphincter relaxation with the important cooperative contraction of pelvic and abdominal muscles. Urinary **retention** is a condition in which no urine is voided. The kidneys produce urine, but the bladder for one reason or another cannot empty itself. In urinary **suppression** the opposite is true. The kidneys do not produce any urine, but the bladder retains the ability to empty itself.

Incontinence (in-KON-ti-nens) is a condition in which the patient voids urine involuntarily. It frequently occurs in patients who have suffered a stroke or spinal cord injury. If the sacral segments of the spinal cord are injured; some loss of bladder function will always occur. Although the voiding reflex may be reestablished to some degree, the bladder does not empty completely. In these individuals the residual urine is often the cause of repeated bladder infections or **cystitis** (sis-TI-tis). Complete destruction or transection of the sacral cord produces what is called an "automatic bladder." Totally cut off from spinal innervation, the bladder musculature acquires some automatic action and periodic but unpredictable voiding occurs.

OUTLINE SUMMARY

Kidneys

A Location—under back muscles, behind parietal peritoneum, just above waistline; right kidney usually a little lower than left

B Internal structure
  1 Cortex—outer layer of kidney substance
  2 Medulla—inner portion of kidney
  3 Pyramids—triangular-shaped divisions of medulla
  4 Papillae—narrow, innermost ends of pyramids
  5 Pelvis—expansion of upper end of ureter
  6 Calyces—divisions of renal pelvis

C Microscopic structure—nephrons are microscopic units of kidneys; consist of following parts
  1 Renal corpuscle—Bowman's capsule with its glomerulus
    a Bowman's capsule—cup-shaped top of nephron
    b Glomerulus—network of blood capillaries tucked into the Bowman's capsule
  2 Renal tubule
    a Proximal convoluted tubule—first segment of a renal tubule
    b Loop of Henle—extension of proximal tubule; consists of descending limb, loop, and ascending limb
    c Distal convoluted tubule—extension of ascending limb of Henle
    d Collecting tubule—straight extension of distal tubule

D Function—urine formation

E How kidneys form urine—by three processes that take place in successive parts of nephron
  1 Filtration—goes on continually in renal corpuscles; glomerular blood pressure causes water and dissolved substances to filter out of glomeruli into Bowman's capsule; normal glomerular filtration rate 125 milliliters per minute
  2 Resorption—movement of substances out of renal tubules into blood in peritubular capillaries; substances resorbed are water, nutrients, and various ions; water resorbed by osmosis from proximal tubules
  3 Secretion—movement of substances into urine in the distal and collecting tubules from blood in peritubular capillaries; hydrogen ions, potassium ions, and certain drugs are secreted by active transport; ammonia is secreted by diffusion

F Control of urine volume—mainly by posterior pituitary hormone ADH, which acts to decrease urine volume

Ureters

A Structure—narrow long tubes with expanded upper end (renal pelvis) located inside kidney and lined with mucous membrane

B Function—drain urine from renal pelvis to urinary bladder

Urinary bladder

A Structure
  1 Elastic muscular organ, capable of great expansion

2 Lined with mucous membrane arranged in rugae, like stomach mucosa

B Functions
    1 Stores urine before voiding
    2 Voiding

Urethra

A Structure
    1 Narrow short tube from urinary bladder to exterior
    2 Lined with mucous membrane
    3 Opening of urethra to exterior called urinary meatus

B Functions
    1 Serves as passageway by which urine leaves bladder exterior
    2 Passageway by which male reproductive fluid leaves body

Micturition

A Refers to passage of urine from body (urination/voiding)

B Regulatory sphincters
    1 Internal urethral sphincter (involuntary)
    2 External urethral sphincter or compressor urethrae (voluntary)

C Bladder wall permits storage of urine with little increase in pressure

D Emptying reflex
    1 Initiated by stretch reflex in bladder wall
    2 Bladder wall contracts
    3 Internal sphincter relaxes
    4 External sphincter relaxes and urination occurs

E Urinary retention—urine produced but not voided

F Urinary suppression—no urine produced but bladder is normal

G Incontinence—urine is voided involuntarily
    1 May be caused by spinal injury or stroke
        a Retention of urine may cause cystitis

NEW WORDS

anuria	glycosuria	polyuria	urination
Bowman's capsule	incontinence	pyramid	urinary retention
calyx	lithotriptor	renal colic	urinary suppression
catheterization	micturition	renal ptosis	void
cystitis	oliguria	trigone	
glomerulus	papilla	uremia	

CHAPTER TEST

1. Failure of the urinary system results in rapid accumulation of toxic wastes—a condition called _____ .
2. Normally a little over _____ of the total blood pumped by the heart each minute will enter the kidneys.
3. The outer portion of the kidney is called the _____ and the inner portion is called the _____ .

4. The functional unit of the kidney is called the _____ .
5. The renal corpuscle is composed of the cup-shaped _____ and a network of blood capillaries called the _____ .
6. The first segment of the renal tubule is called the _____ convoluted tubule.
7. Urine formation begins with the process of _____ of wastes from the blood.
8. Water is resorbed from the proximal convoluted tubules by _____ .
9. Glucose is entirely resorbed from the _____ tubules.
10. Glucose in the urine is called _____ .
11. Production of unusually large amounts of urine is called _____ .
12. The basinlike upper end of the ureter located inside the kidney is called the renal _____ .
13. Attacks of pain caused by the passage of a kidney stone are called _____ _____ .
14. Urine passes from the bladder down the _____ and out its external opening called the urinary _____ .
15. Kidney stones are referred to as renal _____ .
16. Passage of a tube into the bladder for withdrawal of urine is called urinary _____ .
17. The condition in which no urine is voided is called urinary _____ .
18. Voiding urine involuntarily is called _____ .
19. The medical term for bladder infection is _____ .
20. Glomerular filtration normally occurs at the rate of about _____ dl per minute.

Select the most correct answer from Column B for each statement in Column A. (Only one answer is correct.)

	Column A		**Column B**
21.	____ high waste levels	a.	Renal ptosis
22.	____ abnormally low kidney	b.	Glycosuria
23.	____ contain renal corpuscles	c.	Glomerulus
24.	____ functional unit of kidney	d.	ADH
25.	____ tuft of capillaries	e.	Nephron
26.	____ glucose in urine	f.	Uremia
27.	____ absence of urine	g.	Renal calculi
28.	____ water-retaining hormone	h.	Anuria
29.	____ kidney stones	i.	Micturition
30.	____ urination	j.	Cortex

REVIEW QUESTIONS

1 What organs form the urinary system?

2 To operate on a kidney, does a surgeon have to cut through the peritoneum? Explain your answer.

3 Which kidney usually lies a little lower than the other?

4 Name the parts of a nephron.

5 What and where are the glomeruli and Bowman's capsules?

6 Explain briefly the functions of the glomeruli and Bowman's capsules.

7 Explain briefly the function of the renal tubules.

8 What kind of membrane lines the urinary tract?

9 Explain briefly the function of ADH. What is the full name of this hormone? What gland secretes it?

10 Suppose that ADH secretion increases noticeably. Would this increase or decrease urine volume? Why?

11 What hormone might appropriately be nicknamed the "water-retaining hormone"?

12 What hormone might appropriately be nicknamed the "salt- and water-retaining hormone"?

13 What hormone might be called the urine-decreasing hormone?

14 What is the urinary meatus?

15 What and where are the ureters and the urethra?

16 Explain the process of micturition.

17 Define: urinary retention; urinary suppression; incontinence.

UNIT SIX

The Cycle of Life

CHAPTER

16 The Male Reproductive System

CHAPTER OUTLINE

Structural plan

External genitals

Testes
Structure
Functions
Testosterone
Spermatozoa

Ducts
Epididymis, vas deferens, ejaculatory duct, and urethra

Accessory male reproductive glands

BOXED ESSAY

Cryptorchidism

OBJECTIVES

After you have completed this chapter, you should be able to:

1 List the essential and accessory organs of the male reproductive system and give the generalized function of each.

2 Describe the gross and microscopic anatomy of the testis.

3 Discuss the primary functions of testosterone and identify the cell type responsible for its secretion.

4 Trace in sequence the passage of an individual sperm cell from its point of formation through the genital ducts to the exterior of the body.

5 Compare the structure, location, and function of the accessory reproductive glands in the male and identify the components of the external genitals.

We truly are "fearfully and wonderfully made." Almost any one of the body's structures or functions might have inspired this statement, but of them all, perhaps the reproductive systems best deserve such praise. Their achievement is the miracle of reproducing the human body; their goal is the survival of the human species.

While the other body systems are concerned primarily with survival of the individual, the reproductive systems are important not only to the individual but to the entire human race as well.

As you probably noticed, the plural "reproductive systems" was used in the preceding paragraph. The male reproductive system consists of one group of organs and the female reproductive system consists of another group.

These two systems differ in structure. However, they do share a common function—that of reproducing the human body. This chapter discusses the male reproductive system, and Chapter 17 discusses the female reproductive system.

STRUCTURAL PLAN

So many organs make up the male reproductive system that we need to look first at the structural plan of the system as a whole. Reproductive organs can be classified as either **essential** or **accessory.**

The essential organs of reproduction in both males and females are called the **gonads.** The word "gonad" comes from the Greek word *gonos*, meaning "seed." The gonads of the male consist of a pair of main sex glands called the **testes** (TES-tez). The seeds produced are the male sex cells or **spermatozoa.**

The accessory organs of reproduction in the male consist of a series of ducts that carry the male sex cells from the testes to the exterior, additional sex glands, and the external reproductive organs.

Table 16-1 lists the names of all of these structures, and Figure 16-1 shows the location of most of them.

Table 16-1 Male reproductive organs

Essential organs of reproduction or main sex glands (gonads)	Accessory organs of reproduction
Testes (right testis and left testis)	Ducts: epididymis, vas deferens, ejaculatory duct (two each of the preceding), and urethra
	Supportive sex glands: seminal vesicle, bulbourethral or Cowper's gland (two each of the preceding), and prostate gland
	External genitals: scrotum and penis

EXTERNAL GENITALS

The male external reproductive organs or **genitals** consist of two organs: the **scrotum** and the **penis.** The scrotum is a skin-covered pouch suspended from the groin region. Internally, it is divided into two sacs by a septum; each sac contains a testis, epididymis, the lower part of the vas deferens, and the beginning of the spermatic cords. The **spermatic cords** are cylindrical tubes that cover the vas deferens, numerous blood vessels, nerves, and lymphatics as they pass from the scrotum into the abdominal cavity (Figures 16-1 and 16-2). The penis (Figure 16-3) is made up of three separate structures: one corpus cavernosum urethrae and two corpora cavernosa penis. Many anatomic and medical terms are not too difficult to remember if you know what they mean. *Corpos* means "body"; *cavernosum* means "full of small cavities"—a good description of erectile tissue.

Skin covers both the scrotum and the penis. At the distal end of the penis is the **glans,** over which the skin is folded doubly to form a loose-fitting retractable casing called the foreskin or prepuce. If the foreskin fits too tightly about the glans, a **circumcision** or surgical removal of the

Figure 16-1 Longitudinal section of the male pelvis showing the location of the male reproductive organs.

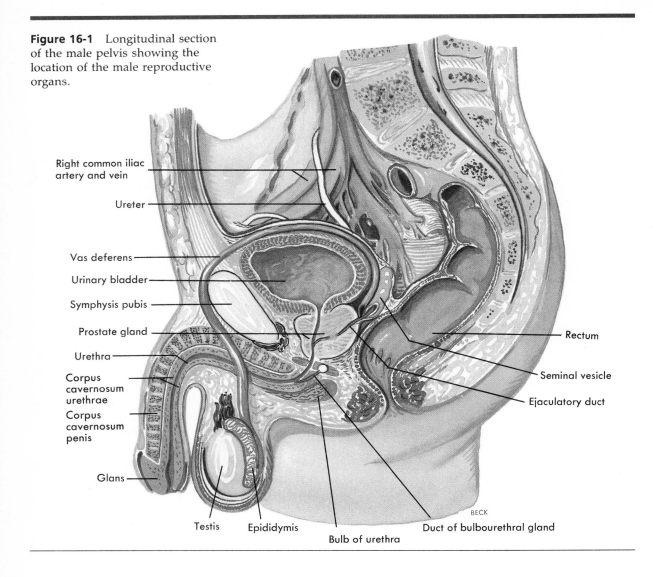

Right common iliac artery and vein

Ureter

Vas deferens

Urinary bladder

Symphysis pubis

Prostate gland

Urethra

Corpus cavernosum urethrae

Corpus cavernosum penis

Glans

Testis   Epididymis

Bulb of urethra

Rectum

Seminal vesicle

Ejaculatory duct

Duct of bulbourethral gland

BECK

foreskin is usually performed to prevent irritation. The external urethral orifice is the opening of the urethra at the tip of the glans.

The penis is the male organ of *coitus* or sexual intercourse. Under the stimulus of sexual emotion, blood floods the spaces in the erectile tissue of the three corpora, distending them enough to produce erection of the penis. The climax of coitus is known as **orgasm.** It is characterized by the ejaculation (expulsion) of semen from the penis and by various other responses—notably

a feeling of intense sexual excitement, accelerated heart and respiratory rates, and a moderate increase in blood pressure. Orgasm and the ejaculation of semen from the penis is the result of the same stimuli that initiate erection.

TESTES
STRUCTURE

The **testes** are the gonads of the male. They are small oval-shaped glands about 1½ inches

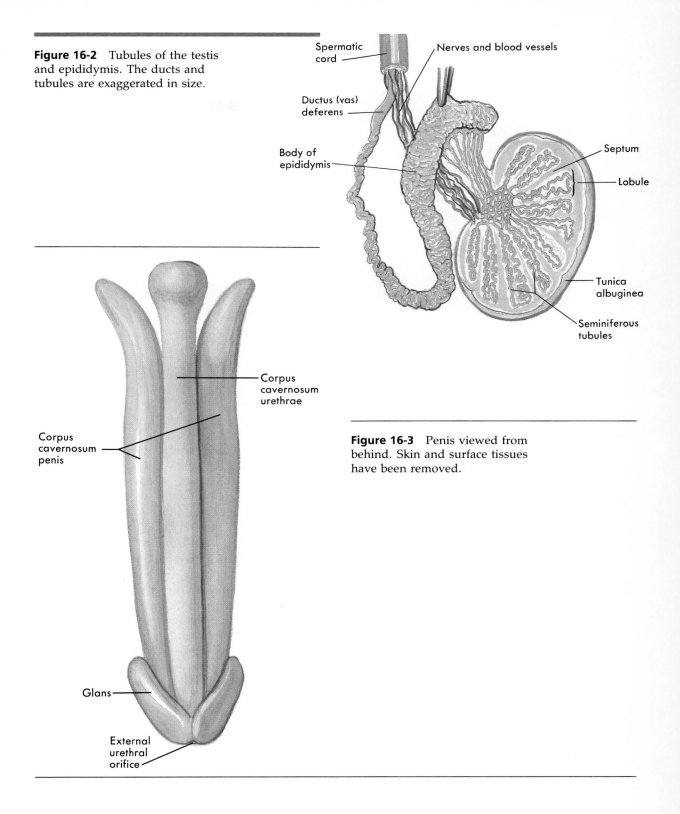

Figure 16-2 Tubules of the testis and epididymis. The ducts and tubules are exaggerated in size.

Spermatic cord

Nerves and blood vessels

Ductus (vas) deferens

Body of epididymis

Septum

Lobule

Tunica albuginea

Seminiferous tubules

Corpus cavernosum urethrae

Corpus cavernosum penis

Glans

External urethral orifice

Figure 16-3 Penis viewed from behind. Skin and surface tissues have been removed.

Figure 16-4 Testis. Low-power view showing several seminiferous tubules surrounded by septa containing interstitial (Leydig) cells. (×70.)

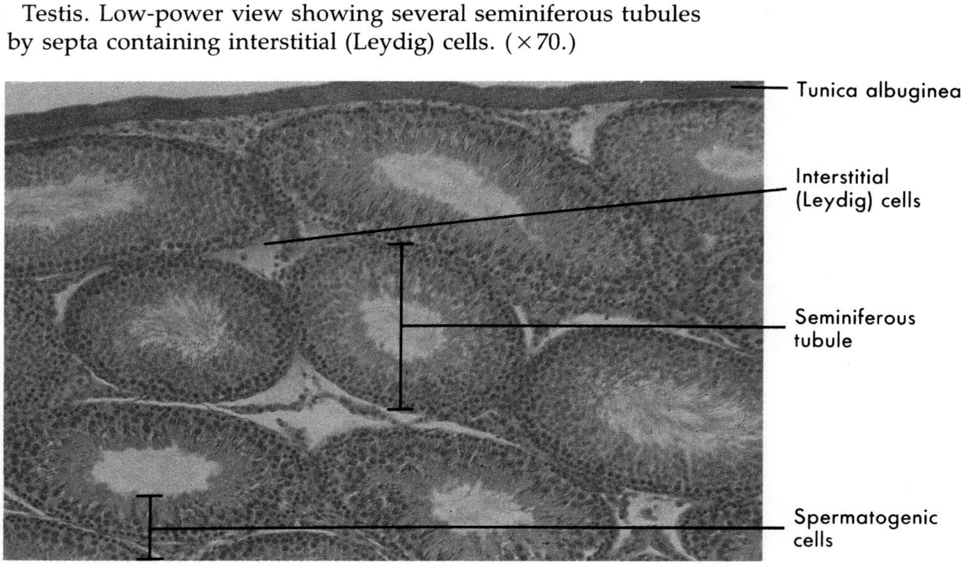

Tunica albuginea

Interstitial (Leydig) cells

Seminiferous tubule

Spermatogenic cells

(3.8 centimeters) long and 1 inch (2.5 centimeters) wide. They are shaped somewhat like an egg that has been flattened slightly from side to side. Note in Figure 16-2 that each testis is surrounded by a tough whitish membrane called the **tunica albuginea** (TU-ni-kah al-bu-JIN-e-ah). This membrane covers the testicle and then enters the gland to form the many septa that divide it into sections or *lobules*. As you can see in Figure 16-2, each lobule consists of a narrow but long and coiled **seminiferous** (se-mi-NIF-er-us) **tubule.** Small clusters of specialized cells lie near the septa that separate the lobules. These are the *interstitial cells* of the testes.

Histological Structure

The microscopic anatomy of testicular tissue is shown in Figures 16-4 and 16-5. If a thin section of tissue is cut from the testicle and viewed under a microscope, individual seminiferous tubules and clusters of interstitial cells can be studied.

In Figure 16-4 the tough saclike covering of the testicle (tunica albuginea) can be seen at the top of the picture. Immediately below the covering membrane are a number of cuts or sections through seminiferous tubules. Connective tissue septa and interstitial cells can be seen between the tubules. Each tubule is a ductlike structure with a central lumen or passageway. Sperm cells develop in the walls of the tubule and are then released into the lumen and begin their journey to the exterior of the body. Figure 16-5 is an enlarged view of a portion of one seminiferous tubule outlined in Figure 16-4. Can you identify the developing sperm that are attached to the inner surface of the seminiferous tubule in Figure 16-5? Note that the sperm tails extend into the lumen of the tube. At puberty, when sexual maturity begins, sperm cells in various stages of development can be identified in the walls of the seminiferous tubules, and the interstitial cells become larger and more noticeable in the surrounding septa.

FUNCTIONS

To judge the testes by their size would be to underestimate their importance. Each testis

Figure 16-5 Testis. High-power view showing portions of three seminiferous tubules with interstitial (Leydig) cells and capillaries between the tubules (×350.)

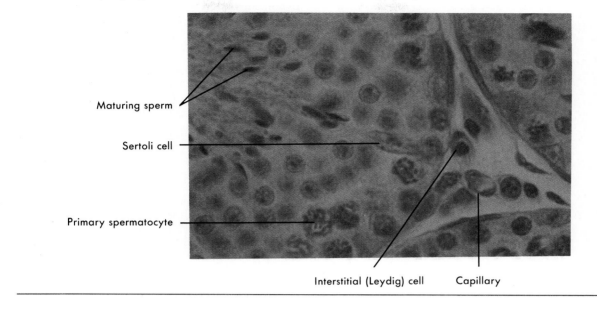

Maturing sperm

Sertoli cell

Primary spermatocyte

Interstitial (Leydig) cell   Capillary

forms many millions of sex cells (*spermatozoa* or *sperm*) every month after puberty. Any one of these tiny cells may join with a female sex cell (*ovum*) to become a new human being (Figure 16-6). Each testis also secretes the male sex hormone testosterone, the hormone that in a few short months transforms a boy into a man. Testosterone lowers the pitch of his voice, makes his muscles grow large and strong, and even changes the size and shape of his bones.

From the age of puberty on, the seminiferous tubules are almost continuously forming spermatozoa or sperm. **Spermatogenesis** (sper-mah-to-JEN-e-sis) is the name of this process. The other function of the testes is to secrete the male hormone testosterone. This function, however, is carried on by the **interstitial cells** of the testes, not by its seminiferous tubules.

The testis, then, is a beautiful example of the familiar principle that "structure determines function" (specifically, the production of sperm

by the seminiferous tubules and secretion of testosterone by the interstitial cells).

TESTOSTERONE

Testosterone serves the following general functions:

1 It masculinizes. The various characteristics that we think of as "male" develop because of testosterone's influence. For instance, when a young boy's voice changes, it is testosterone that brings this about.

2 It promotes and maintains the development of the male accessory organs (prostate gland, seminal vesicles, and so on).

3 It has a stimulating effect on protein anabolism. Testosterone thus is responsible for the greater muscular development and strength of the male.

A good way to remember testosterone's functions is to think of it as "the masculinizing hormone" and the "anabolic hormone."

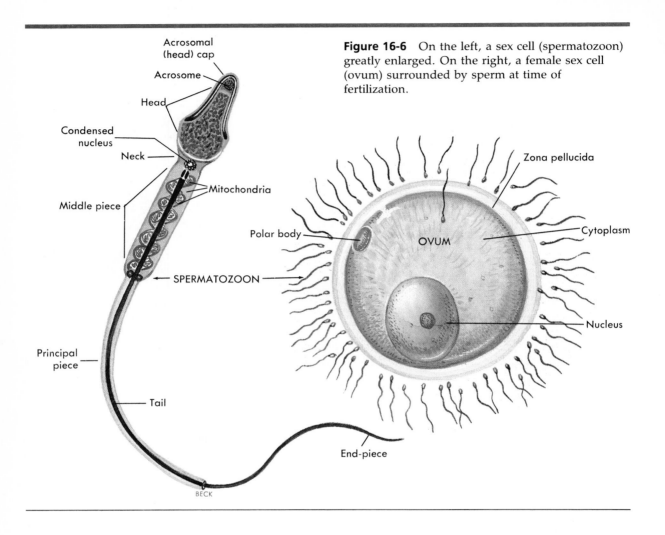

Figure 16-6 On the left, a sex cell (spermatozoon) greatly enlarged. On the right, a female sex cell (ovum) surrounded by sperm at time of fertilization.

How much testosterone the interstitial cells of the testes secrete depends on how much of another hormone—the interstitial cell stimulating hormone (ICSH)—the anterior pituitary gland secretes. The more ICSH secreted, the more it stimulates the interstitial cells to secrete more testosterone. In short, a high blood concentration of ICSH stimulates testosterone secretion. The resulting high blood concentration of testosterone then feeds back via the circulating blood to influence ICSH secretion by the anterior pituitary gland, but here the effect is quite different. Testosterone exerts a negative

effect on ICSH secretion. A high blood concentration of testosterone inhibits ICSH secretion instead of stimulating it. This is an example of a "negative feedback control mechanism"—a term used frequently in our computer-conscious world.

SPERMATOZOA

Spermatozoa are among the smallest and most highly specialized cells in the body. Ejaculation of sperm into the female vagina during sexual intercourse is only one step in the long journey that these sex cells must make before

Cryptorchidism

Early in fetal life the testes are located in the abdominal cavity but normally descend into the scrotum about 2 months before birth. Occasionally a baby is born with undescended testes, a condition called **cryptorchidism** (krip-TOR-ki-dizm), which is readily observed by palpation of the scrotum at the time of delivery. The word "cryptorchidism" is from the Greek words *kryptikos* (hidden) and *orchis* (testis). Failure of the testes to descend may be caused by hormonal imbalances in the developing fetus or by a physical deficiency or obstruction. Regardless of cause, in the cryptorchid infant the testes remain "hidden" in the abdominal cavity. Because the higher temperature inside the body cavity inhibits spermatogenesis, measures must be taken to bring the testes down into the scrotum to prevent permanent sterility. Early treatment of this condition by surgery or by injection of testosterone, which stimulates the testes to descend, will result in normal testicular and sexual development.

they can meet and fertilize an ovum. Sperm cells are, in effect, specialized packages of genetic information. All of the characteristics that a baby will inherit from its father at fertilization are contained in the condensed nuclear (genetic) material found in each sperm head. All other parts of the sperm cell are designed to provide motility. Fertilization normally occurs in the fallopian tubes or oviducts of the female. Therefore, in order for fertilization to occur, sperm must "swim" for relatively long distances through the female reproductive ducts. Figure 16-6 shows the characteristic parts of a single spermatozoon: head, neck, middle piece, and elongated, lash-like tail.

DUCTS

EPIDIDYMIS, VAS DEFERENS, EJACULATORY DUCT, AND URETHRA

The role of the vas deferens in birth control has recently become very important. To find out why, we need first to trace the route by which sperm leave the male reproductive tract in order to enter the female tract. As you read the description of this route in the next sentences, follow it in Figures 16-1 and 16-2. Sperm are formed in the testes by the seminiferous tubules. From the seminiferous tubules, sperm by the millions stream into a narrow but long and tightly coiled duct, the **epididymis** (ep-i-DID-i-mis). From the duct of the epididymis sperm continue on their way through one of the vasa deferentia into an ejaculatory duct from which they move down the urethra and out of the body. Note in Figure 16-1 the location of the vas deferens in the scrotum. There are a pair of these small tubes and here, in the scrotum, they lie near the surface. This fact makes it possible for a surgeon to make a small incision in the scrotum and quickly and easily cut out a section of each vas and tie off each of its separated ends. The technical name for this minor surgery is a bilateral partial **vasectomy** (vah-SEK-to-me). Although a man's seminiferous tubules may continue to form sperm after he has had a vasectomy, they can no longer leave his body. A part of their exit route has been cut away and no detour route has been provided. Therefore the man has become sterile; that is, he can no longer father children. If at a later date the man changes his mind and wishes to be fertile again, it is technically possible in some cases to rejoin the separated ends of the vas deferens. Unfortunately, however, this does not guarantee that the man's fertility will be restored; therefore, an individual should consider carefully before having a vasectomy performed.

It is important to note that a vasectomy does not change a man's ability to have an erection or an ejaculation. As explained in the next paragraph, structures that lie beyond the vas deferens produce a majority of the fluid volume of the ejaculate.

ACCESSORY MALE REPRODUCTIVE GLANDS

The two seminal vesicles, one prostate gland, and two bulbourethral (Cowper's) glands are accessory male glands that produce alkaline secretions. These secretions constitute the gelatinous fluid part of the **semen.** Usually 3 to 5 milliliters (about 1 teaspoonful) of semen is ejaculated at one time, and each milliliter normally contains over 60 million sperm. After a successful bilateral vasectomy, about the same amount of fluid may be ejaculated as before but it contains no sperm. In short, a vasectomy makes a man sterile but not impotent.

The prostate gland claims importance not so much for its function as for its troublemaking.

In older men it often becomes inflamed and enlarged, squeezing on the urethra, which runs through the center of the doughnut-shaped prostate. Sometimes, in fact, the prostate enlarges so much that it closes off the urethra completely. Urination then becomes impossible. (Should you refer to this as urinary retention or urinary suppression? If you are not sure, check your answer on p. 373.)

The small **bulbourethral** (Cowper's) **glands** lie one on either side of the urethra just below the prostate gland. Like the seminal vesicles and the prostate, the bulbourethral glands add an alkaline secretion to the semen. Sperm survive and remain fertile longer in an alkaline fluid than in an acid one.

OUTLINE SUMMARY

Structural Plan

Reproductive organs are classified as either *essential* or *accessory.* The essential organs (gonads) in the male are the testes. The accessory organs of reproduction consist of a series of ducts that carry the male sex cells from the testes to the exterior, additional sex glands, and the external organs (genitals)—see Table 16-1.

External Genitals

A Scrotum—a skin-covered pouch; each half of scrotum containing a testis, epididymis, and lower part of a vas deferens

B Penis—composed of one corpus cavernosum urethrae and two corpora cavernosa penis; erectile tissue makes up corpora; glans penis, bulging distal end of the organ; foreskin, a double fold of skin, covers glans in uncircumcised male; penis functions as organ of coitus; erection produced by blood distending spaces in erectile tissue

Testes

A Structure—pair of small oval glands in the scrotum; tough membrane (tunica albuginea) encloses each testis and forms partitions (septa) inside, dividing it into lobules

B Functions—serve as essential male gland (gonad); seminiferous tubules form male sex cells (process of spermatogenesis); interstitial cells secrete male hormone (testosterone)

C Testosterone

  1 Functions—masculinizes, promotes and maintains development of accessory male organs, stimulates protein anabolism

  2 Control of secretion—high blood level of ICSH stimulates testosterone secretion; conversely, high blood level of testosterone inhibits ICSH secretion—example of "negative feedback control mechanism"

D Spermatozoa—Figure 16-6 shows the characteristic parts of a single spermatozoon: head, neck, middle piece, and elongated, lashlike tail

Ducts

A Epididymis—narrow tube attached to each testis; duct of testis

B Vas deferens—continuation of ducts that start in epididymis

C Ejaculatory duct—continuation of vas deferens

D Urethra—terminal duct in male of both reproductive and urinary tracts

Accessory Male Reproductive Glands

A Seminal vesicles—secrete alkaline substance into semen

B Prostate gland—encircles urethra just below bladder; secretes alkaline fluid into semen

C Bulbourethral glands—pair of small glands located just below prostate; ducts open into urethra where they add alkaline secretion to semen

NEW WORDS

circumcision	genitals	spermatogenesis
coitus	gonads	spermatozoa
cryptorchidism	orgasm	vasectomy
ejaculation	semen	

CHAPTER TEST

1. The gonads of the male are the _____ .
2. In the male the sex cells are called _____ .
3. The essential organs of reproduction in both sexes are called the _____ .
4. The male organ of sexual intercourse is the _____ .
5. The male genitals consist of two organs: the _____ and the _____ .
6. The testis is surrounded by a tough whitish membrane called the tunica _____ .
7. Each lobule of the testis consists of a narrow but long and coiled tube called the _____ tubule.
8. The process of sperm formation is called _____ .
9. Testosterone formation is the function of the _____ cells of the testis.
10. If the testes fail to descend into the scrotum the condition is called _____ .
11. From the seminiferous tubules sperm enter the tightly coiled duct called the _____ .

12. Permanent sterility in the male follows surgical section of both vas deferens—a procedure called _____ .
13. In the male the urethra runs through the center of the doughnut-shaped _____ gland.
14. The bulbourethral glands are also referred to as _____ glands.
15. The pH of both seminal vesicle and prostate secretions is _____ .

Select the most correct answer from Column B for each statement in Column A. (Only one answer is correct.)

Column A	Column B
16. _____ male gonad	a. Spermatozoa
17. _____ male sex cells	b. Testosterone
18. _____ surgical removal of fore-skin	c. Orgasm
	d. Cryptorchidism
19. _____ climax of coitus	e. Prostate
20. _____ sperm formation	f. Testis
21. _____ male sex hormone	g. Interstitial cells
22. _____ undescended testis	h. Penis and scrotum
23. _____ produce testosterone	i. Spermatogenesis
24. _____ accessory reproductive gland	j. Circumcision
25. _____ external genitals	

REVIEW QUESTIONS

1 Identify the essential and accessory organs of reproduction in the male.

2 What organs are included in the *external genitals* of the male?

3 Discuss the structure of the testes. Explain the function of the seminiferous tubules and the interstitial cells of the testes.

4 What is the name of the masculinizing hormone? What glands secrete it? What are its general functions? How is its secretion level controlled?

5 Discuss the anatomy of a sperm cell. Why is it motile? What parts of the sperm cell are designed to provide motility?

6 What structures form semen?

7 Trace a sperm cell from its point of formation in the testes through the male reproductive ducts to ejaculation.

8 Explain why a vasectomy sterilizes a male.

9 How many sperm are normally present in one ejaculation of semen?

10 What is the usual volume of semen ejaculated at one time?

11 Castration is an operation that removes the testes. Would this sterilize the male? Why? What other effects would you expect to see as a result of castration? Why?

12 Discuss cryptorchidism.

CHAPTER
17

The Female Reproductive System

OBJECTIVES

After you have completed this chapter, you should be able to:

1 List the essential and accessory sex organs of the female reproductive system and give the generalized function of each.

2 Describe the structure of the female gonads and explain the steps in the development of mature ova from ovarian follicles.

3 Identify the structures that together constitute the female external genitals.

4 Identify the phases of the endometrial or menstrual cycle and correlate each phase with its occurrence in a typical 28-day cycle.

5 Define the following terms: *hysterectomy, oophorectomy, menopause, menarche, salpingitis.*

The structural plan of the female reproductive system resembles that of the male system. Like the male reproductive system, the female reproductive system consists of **essential** and **accessory** organs of reproduction.

STRUCTURAL PLAN

The essential organs of reproduction in both sexes are called **gonads.** The gonads of the female consist of a pair of main sex glands called **ovaries.** The female sex cells or **ova** are produced here.

The accessory organs of reproduction in the female consist of ducts from the main sex glands to the exterior, additional sex glands, and the external reproductive organs or genitals. To find out the names of these structures in the female, consult Table 17-1 and Figures 17-1 and 17-8.

EXTERNAL GENITALS

The scientific name for the female external genitals is the **vulva** (VUL-vah). Identify the following structures of the vulva in Figure 17-8:

1 Mons pubis
2 Clitoris
3 Orifice of urethra
4 Labia minora (small lips)
5 Hymen
6 Orifice, duct of Bartholin's gland
7 Orifice of vagina
8 Labia majora (large lips)

OVARIES

Although male and female reproductive systems resemble each other in plan, they differ from one another in details. For example, the testes, the main sex glands of the male, are not located inside a body cavity but lie in an external skin-covered pouch, the scrotum. In the female, however, the main sex glands, the ovaries, lie within the pelvic cavity. Also, there are two differences between male and female reproductive systems that sometimes assume clinical importance. One is that the ovarian ducts in the female open into the abdominal cavity; they do not attach directly to the ovaries. Later we shall explain the importance of this fact (p. 398). The other clinically significant difference is that the male urethra serves as a duct for both the urinary tract and reproductive tract. This is not the case in the female. In a woman's body the urethra serves as a duct for the urinary tract only. A separate tube, the vagina, serves the reproductive tract as its duct to the outside. This difference can be important. For example, if a man contracts gonorrhea, the infection can easily spread from the urethra throughout both his reproductive and urinary tracts. Although the urinary and reproductive systems are separated in the female, vaginal infections can spread through the uterine tubes directly into the abdomen. In the male there is no such direct route by which microorganisms can reach the abdominal cavity from the exterior.

Table 17-1 Female reproductive organs	
Essential organs of reproduction or main sex glands (gonads)	**Accessory organs of reproduction**
Ovaries (right and left ovaries)	Ducts: uterine tubes (two), uterus, vagina
	Accessory sex glands: Bartholin's glands, breasts
	External genitals: vulva (pudenda) (see Figure 17-8)

Several thousand sacs, too small to be seen without a microscope, make up the bulk of each ovary. They are called **ovarian follicles** (Figure 17-2). Within each follicle lies an immature *ovum,* the female sex cell. A mature ovum in its sac is called a **graafian** (GRAF-e-an) **follicle,** in honor of the Dutch anatomist who discovered them some 300 years ago.

Like the testes, the ovaries produce both sex cells and sex hormones. However, the ovaries

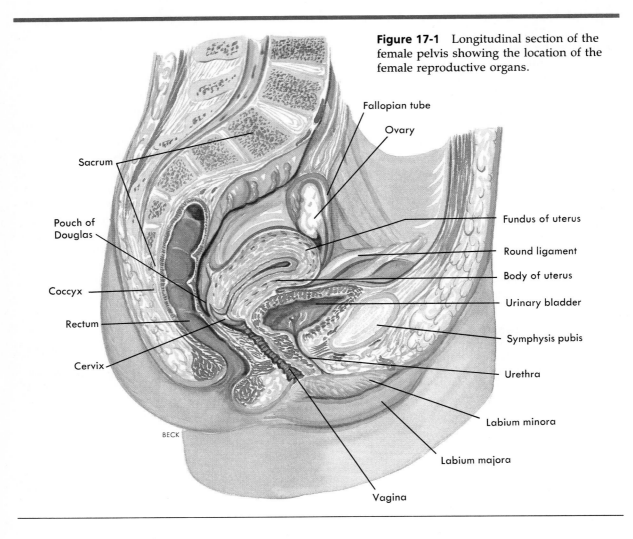

Figure 17-1 Longitudinal section of the female pelvis showing the location of the female reproductive organs.

Fallopian tube

Ovary

Sacrum

Pouch of Douglas

Coccyx

Rectum

Cervix

Fundus of uterus

Round ligament

Body of uterus

Urinary bladder

Symphysis pubis

Urethra

Labium minora

Labium majora

Vagina

BECK

usually produce only one mature sex cell—mature ovum, that is—per month, whereas the testes produce many millions of mature sperm. The ovaries secrete two kinds of female hormones—estrogens and progesterone. The testes, on the other hand, secrete only one kind of male hormone—*androgens;* testosterone is the only important androgen. The only endocrine gland cells of the testes are the interstitial cells. The ovaries contain two endocrine glands—the graafian follicles and the **corpus luteum.** Graafian follicles secrete estrogens. The corpus luteum secretes chiefly progesterone but also some estrogens.

HISTOLOGICAL STRUCTURE

The microscopic detail of a mature ovary is illustrated in Figure 17-2. Note that the graafian follicles are in different stages of development. The youngest follicles are located in the egg nest just above the ovarian ligament. Begin at this point and follow the clockwise movement of a single follicle as it matures. Actual photomicrographs of a developing ovum are shown in Figures 17-3 and 17-4. The maturing ovum can be clearly seen in the center of the developing follicle. It is the body's largest cell at the time of ovulation. The photomicrograph shown in Figure 17-3 clearly illustrates that a majority of the

Figure 17-2   Mammalian ovary showing successive stages of ovarian (graafian) follicle and ovum development. Begin with the first stage (egg nest) and follow around clockwise to the final stage (corpus albicans).

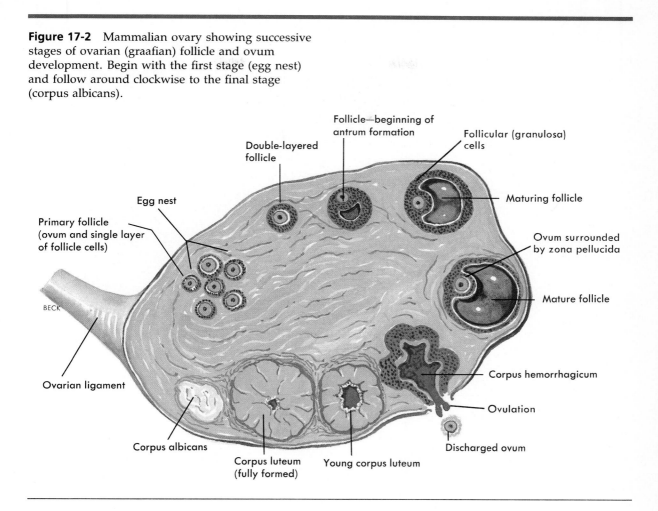

whole follicle is composed of hormone secreting granulosa or follicular cells. The shell-like **zona pellucida** (pel-LU-sid-a) can be seen surrounding the ovum in both photomicrographs. A sperm cell must be able to penetrate this covering layer for fertilization to occur (see Figure 18-1).

Just as androgens are the masculinizing hormone, so estrogens are the feminizing hormone. To discover estrogens' two chief functions, see Figure 17-5. Look next to Figure 16-6; it will tell you the two chief functions progesterone performs.

DUCTS

UTERINE TUBES (FALLOPIAN TUBES)

The **uterine tubes** serve as ducts for the ovaries even though they are not attached to them. The outer end of each tube curves over the top of each ovary and opens into the abdominal cavity. The inner end of each uterine tube attaches to the uterus, and the cavity inside the tube opens into the cavity in the uterus. Each tube measures about 4 inches (10 centimeters) in length.

After ovulation the discharged ovum first enters the abdominal cavity and then finds its way into the uterine tube. In Chapter 18 the details

Figure 17-3 Ovarian follicle. At this stage of development an open area or antrum is present. Hormone-secreting granulosa cells fill the lower portion of the follicle. (×70.)

Antrum

Developing ovum

Follicular (granulosa) cells

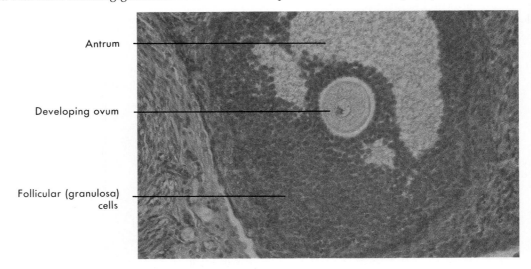

Figure 17-4 Early stage of a developing ovarian follicle. No antrum is present. (×140.)

Follicular (granulosa) cells

Zona pellucida

Developing ovum

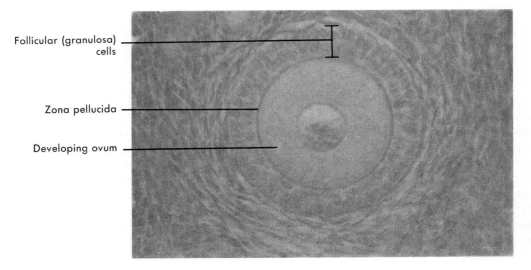

Figure 17-5   Functions of estrogens.

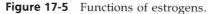

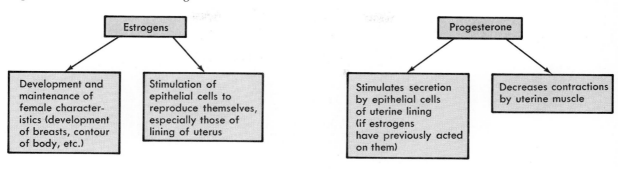

Figure 17-6   Cross section of uterine tube. (×70.)

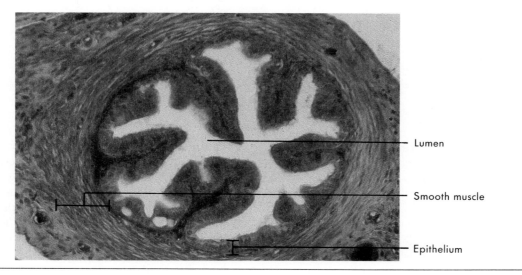

Lumen

Smooth muscle

Epithelium

of fertilization, which normally occurs in the outer one third of the oviduct, will be discussed. Occasionally, because the outer ends of the uterine tubes open into the pelvic cavity and are not actually connected to the ovaries, an ovum does not enter a tube but becomes fertilized in the abdominal cavity. The term "ectopic pregnancy" means a pregnancy that develops outside its proper place in the uterine cavity.

Figure 17-6 is a cross section of a uterine tube. Two ova, which were released from the ovary only a few hours before this tube was removed surgically, can be seen in the center of the photomicrograph. Note the columnar epithelium that forms the mucosal lining of the tube. This lining is directly continuous with the lining of the body cavity on one end and with the lining of the uterus and vagina on the other. This has great

Figure 17-7 Sectioned view of the uterus showing relationship to ovaries and vagina.

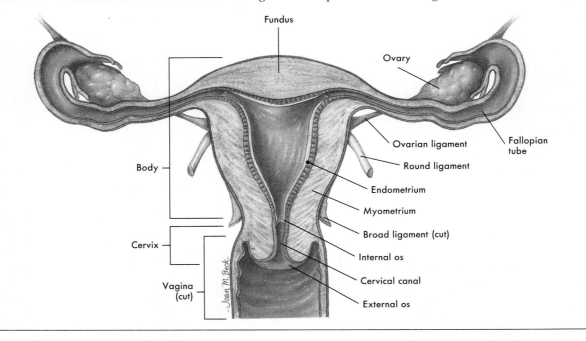

clinical significance, because infections of the vagina or uterus such as gonorrhea may pass into the abdominal cavity, where they may become life-threatening.

UTERUS

The uterus is a small organ—only about the size of a pear—but it is extremely strong. It is almost all muscle, with only a small cavity inside. During pregnancy the uterus grows many times larger so that it becomes big enough to hold a baby plus a considerable amount of fluid. The uterus is composed of two parts: an upper portion, the **body,** and a lower narrow section, the **cervix.** Just above the level where the uterine tubes attach to the body of the uterus, it rounds out to form a bulging prominence called the **fundus.** (See Figure 17-7.) Except during pregnancy, the uterus lies in the pelvic cavity just behind the urinary bladder. By the end of pregnancy it becomes large enough to extend up to the top of the abdominal cavity. It then presses against

the underside of the diaphragm—a fact that explains such a comment as "I can't seem to take a deep breath since I've gotten so big," made by many women late in their pregnancies.

The uterus functions in three processes—menstruation, pregnancy, and labor. The corpus luteum stops secreting progesterone and decreases its secretion of estrogens about eleven days after ovulation. Three days later, when the progesterone and estrogen concentrations in the blood are at their lowest, menstruation starts. Small pieces of **endometrium** (en-do-ME-tre-um) (mucous membrane lining of the uterus) pull loose, leaving torn blood vessels underneath. Blood and bits of endometrium trickle out of the uterus into the vagina and out of the body. Immediately after menstruation the endometrium starts to repair itself. It again grows thick and becomes lavishly supplied with blood in preparation for pregnancy. If fertilization does not take place, the uterus once more sheds the lining made ready for a pregnancy that did not occur. Because these changes in the uterine lining continue to repeat themselves, they are spoken of as the **menstrual cycle.** For a description of this cycle in the form of a diagram, see Figures 16-10 and 16-11.

Menstruation first occurs at puberty, often around the age of 12 years. Normally it repeats itself about every 28 days or thirteen times a year for some 30 to 40 years before it ceases at *menopause* (MEN-o-pawz), when a woman is somewhere around the age of 45 years.

VAGINA

The **vagina** is a distensible tube made mainly of smooth muscle and lined with mucous membrane. It lies in the pelvic cavity between the urinary bladder and the rectum. As the part of the female reproductive tract that opens to the exterior, the vagina is the organ that sperm enter on their journey to meet an ovum, and it is also the organ from which a baby emerges to meet its new world.

ACCESSORY FEMALE REPRODUCTIVE GLANDS

BARTHOLIN'S GLANDS

One of the small Bartholin's glands lies to the right of the vaginal outlet and one to the left of it. Secreting a mucuslike lubricating fluid is the function of Bartholin's glands. Their ducts open into the space between the labia minora and the hymen (Figure 17-8). Bartholinitis, an infection of these glands, occurs frequently. It often develops, for example, when a woman contracts gonorrhea.

BREASTS

The **breasts** lie over the pectoral muscles and are attached to them by connective tissue ligaments (of Cooper) (Figure 17-9). Breast size is determined more by the amount of fat around the glandular (milk-secreting) tissue than by the amount of glandular tissue itself. Hence the size of the breast has little to do with its ability to secrete adequate amounts of milk after the birth of a baby.

Each breast consists of fifteen to twenty di-

Hysterectomy

The word **hysterectomy** (his-te-REK-to-me) comes from the combination of two Greek words: *hystera,* meaning "uterus," and *ektome,* meaning "to cut out." By definition it is the surgical removal of the uterus. Hysterectomy is a term that is often misused, however, by incorrectly expanding its definition to include the removal of the ovaries or other reproductive structures. Only the uterus is removed in a hysterectomy. If the total uterus, including the cervix, is removed, the terms *total hysterectomy* or *panhysterectomy* may be used. If the cervical portion of the uterus is left in place and only the body of the organ is removed, the term *subtotal hysterectomy* is appropriate. The actual removal of the uterus may be performed through an incision made in the abdominal wall—*abdominal hysterectomy,* or through the vagina—*vaginal hysterectomy.* The term **oophorectomy** (o-of-o-REK-to-me) is used to describe removal of the ovaries. Although the two surgical procedures may take place during the same operation—for a woman with uterine or ovarian cancer, for example—the terms used to describe them should not be used interchangeably.

visions or lobes that are arranged radially. Each lobe consists of several lobules, and each lobule consists of milk-secreting glandular cells. The milk-secreting cells are arranged in grapelike clusters called *alveoli.* Small **lactiferous** (lak-TIF-er-us) **ducts** drain the alveoli and converge toward the nipple like the spokes of a wheel. Only one lactiferous duct leads from each lobe to an opening in the nipple. The colored area around the nipple is the **areola** (ah-RE-o-lah). It changes from a delicate pink to brown early in pregnancy and never quite returns to its original color.

A knowledge of the lymphatic drainage of the breast is important because cancerous cells from breast tumors often spread to other areas of the

Figure 17-8 External female genitals (the vulva).

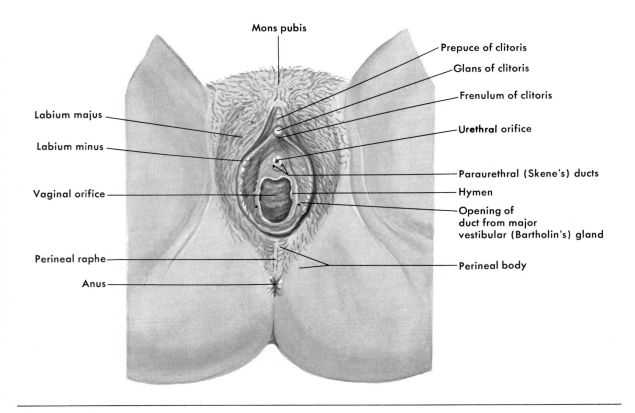

body through the lymphatics. The lymphatic drainage of the breast is discussed in Chapter 11. (See also Figure 11-15.)

MENSTRUAL CYCLE
PHASES AND EVENTS

The menstrual cycle consists of a great many changes—in the uterus, ovaries, vagina, and breasts and in the anterior pituitary gland's secretion of hormones. In the majority of women these changes occur with almost precise regularity throughout their reproductive years. The first indication of changes comes with the event of the first menstrual period. (**Menarche** (me-NAR-ke) is the scientific name for the beginning of the menses.)

A typical menstrual cycle covers a period of 28 days. Each cycle consists of three phases. Although they have been called by several different names, we shall call them the menstrual period, the postmenstrual phase, and the premenstrual phase. Now examine Figure 17-11 to find out what changes take place in the lining of the uterus (endometrium) and in the ovaries during each phase of the menstrual cycle. Be sure you do not overlook the event that occurs on day 14 of a 28-day cycle.

As a general rule, only one ovum matures each month during the 30 or 40 years that a woman has menstrual periods. But there are exceptions to this rule. Some months more than one matures, and some months no ovum matures. Ovulation occurs 14 days before the next

Figure 17-9 Lateral view of the breast (sagittal section). The gland is fixed to the overlying skin and the pectoral muscles by the suspensory ligaments of Cooper. Each lobule of secretory tissue is drained by a lactiferous duct that opens through the nipple.

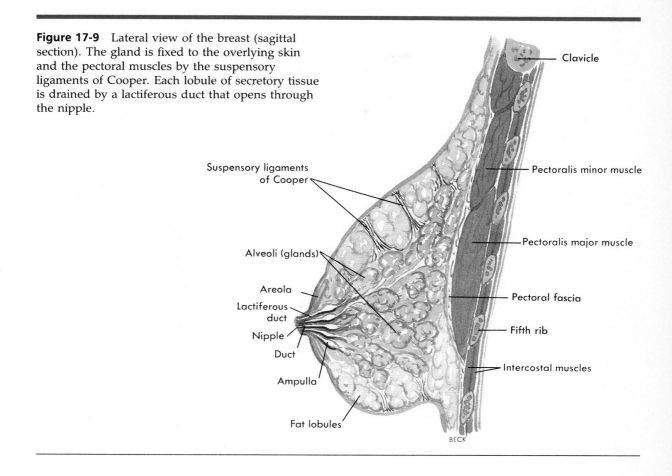

menstrual period begins. In a 28-day menstrual cycle, this means that ovulation occurs on the fourteenth day of the cycle, as shown in Figure 17-10. (Note that the first day of the menstrual period is considered the first day of the cycle.) In a 30-day cycle, however, the fourteenth day before the beginning of the next menses is not the fourteenth cycle day, but the sixteenth. And in a 25-day cycle, the fourteenth day before the next menses begins is the eleventh cycle day.

This matter of the time of ovulation has great practical importance. An ovum lives only a short time after it is ejected from its follicle, and sperm live only a short time after they enter the female body. Fertilization of an ovum by a sperm therefore can occur only around the time of ovulation. In other words, a woman's fertile period lasts

only a few days out of each month. Simply knowing the length of previous cycles, however, cannot ensure with any degree of accuracy the length of a current cycle or some future cycle. The reason is that the vast majority of women show some month-to-month variation in the length of their cycles. This physiological fact probably accounts for most of the unreliability of the old "calendar rhythm method" of contraception. Calendar rhythm has now been replaced by other, far more reliable "natural family planning" methods. Such natural methods base their judgments about fertility on information other than previous cycle lengths—for example, measurement of body temperature and recognition of changes each month in the amount and consistency of cervical mucus.

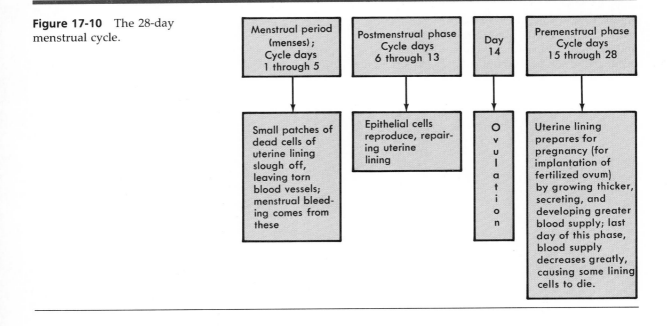

Figure 17-10   The 28-day menstrual cycle.

Changes in the blood levels of the hormones that control ovulation also produce changes in body temperature and in the amount and consistency of the cervical mucus.

Knowledge about the body's method of controlling the events of the menstrual cycle has made possible most of our modern methods of birth control. Knowledge about the male reproductive system has led to other methods—vasectomy, for example. Current and proposed research give promise of still other methods that may prove even more practical.

CONTROL OF MENSTRUAL CYCLE CHANGES

Now that overpopulation looms so large as a threat to man's healthy survival on this planet, knowledge about the body's methods for controlling menstrual cycle events has taken on new and enormous importance. Dominating this control is the anterior pituitary gland, the master gland of the body.

From the first to about the seventh day of the menstrual cycle, the anterior pituitary gland se-

cretes increasing amounts of follicle-stimulating hormone (FSH). A high blood concentration of FSH stimulates several immature ovarian follicles to start growing and secrete estrogens (Figure 17-11). As the estrogen content of blood increases, it stimulates the anterior pituitary gland to secrete another hormone, namely luteinizing hormone (LH). LH causes maturing of a follicle and its ovum, ovulation (rupturing of mature follicle with ejection of ovum), and luteinization (formation of a yellow body, the corpus luteum, in the ruptured follicle).

Which hormone—FSH or LH—would you call the "ovulating hormone"? Do you think ovulation could occur if the blood concentration of FSH remained low throughout the menstrual cycle? If you answered LH to the first question and no to the second, you answered both questions correctly. Ovulation cannot occur if the blood level of FSH stays low, because a high concentration of this hormone is essential to stimulate ovarian follicles to start growing and maturing. With a low level of FSH, no follicles start to grow, and therefore none become ripe

Figure 17-11 Diagram illustrating the interrelationships among the cerebral, hypothalamic, pituitary, ovarian, and uterine functions throughout a usual 28-day menstrual cycle. The variations in basal body temperature are also illustrated.

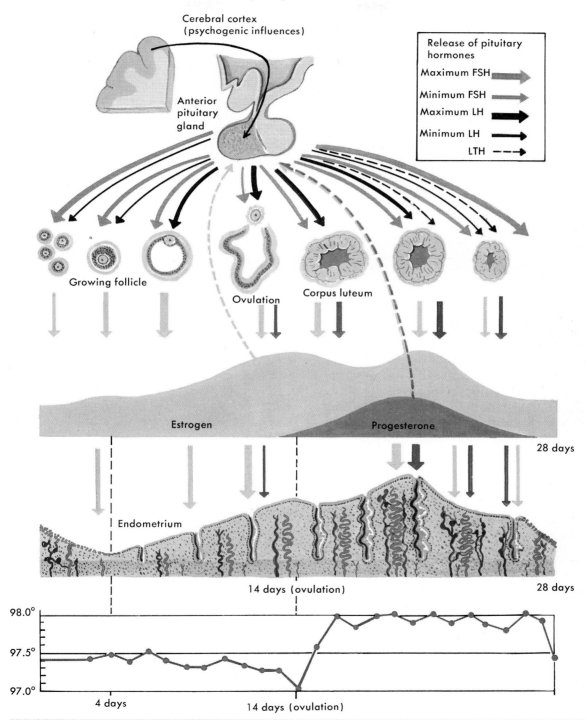

enough to ovulate. Ovulation is caused by the combined actions of FSH and LH. Birth control pills that contain estrogen substances suppress FSH secretion. This indirectly prevents ovulation.

Ovulation occurs, as we have said, because of the combined actions of the two anterior pituitary hormones, FSH and LH. The next question is: what causes menstruation? A brief answer is this: a sudden, sharp decrease in estrogen and progesterone secretion toward the end of the premenstrual period causes the uterine lining to break down and another menstrual period to begin. But exactly what causes the sudden decrease in estrogen and progesterone secretion we do not yet know.

DISORDERS OF REPRODUCTIVE SYSTEM

Infections of the mucous membrane lining of the reproductive tract occur in both sexes. In women the danger of this becomes especially great after childbirth. Physicians and nurses therefore take care not to introduce infectious organisms into the reproductive tract during or after delivery. The name given the inflammation indicates the part of the tract inflamed. For example, **vaginitis** (vaj-i-NI-tis), is inflammation of the vagina; **cervicitis** (ser-vi-SI-tis), inflammation of the cervix; **endometritis** (en-do-me-TRI-tis), inflammation of the endometrium; **salpingitis** (sal-pen-JI-tis), inflammation of the uterine tubes. Tumors frequently develop in the uterus, ovaries, and breasts of women and in the prostate glands of men.

OUTLINE SUMMARY

Structural Plan

Reproductive organs are classified as either *essential* or *accessory*. The essential organs (gonads) in the female are the ovaries. The accessory organs of reproduction consist of ducts from the main sex glands to the exterior, additional sex glands, and the external reproductive organs (genitals)—see Table 17-1.

A External genitals, vulva—see Figure 17-8

B Ovaries—main sex glands of female; located in pelvic cavity; female sex cells (ova) form in graafian follicles in ovaries; graafian follicles secrete estrogens (see Figure 17-5 for estrogen functions); corpus luteum secretes estrogens and progesterone

C Uterine tubes (fallopian tubes)—ducts for ovaries but not attached to them—see Figure 17-6

D Uterus (womb)
   1 Parts—fundus, body, cervix—see Figure 17-7
   2 Structure—strong muscular organ with mucous lining (endometrium)

   3 Functions—menstruation, pregnancy, labor

E Vagina—muscular tube lined with mucous membrane; terminal part of female reproductive tract

F Bartholin's glands—pair of small glands near vaginal orifice; secrete mucuslike lubricating fluid

G Breasts—glands present in both sexes but normally function only in female—see Figure 17-9

Menstrual Cycle

A Length—about 28 days, varies somewhat in different individuals and in the same individual but at different times

B Phases
   1 Menstrual period or menses—about the first 4 or 5 days of the cycle, varies somewhat; characterized by sloughing of bits of endometrium (uterine lining) with bleeding

2 Postmenstrual phase—days between the end of menses and ovulation; varies in length; the shorter the cycle, the shorter the postmenstrual phase; the longer the cycle, the longer the postmenstrual phase; examples: in 28-day cycle postmenstrual phase ends on thirteenth day, but in 26-day cycle it ends on the eleventh day, and in 32-day cycle, it ends on seventeenth day; characterized by repair of endometrium

3 Premenstrual phase—days between ovulation and beginning of next menses; ovulation about 14 days before next menses; characterized by further thickening of endometrium and secretion by its glands in preparation for implantation of fertilized ovum, combined actions of the anterior pituitary hormones FSH and LH cause ovulation; sudden sharp decrease in estrogens and progesterone bring on menstruation if pregnancy does not occur.

Disorders of Reproductive Systems

A Infections—common in both sexes; for example, in women, salpingitis, vaginitis, and endometritis, and in men, prostatitis

B Tumors—occur in both sexes, for example, in women, in uterus, ovaries, and breasts, and in men, prostate gland

NEW WORDS

cervicitis	menopause	prostatitis
climacteric	menses	salpingitis
endometritis	ovulation	vaginitis
menarche		

CHAPTER TEST

1. The essential organs of reproduction in the female are the _____ .
2. The scientific name for the female external genitals is the _____ .
3. The female sex cells are called _____ .
4. A mature ovum in its sac is called a graafian _____ .
5. The ovaries contain two endocrine glands: the follicles and the corpus _____ .
6. For fertilization to occur a sperm must penetrate the shell-like covering of the ovum called the _____ _____ .
7. The female sex hormones are _____ and _____ .
8. A pregnancy that develops outside the uterine cavity is called an _____ pregnancy.
9. The lower narrow section of the uterus that opens into the vagina is called the _____ .

10. The corpus luteum secretes _____ .
11. The mucous membrane lining the uterus is called
    the _____ .
12. Surgical removal of the uterus is called a _____ .
13. The colored area around the nipple is called the _____ .
14. The scientific name for the beginning of the menses
    is called _____ .
15. The average length of a typical menstrual cycle is
    _____ days.

Select the most correct answer from Column B for each statement in Column A. (Only one answer is correct.)

Column A	Column B
16. ___ female gonads	a. Ovarian follicles
17. ___ external genitals	b. Zona pellucida
18. ___ secrete estrogen	c. LH
19. ___ corpus luteum	d. Secrete progesterone
20. ___ surrounds ovum	e. Salpingitis
21. ___ removal of ovaries	f. Ovaries
22. ___ "ovulating hormone"	g. Estrogen
23. ___ inflammation of uterine	h. Oophorectomy
tubes	i. Progesterone
24. ___ decreases uterine contrac-	j. Vulva
tions	
25. ___ feminizing hormone	

REVIEW QUESTIONS

1 Identify the feminizing hormone by its scientific name. What gland secretes it?

2 Identify the ovulating hormone by its scientific name. What glands secrete it?

3 Identify and locate each of the following:

alveoli of breast	graafian follicle
areola	labia majora
cervix	mons pubis
clitoris	ovum
fundus of uterus	uterine tube

4 Define briefly the words listed under "New words."

5 What causes ovulation?

6 What is menstruation?

7 What causes menstruation?

8 How many female sex cells are usually formed each month? How does this compare with the number of male sex cells formed each month?

CHAPTER
18 Growth and Development

CHAPTER OUTLINE

OBJECTIVES

After you have completed this chapter, you should be able to:

1 Discuss the concept of development as a biological process characterized by continuous modification and change.

2 Discuss the major developmental changes characteristic of the prenatal stage of life from fertilization to birth.

3 List and discuss the major developmental changes characteristic of the four postnatal periods of life.

4 Define the following terms: *histogenesis, organogenesis, zygote, quickening, senescence, gestation, presbyopia.*

5 Discuss the effects of aging on the major body organ systems.

any of your fondest and most vivid memories are probably associated with your birthdays. The day of birth is an important milestone of life. Most people continue to remember their birthday in some special way each year; birthdays serve as pleasant and convenient reference points to mark periods of transition or change in our lives. The actual day of birth marks the end of one phase of life called the **prenatal period** and the beginning of a second called the **postnatal period.** The prenatal period begins at conception and ends at birth; the postnatal period begins at birth and continues until death. Although important periods in our lives such as childhood and adolescence are often remembered as a series of individual and isolated events, they are in reality part of an ongoing and continuous process. In reviewing the many changes that occur during the cycle of life from conception to death, it is often convenient to isolate certain periods such as infancy or old age for study. It is important to remember, however, that life is not a series of stop-and-start events

or individual and isolated periods of time. Instead, it is a biological process that is characterized by continuous modification and change.

This chapter discusses some of the important events and changes that occur in the ongoing development of the individual from conception to death. Study of development during the prenatal period will be followed by a review of changes occurring during infancy and adulthood and finally some of the more important changes that occur in the individual organ systems of the body as a result of aging will be discussed.

PRENATAL PERIOD

The **prenatal stage of development** begins at the time of conception or fertilization—that is, at the moment the female ovum and the male sperm cells unite (Figure 18-1). The period of prenatal development continues until birth of the child about 40 weeks later. The science of the development of the individual before birth

Figure 18-1 Fertilization.

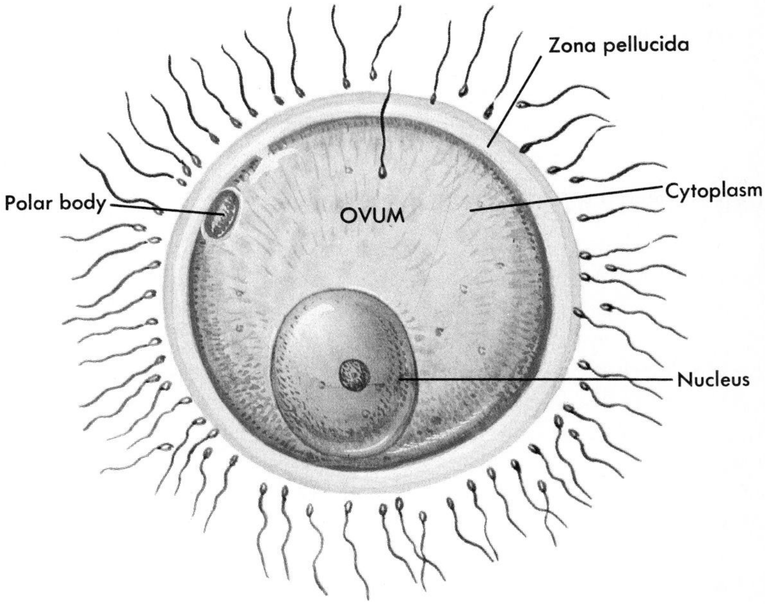

Figure 18-2 Appearance of uterus and uterine tubes, fertilization to implantation. Note that fertilization normally occurs in the outer one third of the uterine tube.

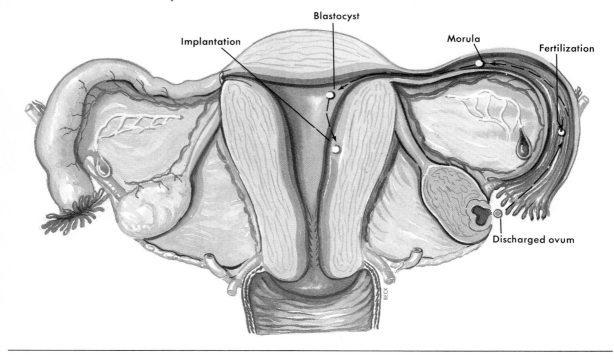

is called **embryology** (em-bre-OL-o-je). It is a story of miracles, describing the means by which a new human life is started and the steps by which a single microscopic cell is transformed into a complex human being.

FERTILIZATION TO IMPLANTATION

After ovulation the discharged ovum first enters the abdominal cavity and then finds its way into the uterine tubes. Sperm cells are required to "swim" up the uterine tubes toward the ovum. Normally fertilization occurs in the outer one third of the oviduct (Figure 18-2). The fertilized ovum or **zygote** (ZI-got) immediately begins to divide, and in about 3 days a solid mass of calls called a **morula** (MOR-u-lah) is formed. Genetically, the zygote is complete. It represents a new single-celled individual. Time and nourishment are all that is needed for expression of those characteristics such as sex, body build, and

skin color that were determined at the time of fertilization. As you can see in Figure 18-3, the cells of the morula continue to divide, and by the time the developing embryo reaches the

On July 25, 1978, the world's first "test-tube" baby, a girl, was born in Oldham, England. Nine months earlier a mature ovum had been removed from the mother's body by laparoscopy ("belly button surgery") and fertilized in a laboratory dish by her husband's sperm. After 2½ days' growth in a controlled environment, the fertilized ovum (now at the eight-cell stage) was returned by the physicians to the mother's uterus. Use of this technique in fertility clinics worldwide now results in the birth of numerous "test-tube" babies each year.

Figure 18-3 Development from fertilization to implantation. Implantation will be complete about 10 days following fertilization.

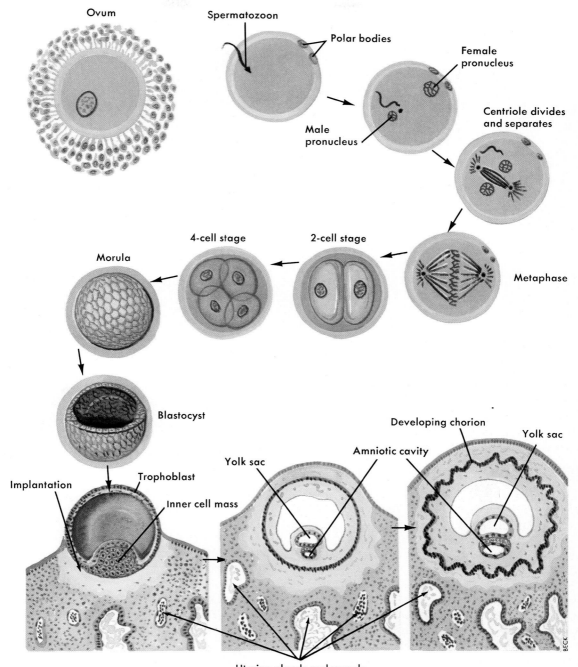

Figure 18-4 Development of the embryo to 4 months of gestation.

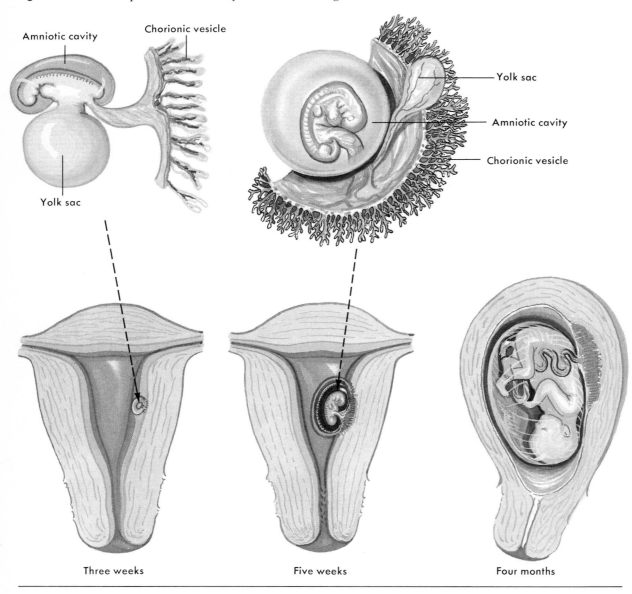

Three weeks

Five weeks

Four months

uterus, it is a hollow ball of cells called a **blastocyst (BLAS-to-sist).** The blastocyst consists of an outer layer of cells and an inner cell mass. In about 10 days from the time of fertilization the blastocyst will be completely implanted in the uterine lining.

Note in Figure 18-3 that as the blastocyst continues to develop it forms a structure with two cavities. The amniotic cavity will become a fluid-filled, shock-absorbing sac in which the embryo will float during development. The **chorion** (KO-re-on) shown in Figure 18-3 will develop into an important fetal membrane in the **placenta** (plah-SEN-tah). The placenta (Figure 18-5) will anchor

Figure 18-5   Sectioned view of placenta showing fetal and maternal blood vessels.

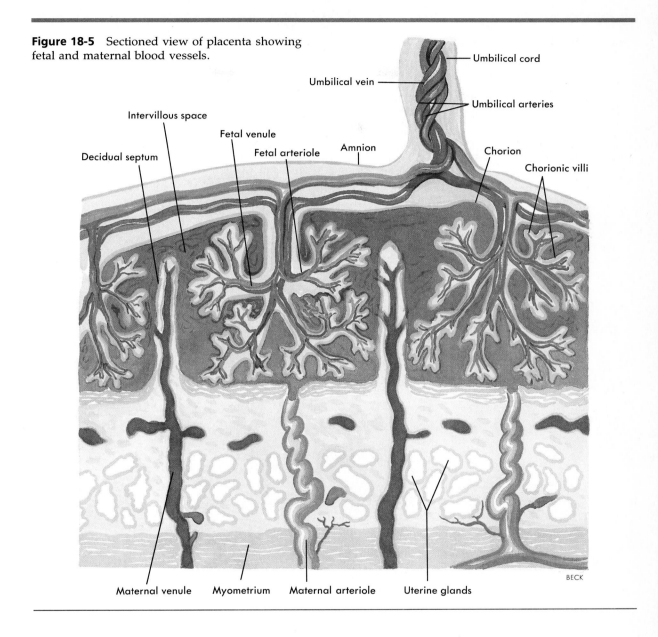

the developing fetus to the uterus and provide a "bridge" for the exchange of nutrients between mother and baby.

As the embryo continues to grow, three layers of specialized cells develop that embryologists call the **primary germ layers**. Each layer will give rise to definite structures such as the skin, mus-cles, or digestive organs. By the fifth week of development the heart is beating and, although the embryo is only ⅜-inch (8 millimeters) long, the organ systems are developing rapidly. By the fourth month all organ systems are formed and functioning. Growth of the embryo to 4 months of age is shown in Figure 18-4.

Figure 18-6 Full-term pregnancy.

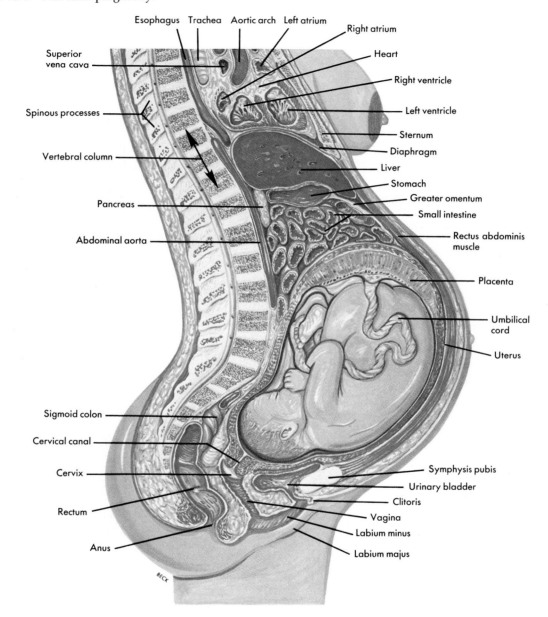

HISTOGENESIS AND ORGANOGENESIS

The study of how the primary germ layers develop into many different kinds of tissues is called **histogenesis** (his-to-JEN-e-sis). How those tissues arrange themselves into organs is called **organogenesis** (or-gah-no-JEN-e-sis). The fascinating story of histogenesis and organogenesis in human development is long and complicated; its telling belongs to the science of embryology. But for the beginning student of anatomy, it seems sufficient to appreciate that life begins when two sex cells unite to form a single-celled zygote and that the new human body evolves by a series of processes consisting of cell multiplication, cell growth, and cell rearrangements, all of which take place in definite, orderly sequence. Development of structure and function go hand in hand, and from 4 months of gestation, when every organ system is in place and functioning, until term (about 280 days), development of the baby is mainly a matter of growth. Figure 18-6 shows the normal intra-uterine placement of a baby just before birth (full-term pregnancy).

POSTNATAL PERIOD

The **postnatal period** begins at birth and lasts until death. It is often divided into major periods for study, but the need for an understanding and appreciation of the fact that growth and development are continuous processes that occur throughout the life cycle cannot be overemphasized. Gradual changes in the physical appearance of the body as a whole and in the relative proportions of the head, trunk, and limbs are quite noticeable between birth and adolescence. Note in Figure 18-7 the obvious changes in the size of bones and in the proportionate sizes between different bones and different body areas. The head, for example, becomes proportionately smaller. Whereas the infant head is approxi-

Figure 18-7 Changes in the proportions of various body parts from birth to maturity. Note the dramatic differences in head size.

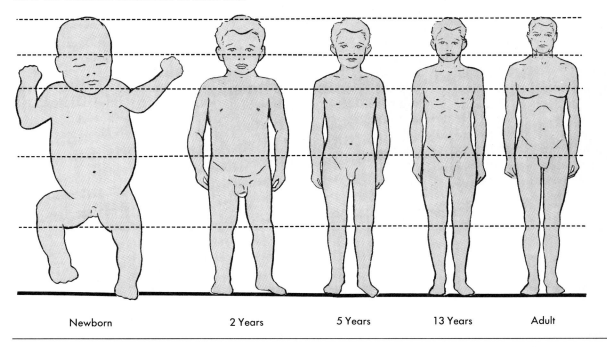

| Newborn | 2 Years | 5 Years | 13 Years | Adult |

Figure 18-8 The neonate infant.

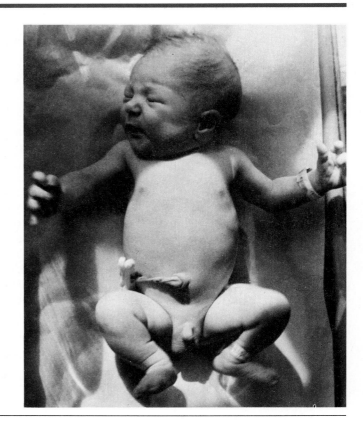

mately one fourth the total height of the body, the adult head is only about one eighth the total height. The facial bones also show several changes between infancy and adulthood. The infant face compared with the entire skull bears the relationship of 1 : 8 that is the face is one eighth of the skull surface; the adult face bears the relationship of 1 : 2 (or one half) to the adult skull. Another change in proportion involves the trunk and lower extremities. The legs become proportionately longer and the trunk proportionately shorter. In addition, the thoracic and abdominal contours change, roughly speaking, from round to elliptical.

Such changes are good examples of the everchanging and ongoing nature of the growth and development process. It is unfortunate that many of the changes that occur in the later years of life do not result in increased function. These degenerative changes are certainly important, however, and will be discussed later in this chapter. The following are the most common postnatal periods: (1) **infancy,** (2) **childhood,** (3) **adolescence** and **adulthood,** and (4) **old age.**

INFANCY

The period of infancy begins abruptly at birth and lasts about 18 months. The first 4 weeks of infancy are often referred to as the **neonatal** (ne-o-NA-tal) **period** (Figure 18-8). Dramatic changes occur at a rapid-fire rate during this short but critical period. **Neonatology** (ne-o-na-TOL-o-je) is the medical and nursing specialty concerned with the diagnosis and treatment of disorders of the newborn. Advances in this area have resulted in dramatically reduced infant mortality.

Many of the changes that occur in the cardio-

Figure 18-9 Normal curvature of the newborn's spine.

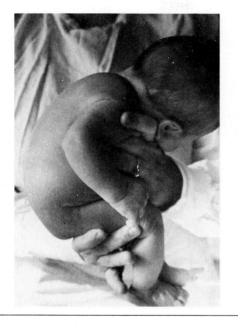

Figure 18-10 Normal lumbar curvature of toddler's spine.

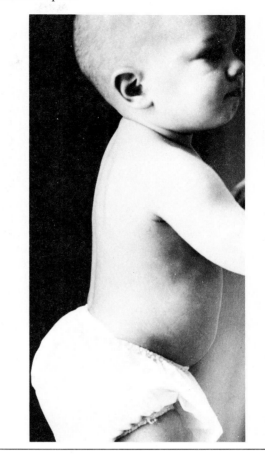

vascular and respiratory systems at the time of birth are necessary for survival. Whereas the fetus was totally dependent on the mother for life support, the newborn infant, to survive, must become totally self-supporting in terms of blood circulation and respiration immediately after birth. A baby's first breath is deep and forceful. It must empty the lungs of amniotic fluid, which partially filled them before birth. The stimulus to breathe results primarily from the increasing amounts of CO_2 that accumulate in the blood after the umbilical cord is cut shortly after delivery.

Many developmental changes occur between the end of the neonatal period and 18 months of age. Birth weight will double during the first 4 months and then triple by 1 year. The baby will also increase in length by 50% by the twelfth month. The "baby fat" that accumulated under the skin during the first year will begin to decrease and the plump infant will become leaner.

Early in infancy the baby has only one spinal curvature (Figure 18-9). The lumbar curvature appears between 12 and 18 months, and the once-helpless infant becomes a toddler who can stand (Figure 18-10). One of the most striking changes to occur during infancy is the rapid development of the nervous and muscular systems. This permits the infant to: follow a moving object with the eyes (2 months); lift the head and raise the chest (3 months); sit when well supported (4 months); crawl (10 months); stand alone (12 months) and actually run (although a bit stiffly) at 18 months of age.

Quickening

Pregnant women will usually notice fetal movement for the first time between the sixteenth and the eighteenth week of pregnancy. The term **quickening** has been used for generations to describe these first recognizable movements of the fetus. From an occasional "kick" during the fourth and fifth months of pregnancy, the frequency of fetal movements will steadily increase as gestation progresses. By the seventh month a normal fetus will be very active—a fact that mothers-to-be have always known. Physicians, however, have only recently realized that the frequency of fetal movements is an excellent indicator of the unborn baby's health. Although diagnostic tests such as x-ray examination and fetal blood sampling provide useful information about the fetus before birth, they carry some risk for either mother or child and are never used unless there is definite evidence of a problem. Noninvasive diagnostic techniques that cause very little risk—such as ultrasound—are inconvenient and require expensive equipment. Recent studies have shown that simply by recording the number of fetal movements each day after the twenty-eighth week of pregnancy, a woman can provide her physician with extremely useful information about the health of her unborn child. Ten or more movements during a daily measurement period are considered normal.

CHILDHOOD

The **childhood period of development** extends from the end of infancy to sexual maturity or puberty—12 to 14 years in girls and 14 to 16 years in boys. Overall, growth during early childhood continues at a rather rapid pace, but month-to-month gains become less consistent. By the age of 6 the child will appear more like a preadolescent than an infant or toddler. The child becomes less chubby, the potbelly becomes flatter, and the face loses its babyish look. The nervous and muscular systems continue to develop rapidly during the middle years of childhood; by 10 years of age the child will have developed numerous motor and coordination skills.

The *deciduous teeth,* which began to appear at about 6 months of age, are lost during childhood beginning at about 6 years of age. The *permanent teeth,* with the possible exception of the third molars or wisdom teeth, have all erupted by age 14.

ADOLESCENCE AND ADULTHOOD

The average age range of **adolescence** will vary but generally the teenage years (13 to 19) are used. The period is marked by rapid and intense physical growth, which ultimately results in sexual maturity. Many of the developmental changes that occur during this period are controlled by the secretion of sex hormones and are classified as **secondary sexual characteristics.** Breast development is often the first sign of approaching puberty in girls, beginning about age 10. Most girls begin to menstruate at 12 to 13 years of age. In boys the first sign of puberty is often enlargement of the testicles, which begins between 10 and 13 years of age. Both sexes show a spurt in height during adolescence. In girls the spurt in height begins between the ages of 10 and 12 and is nearly complete by age 14 or 15. In males the period of rapid growth begins between 12 and 13 and is generally complete by age 16.

Many developmental changes that began early in childhood will not be completed until the early or middle years of adulthood. Examples include the maturation of bone, resulting in the full closure of the growth plates, and changes in the size and placement of other body components such as the sinuses (Figure 18-11). Many body traits do not become apparent for years following birth. Normal balding patterns, for example, are determined at the time of fertilization by heredity but do not appear until maturity (Figure 18-12). As a general rule adulthood is characterized by maintenance of existing body tissues. With the passage of years the on-

Figure 18-11   Development of the frontal and maxillary sinuses. **A,** Infancy; **B,** childhood; **C,** adolescence; **D,** adulthood.

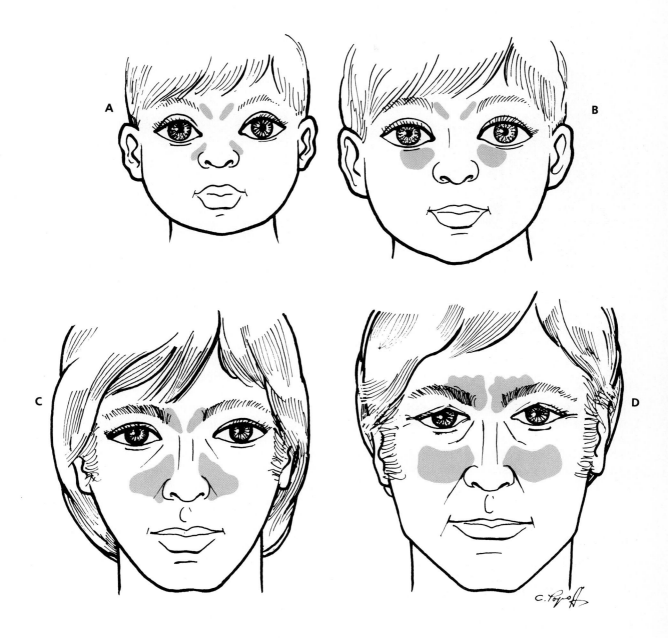

Figure 18-12 Normal balding pattern. This body trait does not appear until adulthood.

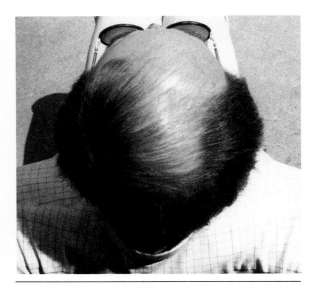

going effort of maintenance and repair of body tissues will become more and more difficult. As a result, degeneration will begin. It is the process of growing old, and it culminates in death.

OLD AGE

Most body systems are in peak condition and function at a high level of efficiency during the early years of adulthood. As a person grows older, a gradual but certain decline takes place in the functioning of every major organ system in the body. The remainder of this chapter will deal with a number of the more common degenerative changes that characterize **senescence** (se-NES-ens) or old age.

EFFECTS OF AGING

SKELETAL SYSTEM

In old age the bones undergo changes in texture, degree of calcification, and shape. Instead of clean-cut margins, old bones develop indistinct and shaggy-appearing margins with

Figure 18-13 Curvature of the spine and shortened stature caused by fracture of vertebrae.

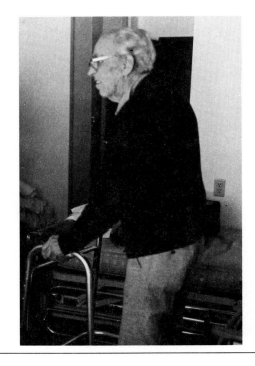

spurs—a process called "lipping." This type of degenerative change restricts movement because of the piling up of bone tissue around the joints. With advancing age, changes in calcification may result in actual reduction of bone size and in bones that are porous and subject to fracture. The lower cervical and thoracic vertebrae are the site of frequent fractures. The result is curvature of the spine and the shortened stature so typical of old age (Figure 18-13). Degenerative joint diseases such as **rheumatoid arthritis** (ROO-mah-toid ar-THRI-tis) are also common in the aged (Figure 18-14).

INTEGUMENTARY SYSTEM (SKIN)

With advancing age the skin becomes dry, thin, and inelastic. It tends to "sag" on the body because of increased wrinkling and skinfolds

Figure 18-14 Rheumatoid arthritis—hand involvement.

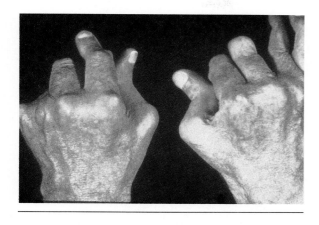

Figure 18-15 Increased wrinkling and skinfolds associated with dry, inelastic skin in old age.

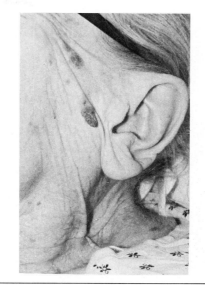

(Figure 18-15). Pigmentation changes, and the thinning or loss of hair are also common problems associated with the aging of skin (Figure 18-16).

URINARY SYSTEM

The number of nephron units in the kidney decreases by almost 50% between the ages of 30 and 75. Also, because less blood flows through the kidneys as an individual ages, there is a reduction in overall function and excretory capacity or the ability to produce urine. In the bladder, significant age-related problems often occur because of diminished muscle tone. Muscle atrophy (wasting) in the bladder wall results in decreased capacity and inability to empty or void completely.

RESPIRATORY SYSTEM

In old age the costal cartilages that connect the ribs to the sternum become hardened or calcified. This makes it difficult for the rib cage to expand and contract as it normally does during inspiration and expiration. In time the ribs gradually become "fixed" to the sternum, and chest movements become difficult. When this occurs the rib cage tends to remain in a more expanded position, respiratory efficiency decreases, and a condition called "barrel chest" results (Figure 18-17). With advancing years a generalized atrophy or wasting of muscle tissue takes place as the contractile muscle cells are replaced by connective tissue. This loss of muscle cells decreases the strength of the muscles associated with inspiration and expiration.

CARDIOVASCULAR SYSTEM

Degenerative heart and blood vessel disease is one of the most common and serious effects of aging. Fatty deposits build up in blood vessel walls and narrow the passageway for the movement of blood, much as the buildup of scale in a water pipe will decrease flow and pressure. The resulting condition, called **atherosclerosis** (ath-er-o-skle-RO-sis), often leads to eventual blockage of the coronary arteries and a "heart attack" (Figure 18-18). If fatty accumulations or other substances in blood vessels calcify, actual hardening of the arteries or **arteriosclerosis** (ar-

Figure 18-16   Marked pigmentation deposits associated with aging.

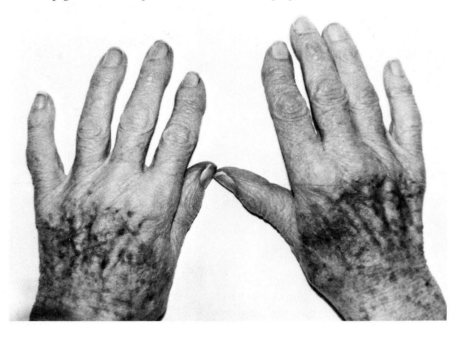

Figure 18-17   Barrel chest. Note the expanded size of the thorax.

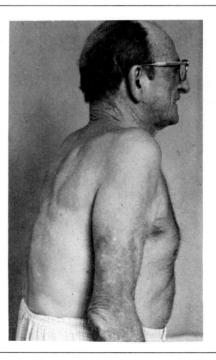

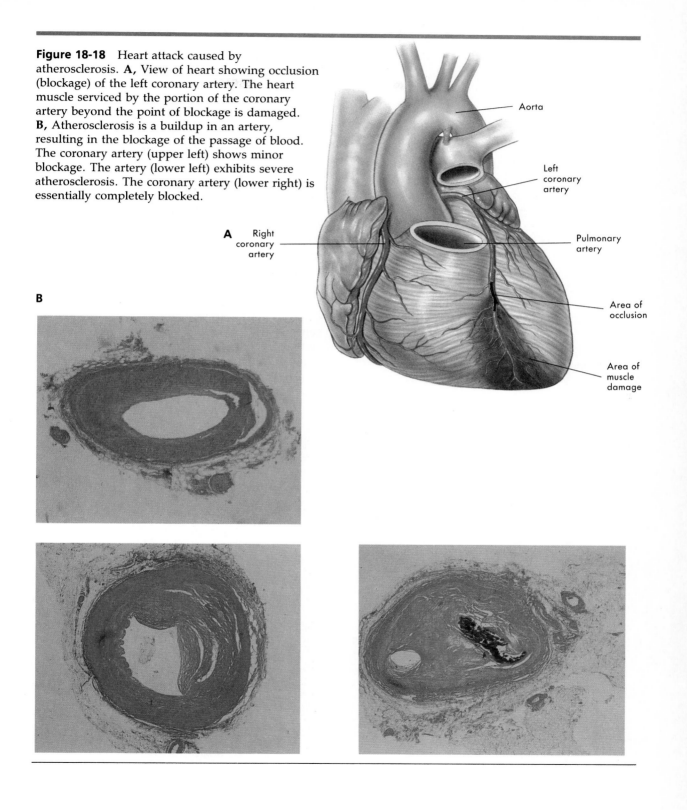

Figure 18-18   Heart attack caused by atherosclerosis. **A,** View of heart showing occlusion (blockage) of the left coronary artery. The heart muscle serviced by the portion of the coronary artery beyond the point of blockage is damaged. **B,** Atherosclerosis is a buildup in an artery, resulting in the blockage of the passage of blood. The coronary artery (upper left) shows minor blockage. The artery (lower left) exhibits severe atherosclerosis. The coronary artery (lower right) is essentially completely blocked.

Aorta

Left
coronary
artery

Pulmonary
artery

A   Right
coronary
artery

Area of
occlusion

Area of
muscle
damage

B

te-re-o-skle-RO-sis) occurs. Rupture of a hardened vessel in the brain (stroke) is a frequent cause of serious disability or death in the aged. **Hypertension** or high blood pressure is also common in the elderly.

SPECIAL SENSES

The sense organs, as a group, all show a gradual decline in performance and capacity as a person ages. Most people are farsighted by age 65 because eye lenses become hardened and lose elasticity; the lenses cannot become curved to accommodate for near vision. This hardening of the lens is called **presbyopia** (pres-be-O-pe-ah), which means "old eye." Many individuals first notice the change at about 40 or 45 years of age when it becomes difficult to do close-up work or read without holding printed material at arm's length. This explains the increased need, with advancing age, for bifocals, or glasses that incorporate two lenses to automatically accommodate for both near and distant vision. Loss of transparency of the lens or its covering capsule

is another common age-related eye change. If the lens actually becomes cloudy and significantly impairs vision, it is called a **cataract** (KAT-ah-rakt) and must be removed surgically. The incidence of **glaucoma** (glaw-KO-mah), the most serious of the age-related eye disorders, increases with age. Glaucoma causes an increase in the pressure within the eyeball and, unless treated, often results in blindness.

In many elderly people a very significant loss of hair cells in the organ of Corti (inner ear) causes a serious decline in ability to hear certain frequencies. In addition, the eardrum and attached ossicles become more fixed and less able to transmit mechanical sound waves. Some degree of hearing impairment is universally present in the aged.

The sense of taste is also decreased in the elderly. Loss of appetite in older people may be caused, at least in part, by the replacement of taste buds with connective tissue cells. Only about 40% of the taste buds present at age 30 remain in an individual at age 75.

OUTLINE SUMMARY

Prenatal Period

A Prenatal period begins at conception and continues until birth (about 40 weeks); see Figure 18-1

B Science of fetal growth and development called embryology

C Fertilization to implantation requires about 10 days
  1 Fertilization normally occurs in outer third of oviduct (Figure 18-2)
  2 Fertilized ovum called a zygote; zygote is genetically complete—all that is needed for expression of hereditary traits is time and nourishment
  3 After 3 days of cell division zygote develops into a solid cell mass called a morula

  4 Continued cell divisions of morula produce a hollow ball of cells called a blastocyst
  5 Blastocyst implants in uterine wall about 10 days following fertilization

D Development of embryo to 4 months
  1 Blastocyst forms amniotic cavity and chorion of placenta (Figure 18-3)
  2 Placenta provides for exchange of nutrients between mother and fetus
  3 Three primary germ layers appear in developing embryo following implantation of the blastocyst
  4 All organ systems are formed and functioning by the fourth month of gestation (Figure 18-5)

E Histogenesis and organogenesis
  1 Formation of new organs and tissues occurs from specific development of the primary germ layers
  2 Each primary germ layer gives rise to such definite structures as skin and muscles
  3 Growth processes include cell multiplication, cell growth, and cell rearrangements
  4 From 4 months of gestation until delivery the development of the baby is mainly a matter of growth (Figure 18-6).

Postnatal Period

A Postnatal period begins at birth and lasts until death
B Divisions of postnatal period into isolated time frames can be misleading; life is a continuous process and growth and development must be understood as continuous
C Obvious changes in physical appearance of body—in whole and in proportion—occur between birth and maturity (Figure 18-7)
D Divisions of postnatal period:
  1 Infancy
  2 Childhood
  3 Adolescence and adulthood
  4 Old age
E Infancy
  1 First 4 weeks called neonatal period (Figure 18-8)
  2 Neonatology—medical and nursing specialty concerned with diagnosis and treatment of disorders of newborn
  3 Many cardiovascular changes occur at time of birth; fetus totally dependent on mother, whereas newborn must immediately become totally self-supporting (respiration and circulation)
  4 Respiratory changes at birth include a deep and forceful first breath to empty lungs of amniotic fluid
  5 Developmental changes between the neonatal period and 18 months include:
    a Doubling of birth weight—4 months; tripled by 1 year
    b 50% increase in body length—12 months
    c Development of normal spinal curvature—15 months (Figure 18-10)
    d Ability to raise head—3 months
    e Ability to crawl—10 months
    f Ability to stand alone—12 months
    g Ability to run—18 months
F Childhood
  1 Extends from end of infancy to puberty—13 years in girls; 15 years in boys
  2 Overall rate of growth remains rapid but is decelerating
  3 Continuing development of motor and coordination skills
  4 Loss of deciduous or baby teeth and eruption of permanent teeth occurs
G Adolescence and adulthood
  1 Average age range of adolescence will vary generally 13 to 19 years
  2 Period of rapid growth resulting in sexual maturity (adolescence)
  3 Appearance of secondary sexual characteristics regulated by secretion of sex hormones
  4 Growth spurt typical of adolescence; begins in girls at about 10 and in boys at about 12
  5 Growth plates fully close in adult; other structures such as the sinuses acquire adult placement (Figure 18-11)
  6 Adulthood characterized by maintenance of existing body tissues
  7 Actual degeneration of body tissue begins in adulthood
H Old age
  1 Degenerative changes characterize old age or sensecence
  2 Every organ system of the body undergoes degenerative changes
  3 Senescence culminates in death

Degenerative Changes in Old Age

A Skeletal system (Figures 18-13 and 18-14)
  1 Aging causes changes in texture, calcification, and shape of bones

2 Bone spurs develop around joints

3 Bones become porous and fracture easily

4 Degenerative joint diseases such as rheumatoid arthritis are common

B Integumentary system (skin)

1 With age skin "sags" and becomes:

a Thin

b Dry

c Wrinkled (Figure 18-15)

2 Pigmentation problems are common (Figure 18-16)

3 Frequent thinning or loss of hair occurs

C Urinary system

1 Nephron units decrease in number by 50% between ages 30 and 75

2 Blood flow to kidney decreases and therefore ability to form urine also decreases

3 Bladder problems such as inability to void completely are caused by muscle wasting in the bladder wall

D Respiratory system

1 Calcification of costal cartilages causes rib cage to remain in expanded position—barrel chest (Figure 18-17)

2 Wasting of respiratory muscles decreases respiratory efficiency

3 Respiratory membrane thickens; movement of oxygen from alveoli to blood is slowed

E Cardiovascular system

1 Degenerative heart and blood vessel disease are among most common and serious effects of aging

2 Fat deposits in blood vessels (atherosclerosis) decreases blood flow to heart and may cause complete blockage of coronary arteries (heart attack: Figure 18-18)

3 Hardening of arteries (arteriosclerosis) may result in rupture of blood vessels, especially in the brain (stroke)

4 Hypertension or high blood pressure is common in old age

F Special senses

1 All sense organs show a gradual decline in performance with age

2 Eye lenses become hard and cannot accommodate for near vision; result is farsightedness in many people by age 45 (presbyopia or "old eye")

3 Loss of transparency of lens or cornea is common in old age (cataract)

4 Glaucoma (increase in pressure in eyeball) is often the cause of blindness in old age

5 Loss of hair cells in inner ear produces frequency deafness in many older people

6 Decreased transmission of sound waves caused by loss of elasticity of eardrum and fixing of the bony ear ossicles is common in old age

7 Some degree of hearing impairment is universally present in the aged

8 Only about 40% of the taste buds present at age 30 remain at age 75

NEW WORDS

arteriosclerosis	embryology	neonate	presbyopia
atherosclerosis	fertilization	neonatology	primary germ layers
barrel chest	glaucoma	organogenesis	rheumatoid arthritis
blastocyst	histogenesis	placenta	zygote
cataract	implantation	postnatal	
chorion	morula	prenatal	

CHAPTER TEST

1. The prenatal period begins at _____ and ends at _____ .

2. The postnatal period begins at _____ and continues until _____ .

3. The science of the development of the individual before birth is called _____ .

4. The fertilized ovum is also called a _____ .

5. The fluid-filled, shock-absorbing sac in which the embryo will float during development is called the _____ cavity.

6. The "bridge" that permits exchange of nutrients between mother and baby before birth is called the _____ .

7. How tissues arrange themselves into organs during development is called _____ .

8. A full term pregnancy lasts about _____ days.

9. Infancy, childhood, adolescence, and adulthood are considered _____ periods.

10. The first 4 weeks of infancy are often referred to as the _____ period.

11. The developmental period that extends from the end of infancy to sexual maturity is called _____ .

12. The scientific name for old age is _____ .

13. Hardening of the arteries is called _____ .

14. If the lens of the eye becomes cloudy and impairs vision the condition is called a _____ .

15. The most serious of the age-related eye disorders is called _____ .

Circle the T before each true statement and the F before each false statement.

T F   16. The prenatal stage of development begins at birth.

T F   17. The fertilized ovum is called the morula.

T F   18. There are 5 primary germ layers present in the developing embryo.

T F   19. In humans all organ systems are formed and functioning by the fourth month of pregnancy.

T F   20. The placenta develops from the fetal membrane called the chorion.

T F   21. The prenatal period in humans lasts about 280 days.

T F   22. Neonatology is the medical and nursing specialty concerned with the diagnosis and treatment of disorders of the newborn.

T F   23. Recognizable fetal movment or "quickening" usually begins about the fourth week of pregnancy.

T F   24. The deciduous teeth usually do not appear until about 1 year of age.

T F   25. Childhood extends from the end of infancy to sexual maturity.

T F   26. The term *senescence* refers to old age.

T F   27. Rheumatoid arthritis is not considered a degenerative joint disease.

T F   28. The kidney is unique in that it does not show degenerative changes with advancing age.

T F   29. The term arteriosclerosis refers to "hardening of the arteries."

T F   30. Cataract is considered the most serious of the age-related eye disorders.

REVIEW QUESTIONS

1 What biological event separates the prenatal from the postnatal period of development?
2 Is it better to think of growth and development as processes or as isolated events in life? Why?
3 Define the following terms: embryology, gestation, fertilization, implantation, morula, blastocyst.
4 What is the difference between an ovum and a zygote?
5 Identify two fetal structures that will develop from the blastocyst after implantation.
6 Discuss the role of the placenta in fetal development.
7 What are the primary germ layers and how are they related to development of the fetus?
8 Briefly define the terms "histogenesis" and "organogenesis."
9 At what point in development of the fetus are all the organ systems formed and functioning?
10 When does the postnatal period begin and how long does it last?
11 List the four subdivisions of the postnatal period.

12 What is the relationship of the neonatal period to infancy? What is neonatology?
13 Why is it necessary for a baby's first breath to be deep and forceful? What is the primary stimulus that causes a newborn baby to take its first breath?
14 List five developmental changes that occur during the infancy period.
15 Is the period of childhood the same length for both males and females? Explain the reason for your answer.
16 What is the average age range of adolescence? Functionally, how would the reproductive system in an individual at the end of adolescence be described?
17 Describe the typical "spurt in height" that occurs during adolescence.
18 Define senescence.
19 Describe the degenerative changes that occur in the skeletal system as a result of aging.
20 How is the skin or integumentary system affected by aging?
21 What is the difference between atherosclerosis and arteriosclerosis?
22 Define the following terms: presbyopia, glaucoma, cataract.

UNIT SEVEN

Fluid, Electrolyte, and Acid-Base Balance

19 Fluid and Electrolyte Balance

OBJECTIVES

After you have completed this chapter, you should be able to:

1 Define the phrase *fluid and electrolyte balance*.

2 List, describe, and compare the body fluid compartments and their subdivisions.

3 Discuss avenues by which water enters and leaves the body and the mechanisms that maintain fluid balance.

4 Discuss the nature and importance of electrolytes in body fluids and explain the aldosterone mechanism of extracellular fluid (ECF) volume control.

5 Explain the interaction between capillary blood pressure and blood proteins and give examples of common fluid imbalances.

Have you ever wondered why sometimes you excrete great volumes of urine and sometimes almost none at all? Why sometimes you feel so thirsty that you can hardly get enough to drink and other times you want no liquids at all? These conditions and many more relate to one of the body's most important functions—that of maintaining its **fluid and electrolyte balance.**

The term *fluid balance* means several things. Of course, it means the same thing as homeostasis of fluids. To say that the body is in a state of fluid balance is to say that the total amount of water in the body is normal and that it remains relatively constant. Health and sometimes even survival itself depend on this complex function.

In this chapter you will find a discussion of body fluids and electrolytes, their normal values, the mechanisms that operate to keep them

normal, and some of the more common types of fluid and electrolyte imbalances.

BODY FLUIDS

If you are a healthy young person and you weigh 120 pounds, there is a good chance that out of the hundreds of compounds present in your body, one substance alone weighs about 72 pounds, or 60% of your total weight! This, the body's most abundant compound, is water. It occupies three main locations known as **fluid compartments.** Look now at Figure 19-1. Note that the largest volume of water by far lies inside cells and that it is called, appropriately **intracellular fluid (ICF).** Note, too, that the water outside of cells—**extracellular fluid (ECF)**—is located in two compartments: in the microscopic spaces between cells, where it is called **intersti-**

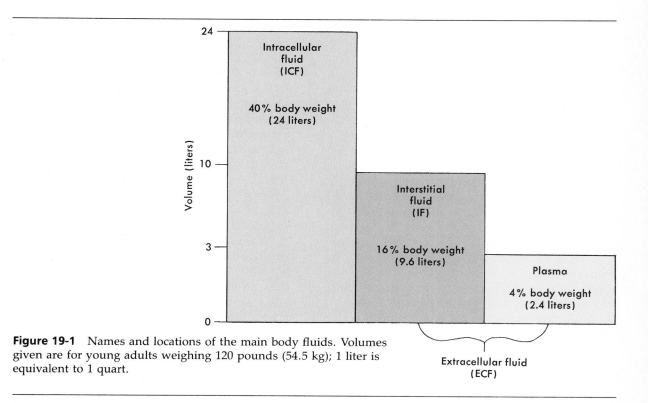

Figure 19-1 Names and locations of the main body fluids. Volumes given are for young adults weighing 120 pounds (54.5 kg); 1 liter is equivalent to 1 quart.

Figure 19-2 Proportion of body weight represented by water.

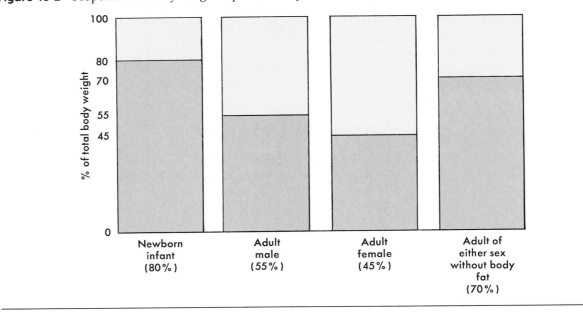

tial fluid (IF), and in the blood vessels, where it is called **plasma.** (Plasma is the liquid part of the blood, constituting a little less than half of the total blood volume; blood cells make up the rest of the volume.)

A normal body maintains fluid balance. The term *fluid balance* means that the volumes of ICF, IF, and plasma and the total volume of water in the body all remain relatively constant. Of course, not all bodies contain the same amount of water. The more a person weighs, the more water his body contains. This is true because, excluding fat or adipose tissue, about 70% of the body weight is water. Since fat is almost water free, the more fat present in the body, the less the total water content will be per unit of weight. In other words, fat people have a lower water content per pound of body weight than slender people. A slender adult body, for instance, typically consists of about 60% water. An obese body, in contrast, may consist of only 50% water or even less.

Sex and age also influence how much of the body's weight consists of water. Infants have more water in comparison to body weight than adults of either sex. In a newborn infant water may account for 80% of the total body weight. There is a rapid decline in the proportion of body water to body weight during the first year of life. Figure 19-2 illustrates the proportion of body weight represented by water in newborn infants (80%); adult males (55%); adult females (45%); and adults of either sex without body fat (70%). The female body contains slightly less water per pound of weight because it contains slightly more fat than the male body. Age and the body's water content are inversely related. In general, as age increases, the amount of water per pound of body weight decreases. Inversely related also (as explained in the preceding paragraph) are the amounts of fat and water in the body. In general, the more fat, the less water per pound of weight.

Table 19-1 Typical normal values for each portal of water entry and exit (with wide variations)

Intake		Output	
Ingested liquids	1,500 ml	Kidneys (urine)	1,400 ml
Water in foods	700 ml	Lungs (water in expired air)	350 ml
Water formed by catabolism	200 ml	Skin	
		By diffusion	350 ml
		By sweat	100 ml
		Intestines (in feces)	200 ml
TOTALS	2,400 ml		2,400 ml

MECHANISMS THAT MAINTAIN FLUID BALANCE

Under normal conditions, homeostasis of the total volume of water in the body is maintained or restored primarily by devices that adjust output (urine volume) to intake and secondarily by mechanisms that adjust fluid intake. There is no question about which of the two mechanisms is more important: the body's chief mechanism, by far, for maintaining fluid balance is to adjust its fluid output so that it equals its fluid intake.

Obviously, as long as output and intake are equal, the total amount of water in the body does not change. Figure 19-3 shows the three sources of fluid intake: the liquids we drink, the water in the foods we eat, and the water formed by catabolism of foods. Table 19-1 gives their normal volumes. However, these can vary a great deal and still be considered normal. Figure 19-3 also indicates that fluid output from the body occurs by way of four organs: the kidneys, lungs, skin, and intestines. The fluid output that changes the most is that from the kidneys. The body maintains fluid balance mainly by changing the volume of urine excreted to match changes in the volume of fluid intake. Everyone knows this from experience. The more liquid one drinks, the more urine one excretes. Conversely, the less the fluid intake, the less the urine volume. How changes in urine volume come about was discussed on p. 368. This would be a good time to review these paragraphs.

It is important to remember from your study of the urinary system that the rate of water and salt resorption by the renal tubules is the most important factor in determining urine volume. In other words, urine volume is regulated chiefly by hormones secreted by the posterior lobe of the pituitary gland (ADH) and the adrenal cortex (aldosterone).

Several factors act as mechanisms for controlling plasma, interstitial fluid, and intracellular fluid volumes. We shall limit our discussion to naming only three of these factors, stating their effects on fluid volumes, and giving some specific examples of these effects. Three of the main factors that influence extracellular and intracellular fluid volumes are:

1 The concentration of electrolytes in the extracellular fluid
2 The capillary blood pressure
3 The concentration of proteins in blood

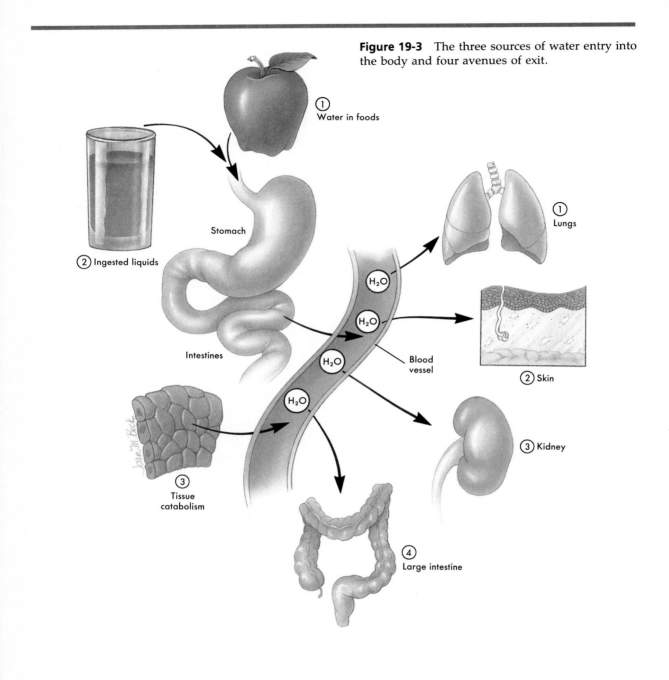

Figure 19-3 The three sources of water entry into the body and four avenues of exit.

Figure 19-4 A basic homeostatic mechanism for adjusting intake to compensate for excess output of body fluid.

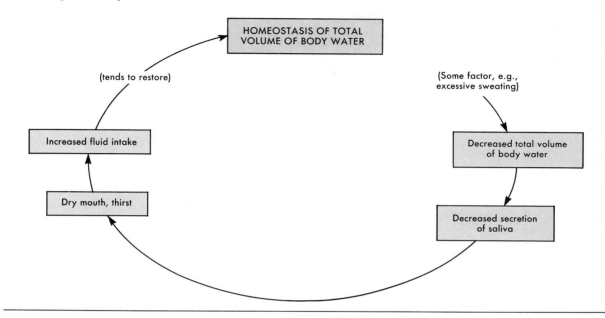

REGULATION OF FLUID INTAKE

Physiologists disagree about the details of the mechanism for controlling and regulating fluid intake to compensate for factors that would lead to dehydration. In general it would appear to operate in this way: when dehydration starts to develop—that is, when fluid loss from the body is exceeding fluid intake—salivary secretion decreases, producing a "dry-mouth feeling" and the sensation of thirst. The individual then drinks water, thereby increasing fluid intake and compensating for previous fluid losses. This tends to restore fluid balance (Figure 19-4). If an individual takes nothing by mouth for days, can his fluid output decrease to zero? The answer—no—becomes obvious after reviewing the information in Table 19-1. Despite every effort of homeostatic mechanisms to compensate for the zero intake, some output (loss) of fluid will occur as long as life continues. Water is continually lost from the body by way of expired air and diffusion through the skin.

IMPORTANCE OF ELECTROLYTES IN BODY FLUIDS

The bonds that hold the molecules of certain organic substances such as glucose together are such that they do not permit the compound to break up or **dissociate** in solution. Such compounds are called **nonelectrolytes.** Compounds such as ordinary table salt or sodium chloride (NaCl) that have molecular bonds that permit them to break up or dissociate in solution into separate particles (Na^+ and Cl^-) are known as **electrolytes.*** The dissociated particles of an electrolyte are called **ions** and carry an electrical charge. Positively charged particles such as Na^+ are called **cations** and negatively charged particles such as Cl^- are called **anions.** A variety of anions and cations serve important nutrient or regulatory roles in the body. Important cations

---

*Electrolytes are compounds that dissociate in solution to yield positively charged particles (cations) and negatively charged particles (anions).

Table 19-2 Electrolyte composition of blood plasma (mEq*)

Cations	Anions
142 mEq Na^+	102 mEq Cl^-
4 mEq K^+	26 mEq HCO_3^-
5 mEq Ca^{++}	17 mEq protein$^-$
2 mEq Mg^{++}	6 mEq other
	2 mEq HPO_4^-
153 mEq/L plasma	153 mEq/L plasma

*The milliequivalent (mEq) is a unit of measurement that indicates how reactive a particular electrolyte is in body fluids.

Diuretics

The word **diuretic** is from the Greek word *diouretikos* meaning "causing urine." By definition a diuretic drug is a substance that promotes or stimulates the production of urine. As a group, diuretics are among the most commonly used drugs in medicine. They are used because of their role in influencing water and electrolyte balance, especially sodium, in the body. Diuretics have their effect on tubular function in the nephron, and the differing types of diuretics are often classified according to their major site of action. Examples would include: (1) *proximal tubule diuretics* such as acetazolamide (Diamox), (2) *loop of Henle diuretics* such as ethacrynic acid (Edecrin) or furosemide (Lasix), and (3) *distal tubule diuretics* such as chlorothiazide (Diuril). Classification can also be made according to the effect the drug has on the level or concentration of sodium (Na^+), chloride (Cl^-), potassium (K^+), and bicarbonate ions in the tubular fluid. Using this classification, ethacrynic acid would be described as a diuretic that acts by inhibiting the resorption of chloride ions (in the loop of Henle). When chloride ion levels are increased in the tubular fluid, resorption of sodium ions is also blocked. The result is retention of NaCl, which must be excreted in the urine, carrying body water with it. Nursing implications for patients receiving diuretics include keeping a careful record of fluid intake and output and assessing the patient for signs and symptoms of electrolyte and water imbalance.

include sodium (Na^+), calcium (Ca^{++}), potassium (K^+), and magnesium (Mg^{++}). Important anions include chloride (Cl^-), bicarbonate (HCO_3^-), phosphate (HPO_4^-), and many proteins. Table 19-2 shows that although blood plasma contains a number of important electrolytes, by far the most abundant one is sodium chloride (ordinary table salt, $Na^+ Cl^-$). To remember how extracellular fluid electrolyte concentration affects fluid volumes, remember this one short sentence: where sodium goes, water soon follows. If, for example, the concentration of sodium in blood increases, the volume of blood soon increases. Conversely, if blood sodium concentration decreases, blood volume soon decreases.

The information in Figure 19-5 is presented in a "flow chart" diagram. Begin in the upper right of the diagram and follow, in sequence, each of the informational steps.

Relate the following facts to the information in Figure 19-6:

1 Overall fluid balance requires that fluid output equal fluid intake.
2 The type of fluid output that changes most is urine volume.
3 Renal tubule regulation of salt and water is the most important factor in determining urine volume.
4 Aldosterone controls salt resorption in the kidney.
5 The presence of sodium "obligates" water (where sodium goes, water soon follows).

Figure 19-5   Aldosterone mechanism that tends to restore normal extracellular fluid (ECF) volume when it decreases below normal. Excess aldosterone, however, leads to excess extracellular fluid volume—that is, excess blood volume (hypervolemia) and excess interstitial fluid volume (edema)—and also to an excess of the total Na$^+$ content of the body.

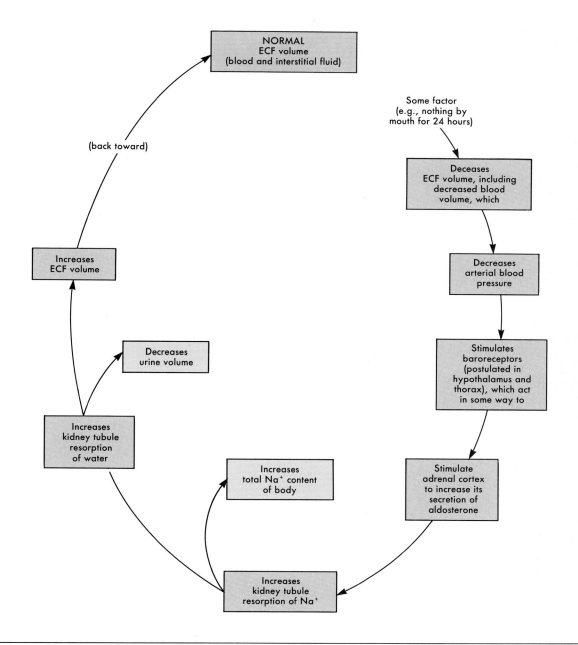

Figure 19-6 Sodium-containing internal secretions. The total volume of these secretions may reach 8,000 or more milliliters in a 24-hour period.

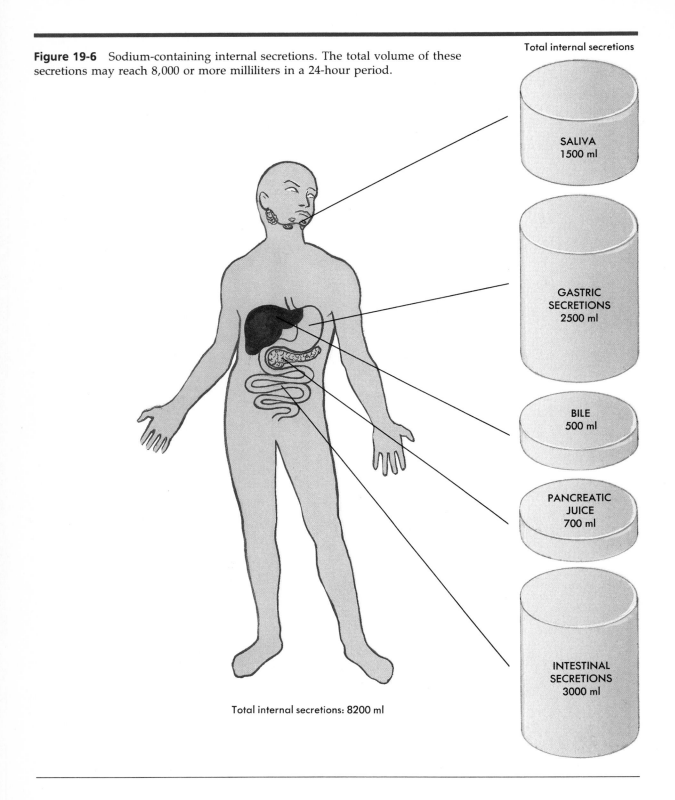

Total internal secretions

SALIVA
1500 ml

GASTRIC
SECRETIONS
2500 ml

BILE
500 ml

PANCREATIC
JUICE
700 ml

INTESTINAL
SECRETIONS
3000 ml

Total internal secretions: 8200 ml

The flow chart diagram in Figure 19-5 explains, in a very concise and brief way, the aldosterone mechanism that helps to restore normal extracellular fluid (ECF) volume when it decreases below normal. Can you construct a similar diagram to show the effect of antidiuretic hormone (ADH) secretion on extracellular fluid volume?

Although wide variations are possible, the average daily diet contains about 100 mEq of sodium. In a healthy individual sodium excretion from the body by the kidney is about the same as intake. The kidney acts as the chief regulator of sodium levels in body fluids. It is important to know that many electrolytes such as sodium not only pass into and out of the body but also move back and forth between a number of body fluids during each 24-hour period. Figure 19-6 shows the large volumes of sodium containing internal secretions that are produced each day. During a 24-hour period, over 8 liters of fluid containing 1,000 to 1,300 mEq of sodium are poured into the digestive system. This sodium, along with most of that contained in the diet, is almost completely resorbed. Very little sodium is lost in the feces. Precise regulation and control of sodium levels are required for survival.

CAPILLARY BLOOD PRESSURE AND BLOOD PROTEINS

Capillary blood pressure is a "water-pushing" force. It tends to push fluid out of the blood that is in capillaries into the interstitial fluid. Therefore, if capillary blood pressure increases, more fluid is pushed—filtered—out of blood into the interstitial fluid. The effect of an increase in capillary blood pressure, then, is to transfer fluid from blood to interstitial fluid. In turn this fluid shift, as it is called, changes both blood and interstitial fluid volumes. It decreases blood volume by increasing interstitial fluid volume. If, on the other hand, capillary blood pressure decreases, less fluid filters out of blood into interstitial fluid.

Water continually moves in both directions through the membranous walls of capillaries. How much water moves out of capillary blood into interstitial fluid depends largely on capillary blood pressure, a water-pushing force. How much water moves in the opposite direction—

Edema

Edema may be defined as the presence of abnormally large amounts of fluid in the intercellular tissue spaces of the body. The condition is a classic example of fluid imbalance and may be caused by disturbances in any of the factors that govern the interchange between blood plasma and interstitial fluid compartments. Examples include: (1) **Retention of electrolytes (especially Na$^+$) in the extracellular fluid** as a result of increased aldosterone secretion or following serious renal disease such as acute glomerulonephritis. (2) **An increase in capillary blood pressure.** Normally fluid is drawn from the tissue spaces into the venous end of a tissue capillary because of the low venous pressure and the relatively high water-pulling force of the plasma proteins. This balance is upset by anything that will increase the capillary hydrostatic pressure. The generalized venous congestion of heart failure is the most common cause of widespread edema. In patients with this condition, blood cannot flow freely through the capillary beds, and therefore the pressure will increase until venous return of blood improves. (3) **A decrease in the concentration of plasma proteins** caused by "leakage" into the interstitial spaces of proteins normally retained in the blood. This may occur as a result of increased capillary permeability caused by infection, burns, or shock.

that is, into blood from interstitial fluid—depends largely on the concentration of proteins present in blood plasma. Plasma proteins act as a water-pulling or water-holding force. They tend to hold water in the blood and to pull it into the blood from interstitial fluid. If, for example, the concentration of proteins in blood decreases appreciably—as it does in some abnormal conditions—less water moves into blood from interstitial fluid. As a result, blood volume decreases and interstitial fluid volume increases. Of the three main body fluids, interstitial fluid volume varies the most. Plasma volume usually fluctuates only slightly and briefly. If a pronounced change in its volume occurs, adequate circulation cannot be maintained.

FLUID IMBALANCES

Fluid imbalances are common ailments. They take several forms and stem from a variety of causes, but they all share a common characteristic—that of abnormally low or abnormally high volumes of one or more body fluids. **Dehydration** is the fluid imbalance seen most often. In this potentially dangerous condition, interstitial fluid volume decreases first, but eventually, if treatment has not been given, intracellular fluid and plasma volumes also decrease below normal levels. Either too small a fluid intake or too large a fluid output causes dehydration. **Overhydration** can also occur but is much less common than dehydration. The grave danger of giving intravenous fluids too rapidly or in too large amounts is overhydration, which can put too heavy a burden on the heart.

OUTLINE SUMMARY

Body Fluids
A Major locations—outside of cells in interstitial spaces (IF) and in blood vessels (plasma) and inside cells (ICF)
B Percentage of body weight and volumes, assuming that weight is 120 pounds:

ECF (extracellular fluid)		
Plasma	4%	2.4 liters
IF (interstitial fluid)	16%	9.6 liters
ICF (intracellular fluid)	40%	24 liters
Total body fluid	60%	36 liters

C Variation in total body water related to:
   1 Total body weight—the more a person weighs, the more water his body contains
   2 Fat content of body—the more fat present, the less total water content per unit of weight (Fat is almost water free.)
   3 Sex—proportion of body weight represented by water is about 10% less in females (45%) than in males (55%)
   4 Age—in newborn infant water may account for 80% of total body weight —see Figure 19-2

Mechanisms that Maintain Fluid Balance
A Fluid output, mainly urine volume, adjusts to fluid intake; ADH from posterior pituitary gland acts to increase kidney tubule reabsorption of water from tubular urine into blood, thereby tending to increase ECF (and total body fluid) by decreasing urine volume
B ECF electrolyte concentration (mainly Na^+ concentration) influences ECF volume; an increase in ECF Na^+ tends to increase ECF volume by increasing osmosis (movement of water) out of ICF and by increasing ADH secretion, which decreases urine volume, and this, in turn, tends to increase ECF volume
C Capillary blood pressure tends to push water out of blood, into IF; blood protein concentration tends to pull water into blood from IF; hence these two forces regulate plasma and IF volume under usual conditions
D Importance of electrolytes in body fluids
   1 Nonelectrolytes—organic substances that do not break up or dissociate when placed in solution; for example, glucose

2 Electrolytes—compounds that break up or dissociate in solution into separate particles called ions, for example, ordinary table salt or sodium chloride

$$(NaCl \rightarrow Na^+ + Cl^-)$$

3 Ions—the dissociated particles of an electrolyte that carry an electrical charge, for example, sodium ion (Na^+)

4 Cations—positively charged ions, for example, potassium (K^+), sodium (Na^+)

5 Anions—negatively charged particles (ions), for example, chloride (Cl^-), bicarbonate (HCO_3^-)

6 Electrolyte composition of blood plasma—see Table 19-2

7 Sodium—most abundant and important plasma cation

  a Normal plasma level—142 mEq/L

  b Average daily intake (diet)—100 mEq

  c Chief method of regulation—kidney

  d Sodium containing internal secretions—see Figure 19-6

E Capillary blood pressure and blood proteins

Fluid Imbalances

A Dehydration—total volume of body fluids smaller than normal; IF volume shrinks first, and then if treatment is not given, ICF volume decreases, and finally, plasma volume; dehydration occurs whenever fluid output exceeds fluid intake for an extended period of time; various factors may cause this; for example, diarrhea, gastrointestinal drainage or suction, or hemorrhage may increase fluid output above intake level; no liquid or food intake may decrease fluid intake below output level

B Overhydration—total volume of body fluids larger than normal; IF volume expands first, and then if treatment is not given, ICF volume increases, and finally, plasma volume increases above normal; overhydration occurs whenever fluid intake exceeds fluid output; various factors may cause this; for example, giving excessive amounts of intravenous fluids or giving them too rapidly may increase intake above output; kidney failure or prolonged hypoventilation (depressed respirations) may decrease output below intake level

NEW WORDS

anions	electrolyte	intracellular fluid (ICF)
cations	extracellular fluid (ECF)	ions
dehydration	fluid balance	nonelectrolytes
dissociate	interstitial fluid (IF)	overhydration
diuretic		

CHAPTER TEST

1. The largest volume of water in the body is classified as _____ water.
2. Extracellular water can be subdivided into the fluid between cells, called the _____ fluid, and the fluid in blood, called _____ .
3. Substances that do not break up or dissociate in solution are called _____ .
4. The dissociated particles of an electrolyte in solution are called _____ .
5. Positively charged particles in an electrolyte solution are called _____ and negatively charged particles are called _____ .
6. The largest quantity of water leaving the body exits as _____ produced by the kidneys.
7. The most important cation of blood plasma is _____ and the most important anion is _____ .
8. Salt resorption in the kidney is regulated largely by the hormone _____ .
9. A drug that promotes or stimulates the production of urine is called a _____ .
10. The presence of abnormally large amounts of fluid in the intercellular tissue spaces of the body is called _____ .

Circle the "T" before each true statement and the "F" before each false statement.

T F   11. About 60% of total body weight is water.
T F   12. Blood plasma is one type of extracellular fluid.
T F   13. There is more interstitial fluid in the body than intracellular fluid.
T F   14. There is a rapid decline in the proportion of body water to body weight during the first year of life.
T F   15. The more fat in the body the more water per pound of weight.
T F   16. The chief mechanism for maintaining fluid balance is to adjust fluid output to equal fluid intake.
T F   17. Capillary blood pressure does not influence either extracellular or intracellular fluid volume.
T F   18. Water is continually lost from the body by way of expired air and diffusion through the skin.
T F   19. Electrolytes dissociate in solution to yield charged particles called ions.
T F   20. Edema is a serious side effect of diuretic drugs.

REVIEW QUESTIONS

1 Suppose a person who had never heard the term "fluid balance" were to ask you what it meant. How would you explain it briefly and simply?
2 Approximately what percentage of a slender adult's body weight consists of water?
3 The volume of blood plasma in a normal-sized adult weighs approximately what percentage of body weight?
4 The proportion of body weight represented by water is about 10% higher in males than in females. Why?
5 ICF makes up approximately what percentage of adult body weight?
6 Interstitial fluid makes up approximately what percentage of adult body weight?
7 To maintain fluid balance, does output usually change to match intake or does intake usually adjust to output?
8 Explain in words and by a diagram how ADH functions to maintain fluid balance.
9 Define the following terms:
    anion   ion
    cation   nonelectrolyte
    electrolyte

10 List the important anions and cations present in blood plasma.
11 Use the phrase, "where sodium goes, water soon follows," to explain how extracellular fluid electrolyte concentration affects fluid volumes.
12 List the sodium-containing internal secretions.
13 Explain by words or diagram how capillary blood pressure and blood protein concentration function to maintain fluid balance.
14 Suppose that an individual has suffered a hemorrhage and that, as a result, his capillary blood pressure has decreased below normal. What change would occur in blood and interstitial fluid volumes as a result of this decrease in capillary blood pressure?
15 Suppose that an individual has a type of kidney disease that allows plasma proteins to be lost in the urine and that, as a result, his plasma protein concentration decreases. How would this tend to change blood and interstitial fluid volumes?
16 If an individual becomes dehydrated, which fluid volume decreases first and which decreases last?

CHAPTER
20 Acid-Base Balance

CHAPTER OUTLINE

pH of body fluids

Mechanisms that control pH of body fluids
Buffers
Respiratory mechanism of pH control
Urinary mechanism of pH control

pH imbalances
Metabolic and respiratory disturbances

BOXED ESSAY

Acid-forming potential of foods

OBJECTIVES

After you have completed this chapter, you should be able to:

1 Discuss the concept of pH and define the phrase *acid-base balance.*

2 Define the term *buffer* and *buffer pair* and contrast strong and weak acids and bases.

3 Contrast the respiratory and urinary mechanisms of pH control.

4 Discuss compensatory mechanisms that may help return blood pH to near normal levels in cases of pH imbalances.

5 Compare and contrast metabolic and respiratory types of pH imbalances.

ne of the requirements for healthy survival is that the body maintain, or quickly restore, the **acid-base balance** of its fluids. Maintaining acid-base balance means keeping the concentration of hydrogen ions in body fluids relatively constant. This is a matter of vital importance. If hydrogen ion concentration veers away from normal even slightly, cellular chemical reactions cannot take place normally and survival is thereby threatened.

pH OF BODY FLUIDS

Water and all water solutions contain both **hydrogen ions (H^+)** and **hydroxyl ions (OH^-).** The term "pH" followed by a number indicates a fluid's hydrogen-ion concentration. More specifically, pH 7.0 means an equal concentration of hydrogen ions and hydroxyl ions. Therefore, pH 7.0 also means that a fluid is neutral in reaction, that is, neither acid nor alkaline (Figure 20-1). The pH of water, for example, is 7.0. A pH higher than 7.0 indicates an alkaline solution, that is, one with a lower concentration of hydrogen ions than hydroxyl ions. The more alkaline a solution, the higher its pH. A pH lower than 7.0 indicates an acid solution, that is, one with a higher hydrogen-ion concentration than hydroxyl-ion concentration. The higher the hydrogen-ion concentration the lower the pH and the more acid a solution is. With a pH of about 1.6, gastric juice is the most acid substance in the body. Saliva has a pH of 6.8, on the acid side. Normally, the pH of arterial blood is about 7.45 and the pH of venous blood is about 7.35. By applying the information given in the last few sentences, you can deduce the answers to the following questions. Is arterial blood slightly acid or slightly alkaline? Is venous blood slightly acid or slightly alkaline? Which is a more accurate statement—venous blood is more acid than arterial blood or venous blood is less alkaline than arterial blood?

Both arterial and venous blood are slightly alkaline because both have a pH slightly higher than 7.0. Venous blood, however, is less alkaline than arterial blood because venous blood's pH

Figure 20-1   The pH range. Note that as concentration of H^+ increases the solution becomes increasingly acidic and the pH value decreases. As OH^- concentration increases, the pH value also increases and the solution becomes more and more basic or alkaline. A pH of 7 is neutral; pH of 1 is very acid; pH of 13 is very basic or alkaline.

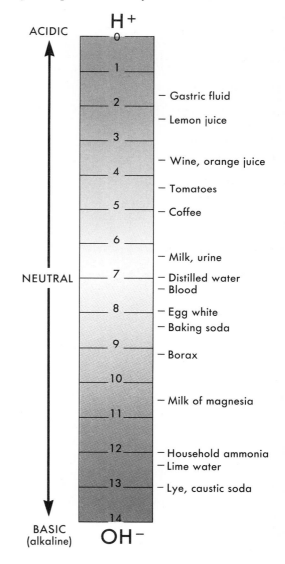

of about 7.35 is slightly lower than arterial blood's pH of 7.45.

MECHANISMS THAT CONTROL pH OF BODY FLUIDS

The body has three mechanisms for regulating the pH of its fluids. They are the buffer mechanism, the respiratory mechanism, and the urinary mechanism. Together they constitute the complex pH homeostatic mechanism—the machinery that normally keeps blood slightly alkaline with a pH that stays remarkably constant. Its usual limits are very narrow, about 7.35 to 7.45.

The slight increase in acidity of venous blood (pH 7.35) compared to arterial blood (pH 7.45) results primarily from carbon dioxide entering venous blood as a waste product of cellular metabolism. As carbon dioxide enters the blood, some of it combines with water and is converted into carbonic acid by **carbonic anhydrase,** an enzyme found in red blood cells:

$$CO_2 + H_2O \xrightarrow{\text{carbonic anhydrase}} H_2CO_3$$

The lungs remove the equivalent of over 20 liters of carbonic acid each day from the venous blood

by elimination of carbon dioxide. This almost unbelievable quantity of acid is so well buffered that a liter of venous blood contains only about $\frac{1}{100,000,000}$ gram more hydrogen ions than does 1 liter of arterial blood. What incredible constancy! The pH homeostatic mechanism does indeed control effectively—astonishingly so.

BUFFERS

Buffers are substances that prevent a sharp change in the pH of a fluid when an acid or base is added to it. Strong acids and bases, if added to blood, would "dissociate" almost completely and release large quantities of hydrogen (H^+) or hydroxyl (OH^-) ions. The result would be drastic changes in blood pH. Survival itself depends on protecting the body from such drastic pH changes. More acids than bases are usually added to body fluids. This is because catabolism, a process that goes on continually in every cell of the body, produces acids that enter blood as it flows through tissue capillaries. Almost immediately, one of the salts present in blood—a buffer, that is—reacts with these relatively strong acids to change them to weaker acids. The weaker acids decrease blood pH only slightly, whereas the stronger acids formed by catabolism would have decreased it greatly if they were not buffered.

Buffers consist of two kinds of substances and are therefore often called **buffer pairs.** One of the main blood "buffer pairs" is ordinary baking soda (sodium bicarbonate or $NaHCO_3$) and carbonic acid (H_2CO_3).

Let us consider, as a specific example of buffer action, how the sodium bicarbonate ($NaHCO_3$)–carbonic acid (H_2CO_3) system works in the presence of a strong acid or base.

Addition of a strong acid, such as hydrochloric acid (HCl), to the sodium bicarbonate–carbonic acid buffer system would initiate the reaction shown in Figure 20-2. Note how this reaction between HCl and sodium bicarbonate ($NaHCO_3$) applies the principle of buffering. As a result of the buffering action of $NaHCO_3$, the weak acid, $H \cdot HCO_3$, replaces the very strong acid, HCl, and therefore the hydrogen ion concentration of the blood increases much less than

Figure 20-2 Buffering of acid HCl by sodium bicarbonate. As a result of buffer action, the strong acid (HCl) is replaced by a weaker acid (H · HCO$_3$). Note that HCl as a strong acid "dissociates" almost completely and releases more hydrogen ions (H$^+$) than carbonic acid. Buffering decreases the number of hydrogen ions in the system.

$$HCl + NaHCO_3 \longrightarrow NaCl + H \cdot HCO_3$$
$$H^+ + Cl^- \qquad\qquad H^+ + HCO_3^-$$
$$\text{(many)} \qquad\qquad\qquad \text{(few)}$$

Figure 20-3 Buffering of base NaOH by carbonic acid. As a result of buffer action, the strong base (NaOH) is replaced by sodium bicarbonate and water. As a strong base, NaOH "dissociates" almost completely and releases large quantities of hydroxyl (OH$^-$) ions. Dissociation of water is minimal. Buffering decreases the number of hydroxyl ions in the system.

$$NaOH + H \cdot HCO_3 \longrightarrow NaHCO_3 + HOH$$
$$Na^+ + OH^- \qquad\qquad H^+ + OH^-$$
$$\text{(many)} \qquad\qquad\qquad \text{(very few)}$$

it would have if HCl were not buffered.

If, on the other hand, a strong base, such as sodium hydroxide (NaOH) were added to the same buffer system, the reaction shown in Figure 20-3 would take place. The hydrogen ion of carbonic acid (H · HCO$_3$), the weak acid of the buffer pair, combines with the hydroxyl ion (OH$^-$) of the strong base sodium hydroxide (NaOH) to form water. Note what this accomplishes. It decreases the number of hydroxyl ions added to the solution, and this in turn prevents the drastic rise in pH that would occur in the absence of buffering.

Figure 20-2 shows how a buffer system works in the presence of a strong acid. Although useful in demonstrating the principles of buffer action, HCl or similar strong acids are never introduced directly into body fluids under normal circumstances. Instead, the sodium bicarbonate buffer system is most often called on to buffer a number of weaker acids produced during catabolism. Lactic acid is a good example. As a weak acid, it does not "dissociate" as completely as HCl. Incomplete dissociation of lactic acid results in fewer hydrogen ions being added to the blood

and a less drastic lowering of blood pH than would occur if HCl were added in an equal amount. In the absence of buffering, however, lactic acid build-up will result in significant hydrogen ion accumulation over a period of time. The resulting decrease of pH can produce a serious acidosis. Ordinary baking soda (sodium bicarbonate or NaHCO$_3$) is one of the main buffers of the normally occurring "fixed" acids in blood. Lactic acid is one of the most abundant of the "fixed" acids, that is, acids that are not volatile and do not break down to form a gas. The following equation shows the compounds formed by the buffering of lactic acid (a "fixed" acid), produced by normal body catabolism:

$$H \cdot lactate + NaHCO_3 \rightarrow H \cdot HCO_3 + Na \cdot lactate$$
$$\text{Lactic acid} \qquad\qquad\qquad \text{Carbonic acid}$$

Figure 20-4 indicates the changes in blood that result from the buffering of fixed or nonvolatile acids in the tissue capillaries.

1 The amount of carbonic acid in blood increases slightly—because the nonvolatile acid (lactic acid in this case) is converted to volatile carbonic acid.

Figure 20-4 Lactic acid (H · lactate) and other nonvolatile or "fixed" acids are buffered by sodium bicarbonate in the blood. Carbonic acid (H · HCO$_3$ or H$_2$CO$_3$, a weaker acid than lactic acid) replaces lactic acid. As a result, fewer hydrogen ions are added to blood than would be if lactic acid were not buffered.

$$H \cdot lactate + NaHCO_3 \longrightarrow Na \cdot lactate + H \cdot HCO_3$$

$$\uparrow\downarrow \qquad\qquad\qquad\qquad\qquad\qquad \uparrow\downarrow$$

$$H^+ + lactate^- \qquad\qquad\qquad\qquad H^+ + HCO_3^-$$

(few) (fewer)

2 The amount of bicarbonate in blood (mainly sodium bicarbonate) decreases because bicarbonate ions become part of the newly formed carbonic acid. Normal arterial blood with a pH of 7.4 contains twenty times more sodium bicarbonate than carbonic acid. If this ratio decreases, blood pH decreases below 7.4.

3 The hydrogen ion concentration of blood increases slightly. Carbonic acid adds hydrogen ions to blood, but it adds fewer of them than lactic acid would have because carbonic acid is a weaker acid than lactic acid. In other words the buffering mechanisms does not totally prevent blood hydrogen-ion concentration from increasing. It simply minimizes the increase.

4 Blood pH decreases slightly because of the small increase in blood hydrogen ion concentration.

Carbonic acid is the most abundant acid in body fluids because it is formed by the buffering of fixed acids and also because carbon dioxide forms carbonic acid by combining with water. Large amounts of carbon dioxide, an end product of catabolism, continually pour into tissue capillary blood from cells. Much of the carbonic acid formed in blood diffuses into red blood cells where it is buffered by the potassium salt of hemoglobin. Carbonic acid is a volatile acid; it breaks down to form the gas, carbon dioxide, and water. This takes place in blood as it moves through the lung capillaries. Read the next paragraphs to find out how this affects blood pH.

RESPIRATORY MECHANISM OF pH CONTROL

Respirations play a vital part in controlling pH. With every expiration, carbon dioxide and water leave the body in the expired air. The carbon dioxide has diffused out of the venous blood as it moves through the lung capillaries. Less carbon dioxide therefore remains in the arterial blood leaving the lung capillaries, so less carbon dioxide is available for combining with water to form carbonic acid. Hence arterial blood contains less carbonic acid and fewer hydrogen ions and has a higher pH (7.45) than does venous blood (pH 7.35).

Let us consider now how a change in respirations can change blood pH. Suppose you were to pinch your nose shut and hold your breath for a full minute or a little longer. Obviously, no carbon dioxide would leave your body by way of the expired air during that time and the blood's carbon dioxide content would necessarily increase. This would increase the amount of carbonic acid and the hydrogen-ion concentration of blood, which in turn would decrease blood pH. Here then are two useful facts to remember. Anything that causes an appreciable decrease in respirations will in time produce **acidosis.** Conversely, anything that causes an excessive increase in respirations will in time produce **alkalosis** (see p. 449).

URINARY MECHANISM OF pH CONTROL

Most people know that the kidneys are vital organs and that life soon ebbs away if they stop functioning. One reason is that the kidneys are the body's most effective regulators of blood pH. They can eliminate much larger amounts of acid than can the lungs and, if it should become necessary, the kidneys can also excrete excess base. The lungs cannot. In short, the kidneys are the body's last and best defense against wide vari-

Figure 20-5 Acidification of urine and conservation of base by distal renal tubule excretion of H ions (see text).

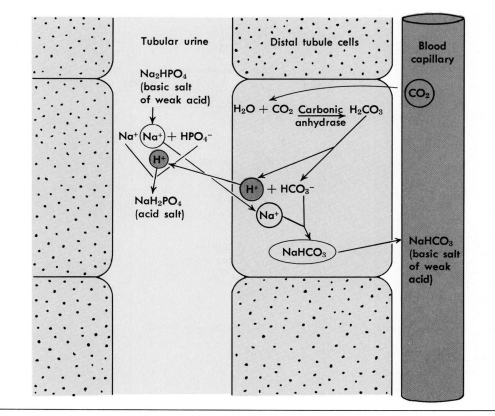

ations in blood pH. If they fail, homeostasis of pH—acid-base balance—fails.

Because more acids than bases usually enter blood, more acids than bases are usually excreted by the kidneys. In other words, most of the time the kidneys acidify urine; that is, they excrete enough acid to give urine an acid pH—frequently as low as 4.8. (How does this compare with normal blood pH?) The distal tubules of the kidneys rid the blood of excess acid and at the same time conserve the base present in it by the two mechanisms illustrated by Figures 20-5 and 20-6. To understand these figures, you need to know only basic chemistry. If you have this knowledge, look at Figure 20-5 and find the carbon dioxide leaving the blood (as it flows through a kidney capillary) and entering one of the cells that helps form the wall of a distal kidney tubule. Note that in this cell the carbon dioxide combines with water to form carbonic acid (H_2CO_3). This occurs rapidly because the cell contains carbonic anhydrase, an enzyme that accelerates this reaction. As soon as carbonic acid forms, some of it dissociates to yield hydrogen ions and bicarbonate ions. Note what happens to these ions. Hydrogen ions diffuse out of the tubule cell into the urine trickling down the tubule. Here it replaces one of the sodium ions in a salt (Na_2HPO_4) to form another salt (NaH_2PO_4), which leaves the body in the urine. Notice next that the Na^+ displaced from Na_2HPO_4 by the hydrogen ion moves out of the tubular urine into

Figure 20-6 Acidification of urine by tubule excretion of ammonia (NH_3). An amino acid (glutamine) leaves blood, enters a tubule cell, and is deaminized to form ammonia, which is excreted into urine. In exchange the tubule cell reabsorbs a basic salt (mainly $NaHCO_3$) into blood from urine.

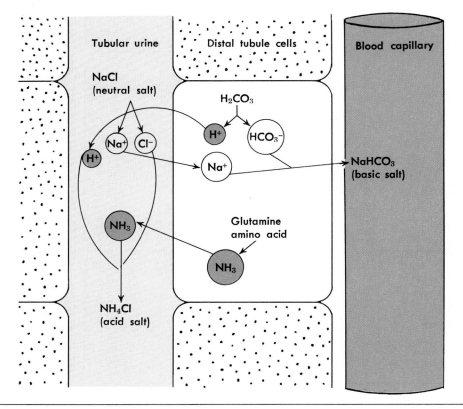

a tubular cell. Here it combines with a bicarbonate (HCO_3^-) ion to form sodium bicarbonate, which then is resorbed into the blood. What this complex of reactions has accomplished is to add hydrogen ions to the urine—that is, acidify it—and to conserve sodium bicarbonate by reabsorbing it into the blood.

Figure 20-6 illustrates another method of acidifying urine, as explained in the legend.

pH IMBALANCES

Acidosis and alkalosis are the two kinds of pH or acid-base imbalance. In acidosis the blood pH falls as hydrogen ion concentration increases. Only rarely does it fall as low as 7.0 (neutrality) and almost never does it become even slightly acid, as death usually intervenes before the pH drops this much. In alkalosis, which develops less often than acidosis, the blood pH is higher than normal.

From a clinical standpoint disturbances in acid-base balance can be considered dependent on the relative quantities (ratio) of carbonic acid and sodium bicarbonate present in the blood. Components of this important "buffer pair" must be maintained at the proper ratio (twenty times more sodium bicarbonate than carbonic acid) if acid-base balance is to remain normal.* It is fortunate that the body can regulate both chemicals in the sodium bicarbonate–carbonic acid buffer system. Blood levels of sodium bicarbonate can be regulated by the kidneys and carbonic acid levels by the respiratory system (lungs).

METABOLIC AND RESPIRATORY DISTURBANCES

Two types of disturbances, metabolic and respiratory, can alter the proper ratio of these components. Metabolic disturbances affect the bicarbonate element of the buffer pair, and respiratory disturbances affect the carbonic acid element, as follows:

1 **Metabolic disturbances**
   a *Metabolic acidosis* (bicarbonate deficit)
   b *Metabolic alkalosis* (bicarbonate excess)
2 **Respiratory disturbances**
   a *Respiratory acidosis* (carbonic acid excess)
   b *Respiratory alkalosis* (carbonic acid deficit)

Vomiting

Vomiting, sometimes referred to as emesis, is the forcible emptying or expulsion of gastric and occasionally intestinal contents through the mouth (p. 454). It occurs as a result of many stimuli, including foul odors or tastes, irritation of the stomach or intestinal mucosa, and some vomitive (emetic) drugs such as ipecac. A "vomiting center" in the brain regulates the many coordinated (but primarily involuntary) steps involved. Severe vomiting such as pernicious vomiting of pregnancy or the repeated vomiting associated with pyloric obstruction in infants can be life threatening. **One of the most frequent**

and serious complications of vomiting is metabolic alkalosis. The bicarbonate excess of metabolic alkalosis results because of the massive loss of chloride from the stomach as hydrochloric acid. It is the loss of chloride that causes a compensatory increase of bicarbonate in the extracellular fluid. The result is metabolic alkalosis. Therapy includes intravenous administration of chloride-containing solutions such as **normal saline.** The chloride ions of the solution replace bicarbonate ions and thus help relieve the bicarbonate excess responsible for the imbalance.

The *ratio* of sodium bicarbonate to carbonic acid levels in the blood is the key to acid base balance. If the normal ratio (20 to 1 sodium bicarbonate to carbonic acid) can be maintained, the acid-base balance and pH will remain normal despite changes in the absolute amounts of either component of the buffer pair in the blood.

As a clinical example, in a person suffering from untreated diabetes, abnormally large amounts of acids enter the blood. The normal 20 to 1 ratio of sodium bicarbonate to carbonic acid will be altered as the sodium bicarbonate component of the "buffer pair" reacts with the acids. Blood levels of sodium bicarbonate decrease rapidly in these patients. The result will be a lower ratio of sodium bicarbonate to carbonic acid (perhaps 10 to 1) and lower blood pH. The condition is called **uncompensated metabolic acidosis.** The body will attempt to correct or *compensate* for the acidosis by altering the *ratio* of $NaHCO_3$ to H_2CO_3. Acidosis in a diabetic patient is often accompanied by rapid breathing or hyperventilation. This compensatory action of the respiratory system results in a "blow-off" of carbon dioxide. Decreased blood levels of CO_2 result in lower carbonic acid levels. A new compensated ratio of sodium bicarbonate to carbonic acid (perhaps 10 to 0.5) may result. In such individuals the blood pH would return to normal or near normal levels. The condition is called **compensated metabolic acidosis.**

---

*Actually, in a state of acid-base balance a liter of plasma contains 27 mEq of sodium bicarbonate ($NaHCO_3$) or ordinary baking soda and 1.3 mEq of carbonic acid (H_2CO_3):

$$\frac{27 \text{ mEq NaHCO}_3}{1.3 \text{ mEq H}_2\text{CO}_3} = \frac{20}{1} = \text{pH } 7.4$$

The vomiting act.

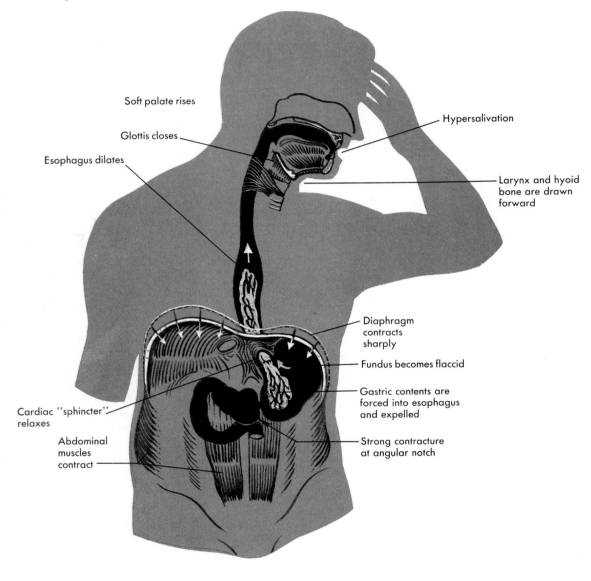

Soft palate rises

Glottis closes

Esophagus dilates

Hypersalivation

Larynx and hyoid bone are drawn forward

Diaphragm contracts sharply

Fundus becomes flaccid

Gastric contents are forced into esophagus and expelled

Cardiac "sphincter" relaxes

Abdominal muscles contract

Strong contracture at angular notch

OUTLINE SUMMARY

pH of Body Fluids

A Definition of pH—a number that indicates the hydrogen-ion concentration of a fluid; pH 7.0 indicates neutrality, pH higher than 7.0 indicates alkalinity, and pH less than 7.0 indicates acidity—see Figure 20-1

B Normal arterial blood pH—about 7.45

C Normal venous blood pH—about 7.35

Mechanisms that Control pH of Body Fluids

A Buffers

  1 Definition—buffers are substances that prevent a sharp change in the pH of a fluid when an acid or base is added to it—see Figures 20-2 and 20-3

  2 Nonvolatile acids are buffered mainly by sodium bicarbonate ($NaHCO_3$)

  3 Changes in blood produced by buffering of nonvolatile acids in the tissue capillaries

    a Amount of carbonic acid (H_2CO_3) in blood increases slightly

    b Amount of sodium bicarbonate in blood decreases; ratio of amount of $NaHCO_3$ to the amount of H_2CO_3 does not normally change; normal ratio is 20:1

    c Hydrogen ion concentration of blood increases slightly

    d Blood pH decreases slightly below arterial level

B Respiratory mechanism of pH control—respirations remove some of the carbon dioxide from blood; as blood flows through lung capillaries, the amount of carbonic acid in blood is decreased and thereby its hydrogen ion concentration is decreased, and this in turn increases blood pH from its venous to its arterial level

C Urinary mechanism of pH control—the body's most effective regulator of blood pH; kidneys usually acidify urine by the distal tubules excreting hydrogen ions and ammonia (NH) into the urine from blood in exchange for sodium bicarbonate being resorbed into the blood

pH Imbalances

A Acidosis and alkalosis are the two kinds of pH or acid-base imbalance

B Disturbances in acid-base balance depend on relative quantities of sodium bicarbonate and carbonic acid in the blood

C Body can regulate both components of the $NaHCO_3$–H_2CO_3 buffer system

  1 Blood levels of $NaHCO_2$ regulated by kidneys

  2 Carbonic acid levels regulated by lungs

D Two basic types of pH disturbances, metabolic and respiratory, can alter the normal 20 to 1 ratio of $NaHCO_3$ to H_2CO_3 in blood

  1 Metabolic disturbances affect the sodium bicarbonate levels in blood

  2 Respiratory disturbances affect the carbonic acid levels in blood

E Types of pH or acid-base imbalances

  1 Metabolic disturbances

    a Metabolic acidosis (bicarbonate deficit)

    b Metabolic alkalosis (bicarbonate excess)

  2 Respiratory disturbances

    a Respiratory acidosis (carbonic acid excess)

    b Respiratory alkalosis (carbonic acid deficit)

F Clinical example

  1 Uncompensated metabolic acidosis

  2 Compensated metabolic acidosis

NEW WORDS

acid solution
acidosis (metabolic and respiratory)
alkaline solution
alkalosis (metabolic and respiratory)
buffer

buffer pairs
carbonic anhydrase
emesis
pH

CHAPTER TEST

1. A fluid having a pH of 7.0 would be described as ——————— in reaction.
2. A pH higher than 7.0 indicates an ——————— solution and one with a pH lower than 7.0 indicates an ——————— solution.
3. As CO_2 enters blood some of it is converted into carbonic acid by an enzyme called ——————— ——————— found in red blood cells.
4. Substances that prevent a sharp change in pH of a fluid when an acid or base is added are called ——————— .
5. The most abundant acid in the body is ——————— acid.
6. Anything that causes an appreciable decrease in respirations will in time produce the condition called ——————— .
7. Acidification of urine can occur by renal tubule excretion of ——————— .
8. One of the most frequent complications of vomiting is metabolic ——————— .
9. A body deficit of sodium bicarbonate would result in metabolic ——————— .
10. The ability of the body to maintain a constant blood pH during changing conditions is an example of ——————— .

Select the most correct answer from Column B for each statement in Column A. (Only one answer is correct.)

Column A	Column B
11. _____ pH of arterial blood	a. Lactic acid
12. _____ neutral in reaction	b. pH 7.0
13. _____ lost as CO_2	c. Excretion of ammonia
14. _____ prevent pH changes	d. Buffers
15. _____ "fixed" acid	e. Metabolic alkalosis
16. _____ hydrogen-ion concentration	f. pH 7.45
17. _____ acidification of urine	g. Respiratory acidosis
18. _____ vomiting	h. pH
19. _____ carbonic acid excess	i. Hyperventilation
20. _____ compensates for acidosis	j. Carbonic acid

REVIEW QUESTIONS

1 Explain briefly what the term "pH" means.

2 What is the normal range for blood pH?

3 What is a typical normal pH for venous blood? for arterial blood?

4 When the body is in acid-base balance, arterial blood contains how many times more base bicarbonate (mainly $NaHCO_3$) than carbonic acid? In other words, what is the normal base bicarbonate/carbonic acid ratio in blood?

5 What function do buffers serve?

6 Explain how respirations affect blood pH.

7 Explain how the kidneys function to maintain normal blood pH.

8 How does prolonged hyperventilation (abnormally increased respirations) affect blood pH?

9 Which is the most important for maintaining acid-base balance—buffers, respirations, or kidney functioning?

10 Briefly, how does compensated acidosis differ from uncompensated acidosis? Define each of these terms briefly.

11 List the types of pH or acid-base imbalances.

Common Medical Abbreviations, Prefixes, and Suffixes

ABBREVIATIONS

āā of each
a.c. before meals
ad lib. as much as desired
alb. albumin
AM before noon
amt. amount
ante before
aq. water
Av. average
Ba barium
b.i.d. twice a day
b.m. bowel movement
BMR basal metabolic rate
BP blood pressure
BRP bathroom privileges
BUN blood urea nitrogen
c̄ with
CBC complete blood count
CCU coronary care unit
CHF congestive heart failure
CNS central nervous system
Co cobalt
CVA cerebrovascular accident, stroke
D&C dilation and curettage
d/c discontinue
DOA dead on arrival
Dx diagnosis
ECG electrocardiogram
EDC expected date of confinement
EEG electroencephalogram
EENT ear, eye, nose, throat
EKG electrocardiogram

ER emergency room
FUO fever of undetermined origin
GI gastrointestinal
GP general practitioner
GU genitourinary
h. hour
HCT hematocrit
Hgb hemoglobin
H₂O water
h.s. at bedtime
ICU intensive care unit
KUB kidney, ureter, and bladder
MI myocardial infarction
non rep. do not repeat
NPO nothing by mouth
OR operating room
p.c. after meals
per by
PH past history
PI previous illness
PM after noon
p.c. after meals
p.r.n. as needed
q. every
q.d. every day
q.h. every hour
q.i.d. four times a day
q.n.s. quantity not sufficient
q.o.d. every other day
q.s. quantity sufficient
RBC red blood cell
℞ prescription

s̄ without
sp. gr. specific gravity
s̄s̄. one half
stat. at once, immediately
T & A tonsillectomy and adenoidectomy
T.b. tuberculosis
t.i.d. three times a day
TPR temperature, pulse, respiration
TUR transurethral resection
WBC white blood cell

PREFIXES

a- without
ab- away from
ad- to, toward
adeno- glandular
amphi- on both sides
an- without
ante- before, forward
anti- against
bi- two, double, twice
circum- around, about
contra- opposite, against
de- away from, from
di- double
dia- across, through
dis- separate from, apart
dys- difficult
e- out, away
ecto- outside
en- in
endo- in, inside
epi- on
eu- well
ex- from, out of, away from
exo- outside
extra- outside, beyond; in addition
hemi- half
hyper- over, excessive, above
hypo- under, deficient
infra- underneath, below
inter- between, among
intra- within, on the side
intro- into, within
iso- equal, like
para- beside
peri- around, beyond
post- after, behind

pre- before, in front of
pro- before, in front of
re- again
retro- backward, back
semi- half
sub- under, beneath
super- above, over
supra- above, on upper side
syn- with, together
trans- across, beyond
ultra- excessive

SUFFIXES

-algia pain, painful
-asis condition
-blast young cell
-cele swelling
-centesis puncture for aspiration
-cide killer
-cyte cell
-ectomy cut out
-emia blood
-genesis production, development
-itis inflammation
-kinin motion, action
-logy study of
-megaly enlargement
-odynia pain
-oid resembling
-oma tumor
-osis condition
-opathy disease
-penia abnormal reduction
-pexy fixation
-phagia eating, swallowing
-phasia speaking condition
-phobia fear
-plasty plastic surgery
-plegia paralysis
-poiesis formation
-ptosis downward displacement
-rhaphy suture
-scope instrument for examination
-scopy examination
-stomy creation of an opening
-tomy incision
-uria urine

Chapter Test Answers

CHAPTER 1

1. anatomy; physiology
2. tissue
3. dorsal; ventral; lateral
4. lateral
5. anterior or front; posterior or back
6. systems
7. anatomical
8. axial, appendicular
9. homeostasis
10. leg
11. breast
12. chest
13. fingers or toes
14. wrist
15. groin
16. chest area over heart
17. buttock
18. skull
19. area between anus and genitals
20. sole of foot
21. a
22. d
23. b
24. c
25. c

CHAPTER 2

1. c
2. d
3. a
4. b
5. a
6. c
7. d
8. a
9. b
10. d
11. e
12. i
13. c
14. a
15. g
16. f
17. d
18. b
19. j
20. h
21. element
22. ribosomes
23. osmosis, dialysis
24. active
25. lyse
26. DNA
27. prophase
28. epithelial
29. neurons
30. epithelial

CHAPTER 3

1. cardiovascular; lymphatic
2. skin
3. bones
4. skeletal/voluntary; smooth/involuntary; cardiac
5. communication; integration; control
6. hormones
7. lymphatic
8. urine
9. accessory
10. alveoli
11. e
12. h
13. c
14. j
15. a
16. i
17. b
18. d
19. f
20. g

CHAPTER 4

1. skin
2. cutaneous
3. epidermis; dermis
4. keratin
5. melanin
6. dermal; papillae
7. root
8. pressure
9. eccrine; apocrine
10. sebaceous
11. acne
12. burns
13. epithelial; connective tissue
14. parietal; visceral
15. synovial
16. f
17. j
18. a
19. h
20. b
21. d
22. c
23. e
24. g
25. i

CHAPTER 5

1. hemopoiesis
2. trabeculae
3. Haversian systems
4. osteoblasts; osteoclasts
5. diaphysis
6. axial; appendicular
7. sinuses
8. scoliosis
9. cervical
10. scapula; clavicle
11. radius; ulna
12. olecranon
13. carpal; phalanges
14. ileum; ischium; pubis
15. patella; tibia
16. feet
17. osteoporosis
18. skull
19. sternum
20. diarthrotic
21. F
22. T
23. F
24. F
25. T
26. F
27. F
28. T
29. T
30. F
31. T
32. F
33. T
34. T
35. F

CHAPTER 6

1. b
2. a
3. d
4. a
5. c
6. b
7. b
8. c
9. d
10. a
11. striated; voluntary
12. smooth or involuntary
13. tendons
14. myosin
15. sarcomere
16. sliding-filament
17. synergists
18. antagonists
19. posture
20. oxygen debt
21. all-or-none
22. isotonic
23. plantar
24. atrophy
25. hamstrings
26. e
27. h
28. a
29. b
30. f
31. c
32. j
33. d
34. g
35. i

CHAPTER 7

1. central; peripheral
2. meninges
3. choroid plexus
4. hydrocephalus
5. neurons; neuroglia
6. oligodendroglia
7. myelin sheath
8. away; axon
9. synapse
10. saltatory conduction
11. neurotransmitters
12. dermatome
13. endorphins; enkephalins
14. brain stem
15. hypothalamus; thalamus
16. cerebral cortex
17. 12; 31
18. cerebellum
19. reflex arc
20. connective tissue
21. e
22. i
23. a
24. b
25. j
26. c
27. h
28. d
29. f
30. g

CHAPTER 8

1. cardiac muscle; smooth muscle; glandular epithelium
2. ganglia
3. effectors
4. sympathetic; parasympathetic
5. thoracolumbar
6. sympathetic
7. increases; decreases
8. "fight-or-flight"
9. acetylcholine
10. adrenergic
11. f
12. h
13. a
14. j
15. c
16. d
17. b
18. g
19. i
20. e

CHAPTER 9

1. hormones
2. exocrine
3. target
4. second messenger
5. negative
6. prostaglandins
7. hypersecretion
8. thyroid stimulating
9. nervous
10. anterior pituitary
11. antidiuretic
12. acromegaly
13. thyroxine
14. goiter
15. parathyroid
16. glomerulosa
17. medulla
18. Cushing's
19. glucagon
20. diabetes mellitus
21. progesterone
22. testosterone
23. thymosin
24. pineal
25. estrogen
26. T
27. F
28. T
29. T
30. F
31. T
32. F
33. T
34. T
35. T
36. F
37. F
38. F
39. T
40. F
41. T
42. T
43. T
44. T
45. F

CHAPTER 10

1. b
2. b
3. d
4. a

5. c
6. d
7. a
8. c

9. a
10. b
11. plasma
12. leukocytes

13. red bone marrow
14. polycythemia
15. hemoglobin
16. clot

17. embolus
18. antigen
19. AB
20. Rh

CHAPTER 11

1. cardiovascular; lymphatic
2. apex
3. cardiopulmonary resuscitation
4. atria; ventricles
5. myocardium
6. endocarditis

7. pericardium
8. atrioventricular
9. tricuspid
10. vena cava
11. pulmonary
12. angina pectoris
13. pacemaker
14. electrocardiogram

15. ventricles
16. veins
17. tunica media
18. systemic
19. capillary
20. foramen ovale
21. cardiac output

22. thoracic duct
23. afferent; efferent
24. leukocytes
25. clot
26. f
27. j
28. a

29. h
30. c
31. e
32. b
33. d
34. g
35. i

CHAPTER 12

1. antibodies; complement
2. immunoglobulins
3. antigens
4. epitopes
5. humoral
6. opsonins

7. complement fixation
8. interferon
9. monoclonal
10. lymphocytes
11. neutrophils; monocytes
12. lymphocytes

13. humoral
14. cell-mediated
15. AIDS
16. T
17. T
18. F

19. F
20. T
21. F
22. T
23. F
24. T

25. T
26. F
27. T
28. T
29. T
30. F

CHAPTER 13

1. mechanical; chemical
2. feces
3. gastroenterology
4. enteritis
5. accessory
6. lumen
7. mucosa
8. palates

9. crown
10. mastication
11. amylase
12. esophagus
13. chyme
14. peristalsis
15. jejunum
16. e

17. h
18. a
19. g
20. c
21. b
22. j
23. d
24. f

25. i
26. T
27. F
28. F
29. F
30. T
31. T
32. F

33. T
34. F
35. F
36. F
37. T
38. T
39. T
40. F

CHAPTER 14

1. distributor; exchanger
2. warms; filters; humidifies
3. diffusion
4. surfactant
5. lower

6. epistaxis
7. paranasal
8. oropharynx
9. nasopharynx
10. larynx

11. epiglottis
12. pneumothorax
13. external
14. oxyhemoglobin
15. vital capacity

16. f
17. j
18. a
19. h
20. c

21. b
22. g
23. d
24. e
25. i

CHAPTER 15

1. uremia
2. 10%
3. cortex; medulla
4. nephron
5. Bowman's capsule; glomerulus
6. proximal
7. filtration
8. osmosis
9. proximal
10. glycosuria
11. polyuria
12. pelvis
13. renal colic
14. urethra; meatus
15. calculi
16. catheterization
17. retention
18. incontinence
19. cystitis
20. 125
21. f
22. a
23. j
24. e
25. c
26. b
27. h
28. d
29. g
30. i

CHAPTER 16

1. testes
2. spermatozoa
3. gonads
4. penis
5. scrotum; penis
6. albuginea
7. seminiferous
8. spermatogenesis
9. interstitial
10. cryptorchidism
11. epididymis
12. vasectomy
13. prostate
14. Cowper's
15. alkaline
16. f
17. a
18. j
19. c
20. i
21. b
22. d
23. g
24. e
25. h

CHAPTER 17

1. ovaries
2. vulva
3. ova
4. follicle
5. graafian; luteum
6. zona pellucida
7. estrogen; progesterone
8. ectopic
9. cervix
10. progesterone
11. endometrium
12. hysterectomy
13. areola
14. menarche
15. 28
16. f
17. j
18. a
19. d
20. b
21. h
22. c
23. e
24. i
25. g

CHAPTER 18

1. conception; birth
2. birth; death
3. embryology
4. zygote
5. amniotic
6. placenta
7. organogenesis
8. 280
9. postnatal
10. neonatal
11. childhood
12. senescence
13. arteriosclerosis
14. cataract
15. glaucoma
16. F
17. F
18. F
19. T
20. T
21. T
22. T
23. F
24. F
25. T
26. T
27. F
28. F
29. T
30. F

CHAPTER 19

1. intracellular
2. interstitial; plasma
3. nonelectrolytes
4. ions
5. cations; anions
6. urine
7. sodium; chloride
8. aldosterone
9. diuretic
10. edema
11. T
12. T
13. F
14. T
15. F
16. T
17. F
18. T
19. T
20. F

CHAPTER 20

1. neutral
2. alkaline; acid
3. carbonic anhydrase
4. buffers
5. carbonic
6. acidosis
7. ammonia
8. alkalosis
9. acidosis
10. homeostasis
11. f
12. b
13. j
14. d
15. a
16. h
17. c
18. e
19. g
20. i

Glossary

abdomen (ab-dōmen; ab-do'men) body area between the diaphragm and pelvis.

abduct (ab-dukt') to move away from the midline; opposite of adduct.

absorption (ab-sorp'shun) passage of a substance through a membrane (for example, skin or mucosa) into blood.

acetabulum (as"e-tab'u-lum) socket in the hipbone (os coxa or innominate bone) into which the head of the femur fits.

acetylcholine (as-e-til-ko'-len) chemical neurotransmitter.

Achilles tendon (ah-kil'ēz ten'dun) tendon inserted on calcaneus; so called because of the Greek myth that Achilles' mother held him by the heels when she dipped him in the river Styx, thereby making him invulnerable except in this area.

acidosis (as"i-do'sis) condition in which there is an excessive proportion of acid in the blood.

acne (ak'-ne) inflammation and blockage of sebaceous glands.

acromegaly (ak"ro-meg'ah-le) condition caused by hypersecretion of growth hormone after puberty.

acromion (ah-kro'me-on) long projection of the scapula; forms point of the shoulder.

actin (ak-tin) contractile protein found in the thin myofilaments of skeletal muscle.

adduct (ah-dukt') to move toward the midline; opposite of abduct.

adenohypophysis (ad"ē-no-hi-pof'i-sis) anterior pituitary gland.

adenoid (ad'ē-noid) literally, glandlike; adenoids, or pharyngeal tonsils, are paired lymphoid structures in the nasopharynx.

adenosine triphosphate (ATP) (ah-den'-o-sen tri-fos'-fate) chemical compound that provides energy for use by body cells.

adipose (ad'-i-pose) fat tissue.

adolescence (ad"o-les'ens) period between puberty and adulthood.

adrenergic fibers (ad"ren-er'jik fi'bers) axons whose terminals release norepinephrine and epinephrine.

afferent neuron (af'er-ent nu'ron) transmitting impulses to the central nervous system.

AIDS acquired immune deficiency syndrome.

albuminuria (al"byu-mĭ-nu're-ah) albumin in the urine.

aldosterone (al-dos'te-rōn) hormone secreted by adrenal cortex.

alkalosis (al"kah-lo'sis) condition in which there is an excessive proportion of alkali in the blood; opposite of acidosis.

alveolus (al-ve'o-lus) literally, a small cavity; alveoli of lungs are microscopic saclike dilation of terminal bronchioles.

amenorrhea (ah-men"o-re'ah) absence of menses.

amino acid (a-mee'no as'id) organic compound having an NH_3 and a COOH group in its molecule; has both acid and basic properties; amino acids are the structural units from which proteins are built.

amphiarthrosis (am"fe-ar-thro'sis) slightly movable joint.

anabolism (ah-nab'o-lizm) synthesis by cells of complex compounds (for example, protoplasm and hormones) from simpler compounds (amino acids, simple sugars, fats, and minerals); opposite of catabolism, the other phase of metabolism.

anaphase (an'a-fāz) stage of mitosis; duplicate chromosomes move to poles of dividing cell.

anastomosis (ah-nas"to-mo'sis) connection between vessels; the circle of Willis, for example, is an anastomosis of certain cerebral arteries.

anatomy (ah-nat'-o-me) study of the structure of an organism and the relationship of its parts.

androgen (an'dro-jen) male sex hormone.

anemia (ah-ne'me-ah) deficient number of red blood cells or deficient hemoglobin.

anesthesia (anes-the'ze-ah) loss of sensation.

aneurysm (an'u-rizm) blood-filled saclike dilation of the wall of an artery.

angina (an-ji'nah) any disease characterized by spasmodic suffocative attacks; for example, angina pectoris and paroxysmal thoracic pain with feeling of suffocation.

Angstrom (ang'strum) unit 0.1 μm ($\frac{1}{10,000,000,000}$ of a meter or about 1/250,000,000 of an inch).

anions (an'-i-unz) negatively charged particles.

ankylosis (ang"ki-lo'sis) abnormal immobility of a joint.

anorexia (an"o-rek'se-ah) loss of appetite.

anoxia (an-ok'se-ah) deficient oxygen supply to tissues.

antagonistic muscles (an-tag'o-nis-tik mus'elz) those having opposing actions; for example, muscles that flex the upper arm are antagonists to muscles that extend it.

anterior (an-te′re-or) front or ventral; opposite of posterior or dorsal.

antibody, immune body (an′tĭ-bod″e, ĭ-myun′ bod′e) substance produced by the body that destroys or inactivates a specific substance (antigen) that has entered the body; for example, diphtheria antitoxin is the antibody against diphtheria toxin.

antigen (an′tĭ-jen) substance that, when introduced into the body, causes formation of antibodies against it.

antrum (an′trum) cavity; for example, the antrum of Highmore, the space in each maxillary bone, or the maxillary sinus.

anus (a′nus) distal end or outlet of the rectum.

appendicitis (ah-pen″-di-si′-tis) inflammation of the appendix.

appendicular (ap″-en-dik′-u-lar) refers to the upper and lower extremities of the body.

apex (a′peks) pointed end of a conical structure.

aphasia (ah-fa′ze-ah) loss of a language faculty, such as the ability to use words or understand them.

apnea (ap-ne′ah) temporary cessation of breathing.

aponeurosis (ap″o-nu-ro′sis) flat sheet of white fibrous tissue that serves as a muscle attachment.

arachnoid (ah-rak′noid) delicate, weblike middle membrane of the meninges.

areola (ah-re′o-lah) small space; the pigmented ring around the nipple.

areolar (ah-re′-o-lar) a type of connective tissue consisting of fibers and a variety of cells embedded in a loose matrix of soft, sticky gel.

arteriole (ar-te′re-ōl) small branch of an artery.

arteriosclerosis (ar-ter′i-o-skle-ro′sis) hardening of the arteries.

artery (ar′ter-e) vessel carrying blood away from the heart.

arthrosis (ar-thro′sis) joint or articulation.

articulation (ar-tik-u-la′shun) joint.

ascites (ah-si′tēz) accumulation of serous fluid in the abdominal cavity.

asphyxia (as-fik′se-ah) loss of consciousness caused by deficient oxygen supply.

aspirate (as′pŭ-rāt) to remove by suction.

ataxia (ah-tak′se-ah) loss of power of muscle coordination.

atherosclerosis (ath-e-ro″-skle-ro′-sis) hardening of arteries; lipid deposits in lining coat.

atrium (a′trē-um) chamber or cavity; for example, atrium of each side of the heart.

astrocyte (as′-tro-sit) a neuroglial cell.

atrophy (at′ro-fe) wasting away of tissue; decrease in size of a part.

auricle (aw′re-kl) part of the ear attached to the side of the head; earlike appendage of each atrium of heart.

autonomic (aw″to-nom′ik) self-governing; independent.

axial (ak′-se-al) refers to the head, neck, and torso or trunk of the body.

axilla (ak-sil′ah) armpit.

axon (ak′son) nerve cell process that transmits impulses away from the cell body.

Bartholin (bar′to-lin) seventeenth century Danish anatomist.

basophil (ba′so-fil) white blood cell that stains readily with basic dyes.

biceps (bi′seps) a muscle having two heads.

bilirubin (bil″e-roo′bin) red pigment in the bile.

biliverdin (bil″e-ver′din) green pigment in the bile.

blastocyst (blas′to-sist) post-morula stage of developing embryo; hollow ball of cells.

bolus (bo′-lus) a small rounded mass of masticated food to be swallowed.

Bowman (bo′man) nineteenth century English physician.

brachial (brak′e-al) pertaining to the arm.

bronchiectasis (brong″ke-ek′tah-sis) dilation of the bronchi.

bronchiole (brong′ke-ōl) small branch of a bronchus.

bronchus (brong′kus) one of the two branches of the trachea.

buccal (buk′al) pertaining to the cheek.

buffer (buf′er) compound that combines with an acid or with a base to form a weaker acid or base, thereby lessening the change in hydrogen-ion concentration that would occur without the buffer.

buffy coat (buff′e kōt) layer of white cells located between plasma and packed red cells in a hematocrit tube.

bursa (bur′sah) fluid-containing sac or pouch lined with synovial membrane.

bursitis (bur-si′tis) inflammation of a bursa.

buttock (but′ok) prominence over the gluteal muscles.

calculus (kal′kyu-lus) stone; formed in various parts of the body; may consist of different substances.

calorie (kal′o-re) heat unit; a large calorie is the amount of heat needed to raise the temperature of 1 kilogram of water 1 degree Celsius.

calyx (ka′liks) cup-shaped division of the renal pelvis.

canaliculi (kan″-ah-lik′-u-li) an extremely narrow tubular passage or channel in compact bone.

capillary (kap′i-la″re) microscopic blood vessel; capillaries connect arterioles with venules; also, microscopic lymphatic vessels.

carbaminohemoglobin (kar-bam″ĭ-no-he″mo-glo′bin) compound formed by union of carbon dioxide with hemoglobin.

carbohydrate (kar″bo-hi′drāt) organic compounds containing carbon, hydrogen, and oxygen in certain specific proportions; for example, sugars, starches, and cellulose.

carboxyhemoglobin (kar-bok″se-he″mo-glo′bin) compound formed by union of carbon monoxide with hemoglobin.

carcinoma (kar″sĭ-no′mah) cancer; a malignant tumor.

cardiopulmonary resuscitation (kar″de-o-pul′mo-ner-e re-sus″i-ta′shun) (CPR) combined cardiac (heart) massage and artificial respiration.

cardiovascular (kar″-de-o-vas′-ku-lar) pertaining to the heart and blood vessels.

caries (kar′ēz) decay of teeth or of bone.

carotid (kah-rot′id) from Greek word *karos*, meaning "deep sleep"; carotid arteries of the neck so called because pressure on them may produce unconsciousness.

carpal (kar'pal) pertaining to the wrist.

casein (ka'se-in) protein in milk.

cast (kast) mold; for example, formed in renal tubules.

castration (kas-tra'shun) removal of testes or ovaries.

catabolism (kah-tab'o-lism) breakdown of food compounds or protoplasm into simpler compounds; opposite of anabolism, the other phase of metabolism.

catalyst (kat'ah-list) substance that accelerates the rate of a chemical reaction.

cataract (kat'ah-rakt) opacity of the lens of the eye.

catecholamines (kat"e-kol-am'inz) norepinephrine and epinephrine.

catheterization (kath"e-ter-i-za'shun) passage of a flexible tube (catheter) into the bladder through the urethra for the withdrawal of urine (urinary catheterization).

cations (cat'i-unz) positively charged particles.

cavity (kav'i-te) hollow place or space in a tooth; dental caries.

cecum (se'kum) blind pouch; the pouch at the proximal end of the large intestine.

celiac (se'le-ak) pertaining to the abdomen.

cell (sel) the basic biological and structural unit of the body consisting of a nucleus surrounded by cytoplasm and enclosed by a membrane.

cellulose (sel'yu-lōs) polysaccharide, the main plant carbohydrate.

centimeter (sen'tĭ-me"ter) 1/100 of a meter, about ⅖ of an inch.

centrioles (sen'trĭ-ōlz) two dots seen with a light microscope in the centrosphere of a cell; active during mitosis.

centromere (sen'-tro-mer) a beadlike structure that attaches one chromatid to another during the prophase stage of mitosis.

cerumen (sĕ-roo'men) earwax.

cervicitis (ser"vi-si'tis) inflammation of the cervix.

cervix (ser'viks) neck; any necklike structure.

chemotactic factor chemical substance that attracts macrophages to "enemy" cells.

chiasm (ki'-azm) crossing; specifically, a crossing of the optic nerves; also **chiasma.**

cholecystectomy (ko"le-sis-tek'to-me) removal of the gallbladder.

cholesterol (ko-les'ter-ol) organic alcohol present in bile, blood, and various tissues.

cholinergic fibers (ko"lin-er'jik fi'bers) axons whose terminals release acetylcholine.

cholinesterase (ko"lin-es'ter-ās) enzyme; catalyzes breakdown of acetylcholine.

chondrocyte (kon'dro-sit) cartilage cell.

chromatid (kro'-mah-tid) a chromosome strand.

chromatin (kro'muh-tin) deep-staining substance in the nucleus of cells; divides into chromosomes during mitosis.

chromosome (kro'mo-sōm) one of the segments into which chromatin divides during mitosis; involved in transmitting hereditary characteristics.

chyle (kīl) milky fluid; the fat-containing lymph in the lymphatics of the intestine.

chyme (kīm) partially digested food mixture leaving the stomach.

cilia (sil"e-ah) hairlike projections of protoplasm.

circumcision (ser"kum-sizh'un) surgical removal of the foreskin or prepuce.

circadian (ser"kah-de'an) daily.

clone a family of many identical cells descended from a single "parent" cell.

cochlea (kok'le-ah) snail shell or structure of similar shape.

coenzyme (ko-en'zīn) nonprotein substance that activates an enzyme.

coitus (ko'i-tus) sexual intercourse.

collagen (kol'ah-jen) principle organic constituent of connective tissue.

colloid (kol'oid) dissolved particles with diameters of 1 to 100 millimicrons (1 millimicron equals about 1/25,000,000 of an inch).

colostrum (ko-los'trum) first milk secreted after childbirth.

columnar (kol'-um-nar) shape in which cells are higher than they are wide.

combining sites (com-bin'ing sīts) antigen-binding sites, antigen receptor regions on antibody molecule; shape of each combining site is complementary to shape of a specific antigen's epitopes.

comedo (kom'-e-do) inflamed plug that blocks sebaceous gland duct in acne.

complement (com'ple-ment) several inactive enzymes normally present in blood, which kill foreign cells by dissolving them.

compound (kom-pownd) substance composed of two or more elements.

concha (kong'kah) shell-shaped structure; for example, bony projections into the nasal cavity.

condyle (kon'dīl) rounded projection at the end of a bone.

congenital (kon-jen'ĭ-tal) present at birth.

contralateral (kon"trah-lat'er-al) on the opposite side.

coracoid (kor'ah-koid) like a raven's beak in form.

corium (ko're-um) true skin or derma.

coronal (ko-ro'nal) like a crown.

coronary (kor'o-na-re) encircling; in the form of a crown.

corpus (kor'pus) body.

corpuscle (kor'pus"l) very small body or particle.

cortex (kor'teks) outer part of an internal organ, for example, of the cerebrum and of the kidneys.

cortisol (kor'ti-sol) the chief hormone secreted by the adrenal cortex; hydrocortisone; compound F.

costal (kos'tal) pertaining to the ribs.

cranial (kra'ne-al) toward the head.

crenation, plasmolysis (kre-na'shun, plaz-mol'ĭ-sis) shriveling of a cell caused by water withdrawal.

cretinism (kre'tin-izm) dwarfism caused by hypofunction of the thyroid gland.

cribriform (krib'rĭ-form) sievelike.

cricoid (kri'koid) ring-shaped; a cartilage of this shape in the larynx.

cryptorchidism (krip-tor'ki-dism) undescended testis.

crystalloid (kris'tal-loid) dissolved particle less than 1 millimicron in diameter.

cuboidal (ku-boi'-dal) cell shape resembling a cube.

Cushing's syndrome (koosh'ingz sin'drom) condition

caused by hypersecretion of glucocorticoids from the adrenal cortex.

cutaneous (kyu-ta′ne-us) pertaining to the skin.

cuticle (ku′-te-kl) skin fold covering root of nail.

cyanosis (si″ah-no′sis) bluish appearance of the skin caused by deficient oxygenation of the blood.

cystitis (sis-ti′tis) inflammation of urinary bladder.

cytology (si-tol′o-je) study of cells.

cytoplasm (si′to-plasm″) the protoplasm of a cell exclusive of the nucleus.

deciduous (de-sid′yu-us) temporary; shedding at a certain stage of growth; for example, deciduous teeth.

decussation (de″kus-sa′shun) crossing over like an X.

defecation (def″e-ka′shun) elimination of waste matter from the intestines.

deglutition (de″gloo-tish′un) swallowing.

deltoid (del′toid) triangular; for example, deltoid muscle.

dehydration (de″hi-dra-shun) excessive loss of body water.

dendrite, dendron (den′dr-it, den′dron) branching or tree-like; a nerve cell process that transmits impulses toward the cell body.

dens (denz) tooth.

dentate (den′tāt) having toothlike projections.

dentine, dentin (den′tēn; den′tin) main part of a tooth, under the enamel.

dentition (den-tish′un) teething; also, number, shape, and arrangement of the teeth.

depilatories (de-pil′ah-toe-res) hair removers.

dermis, corium (der′mis, ko′re-um) true skin.

dextrose (deks′trōs) glucose, a monosaccharide, the principal blood sugar.

diabetes insipidus (di″ah-be′tez in-sip-i-dus) condition characterized by a large urine volume caused by deficiency of antidiuretic hormone (ADH).

diabetes mellitus (di″ah-be′tez mel″li-tus) condition characterized by a high blood glucose level caused by a deficiency of insulin.

dialysis (di-al′i-sis) separation of smaller (diffusible) particles from larger (nondiffusable) particles through a semipermeable membrane.

diaphragm (di′ah-fram) membrane or partition that separates one thing from another; the muscular partition between the thorax and abdomen; the midriff.

diaphysis (di-af′ĭ-sis) shaft of a long bone.

diarrhea (di-a-re′-a) defecation of liquid feces.

diarthrosis (di″ar-thro′sis) freely movable joint.

diastole (di-as′to-le) relaxation of the heart, interposed between its contractions; opposite of systole.

diastolic pressure (di″ah-stol′ik presh′ur) blood pressure in arteries during diastole (relaxation) of heart.

diencephalon (di″en-sef′ah-lon) "tween" brain; parts of the brain between the cerebral hemispheres and the mesencephalon or midbrain.

diffusion (dĭ-fyu′zhun) spreading; for example, scattering of dissolved particles.

digestion (di-jes′chun) conversion of food into assimilable compounds.

diplopia (dĭ-plo′pe-ah) double vision; seeing one object as two.

disaccharide (di-sak′ah-rīd) sugar formed by the union of two monosaccharides; contains twelve carbon atoms.

dissection (di-sek′-shun) cutting technique used to separate body parts for study.

distal (dis′tal) toward the end of a structure; opposite of proximal.

diuresis (di″u-re′sis) increased urine production.

dopamine (do′-pah-men) chemical neurotransmitter.

dorsal, posterior (dor′sal, pos-te′re-or) pertaining to the back; opposite of ventral; in humans posterior is dorsal.

dropsy (drop′se) accumulation of serous fluid in a body cavity or in tissues; edema.

dura mater (du′rah ma′ter) literally strong or hard mother; outermost layer of the meninges.

dyspnea (disp′ne-ah) difficult or labored breathing.

dystrophy (dis′tro-fe) faulty nutrition.

ectopic (ek-top′ik) displaced; not in the normal place; for example, extrauterine pregnancy.

edema (e-de′mah) excessive fluid in tissues; dropsy.

effector (ef-fek′tor) responding organ; for example, voluntary and involuntary muscle, the heart, and glands.

efferent (ef′er-ent) carrying from, as neurons that transmit impulses from the central nervous system to the periphery; opposite of afferent.

electrocardiogram (e-lek″tro-kar″de-o-gram) graphic record of heart's action potentials.

electroencephalogram (e-lek″tro-en-sef′ah-lo-gram) graphic record of brain's action potentials.

electrolyte (e-lek′tro-l-it) substance that ionizes in solution, rendering the solution capable of conducting an electric current.

electron (e-lek′tron) minute, negatively charged particle.

element (el′-e-ment) substance that cannot be broken down into two or more different substances.

elimination (e-lim″ĭ-na′shun) expulsion of wastes from the body.

embolism (em′bo-lizm) obstruction of a blood vessel by foreign matter carried in the bloodstream.

embolus (em′bo-lus) a blood clot or other substance (bubble of air) that is moving in the blood and may block a blood vessel.

embryo (em′bre-o) animal in early stages of intrauterine development; the human fetus the first 3 months after conception.

embryology (em″bre-ol′o-je) study of the development of an individual from conception to birth.

emesis (em′e-sis) vomiting.

emphysema (em″fĭ-se′mah) dilation of pulmonary alveoli.

empyema (em″pi-e′mah) pus in a cavity; for example, in the chest cavity.

encephalon (en-sef′ah-lon) brain.

endocrine (en′do-krin) secreting into the blood or tissue fluid rather than into a duct; opposite of exocrine.

endometritis (en″do-me-tri′tis) inflammation of the uterine lining or endometrium.

endoplasmic reticulum (en'do-plas'mic re-tic'u-lum) network tubules and vesicles in cytoplasm.

endorphin (en-dor'-fins) chemical in central nervous system that influences pain perception—a natural painkiller.

energy (en'er-je) capacity for doing work.

enkephalin (en-kef'-a-lin) peptide chemical in central nervous system that acts as a natural painkiller.

enteritis (en"ter-i'tis) inflammation of the intestines.

enteron (en'ter-on) intestine.

enzyme en'zīm) catalytic agent formed in living cells.

eosinophil, acidophil (e"o-sin'o-fil, a-sid'o-fil") white blood cell readily stained by eosin.

epidermis (ep"ĭ-der'mis) "false" skin; outermost layer of the skin.

epiglottis (ep'i-glot'is) lidlike cartilage overhanging entrance to larynx.

epinephrine (ep"ĭ-nef'rin) adrenaline; secretion of the adrenal medulla.

epiphyses (e-pif'ĭ-sēz) ends of a long bone.

epistaxis (ip-i-stak'-sis) nosebleed.

epitopes (ep'-i-topes) surface areas that interact with antibody combining sites.

erythrocyte (e-rith'ro-sīt) red blood cell.

ethmoid (eth'moid) sievelike.

eupnea (yoop-ne'ah) normal respiration.

Eustachio (yu-stah'ke-o) sixteenth century Italian anatomist.

eustachian canal (yu-sta'ke-an că-nal') tube from inside of ear to throat to equalize air pressure.

exocrine (ek'so-krin) secreting into a duct; opposite of endocrine.

exophthalmos (ek"sof-thal'mos) abnormal protrusion of the eyes.

extension (ek-sten'shun) increasing the angle between two bones at a joint.

extrinsic (eks-trin'sik) coming from the outside; opposite of intrinsic.

fallopian tubes (fal-lo'pe-in toobs) pair of tubes that conduct ovum from ovary to uterus.

Fallopius (fal-lo'pe-us) sixteenth century Italian anatomist.

fascia (fash'e-ah) sheet of connective tissue.

fasciculus (fah-sik'yu-lus) little bundle.

fatigue (fah-teg') loss of muscle power—weakness.

feces (fe'sez) waste material discharged from intestines.

fetus (fe'tus) unborn young, especially in the later stages; in human beings, from third month of intrauterine period until birth.

fiber (fi'ber) threadlike structure.

fibrin (fi'brin) insoluble protein in clotted blood.

fibrinogen (fi-brin'o-jen) soluble blood protein that is converted to insoluble fibrin during clotting.

filtration (fil-tra'-shun) movement of water and solutes through a membrane by a higher hydrostatic pressure on one side.

fimbria (fim'bre-ah) fringe.

fissure (fish'ūr) groove.

flaccid (flak'sid) soft, limp.

flexion (flek'shun) act of bending; decreasing the angle between two bones at a joint.

fluosol (flu'o-sol) artificial blood.

follicle (fol'lĭ-k"l) small sac or gland.

fontanelles (fon"tah-nelz') "soft spots" of the infant's head; unossified areas in the infant skull.

foramen (fo-ra'men), plural **foramina** (fo-ram'in-ah) small opening.

fossa (fos'sah) cavity or hollow.

fovea (fo've-ah) small pit or depression.

fundus (fun'dus) base of a hollow organ; for example, the part farthest from its outlet.

ganglion (gang'gle-on) cluster of nerve cell bodies outside the central nervous system.

gasserian (gas-se're-an) named for Gasser, a sixteenth century Austrian surgeon; trigeminal ganglion.

gastric (gas'trik) pertaining to the stomach.

gastritis (gas-tri'tis) inflammation of the stomach.

gastroenterology (gas'tro-en"ter-ol'o-je) study of the stomach and intestines and their diseases.

gene (jēn) part of the chromosome that transmits a given hereditary trait.

genitals (gen'i-tlz) reproductive organs; genitalia.

gestation (jes-ta'shun) pregnancy.

gland (gland) secreting structure.

glaucoma (glaw-ko'-ma) disorder characterized by elevated pressure in the eye.

glomerulus (glo-mer'yu-lus) compact cluster; for example, of capillaries in the kidneys.

glossal (glos'al) of the tongue.

glucagon (gloo'kah-gon) hormone secreted by alpha cells of the islands of Langerhans.

glucocorticoids (gloo"ko-kor'tĭ-koidz) hormones that influence food metabolism; secreted by adrenal cortex.

gluconeogenesis (gloo"ko-ne"o-jen'e-sis) formulation of glucose or glycogen from protein or fat compounds.

glucose (gloo'kōs) monosaccharide or simple sugar; the principal blood sugar.

gluteal (gloo'te-al) of or near the buttocks.

glycerin, glycerol (glis'er-in, glis'er-ol) product of fat digestion.

glycogen (gli'ko-jen) polysaccharide; animal starch.

glycogenesis (gli"ko-jen'e-sis) formation of glycogen from glucose or from other monosaccharides, fructose or galactose.

glycogenolysis (gli"ko-jĕ-nol'ĭ-sis) hydrolysis of glycogen to glucose-6-phosphate or to glucose.

glyconeogenesis (gli"ko-ne"o-jen'e-sis) See **gluconeogenesis.**

goiter (goi'ter) enlargement of thyroid gland.

gonad (gon'ad) sex gland in which reproductive cells are formed.

gonorrhea (gon"o-re'ah) sexually transmitted disease caused by *Neisseria gonorrhoeae.*

graafian (graf'e-an) named for Graaf, a seventeenth century Dutch anatomist; pertaining to ovarian follicle.

gradient (gra'de-ent) a slope or difference between two levels; for example, blood pressure gradient—a difference between the blood pressure in two different vessels.

gustatory (gus'tah-to"re) pertaining to taste.

gyrus (ji"rus) convoluted ridge.

Haversian (ha-ver′shan) named for Havers, English anatomist of late seventeenth century; pertaining to small blood vessels in bone.

heart block (hart blok) blockage of impulse conduction from atria to ventricles so that heart beats at a slower rate than normal.

Heimlich maneuver lifesaving technique used to free the trachea of objects blocking the airway.

hematocrit (he-mat′o-krit) volume percent of blood cells in whole blood.

hemiplegia (hem″e-ple′je-ah) paralysis of one side of the body.

hemodialysis (he″mo-di-al′i-sis) use of dialysis to separate waste products from the blood.

hemoglobin (he″mo-glo′bin) iron-containing protein in red blood cells.

hemolysis (he-mol′ĭ-sis) destruction of red blood cells with escape of hemoglobin from them into surrounding medium.

hemopoiesis (he″mo-poi-e′sis) blood cell formation.

hemorrhage (hem′or-ij) bleeding.

hepar (he′par) liver.

heparin (hep′ah-rin) substances obtained from the liver; inhibits blood clotting.

heredity (he-red′ĭ-te) transmission of characteristics from a parent to a child.

hernia, "rupture" (her′ne-ah, rup′chur) protrusion of a loop of an organ through an abnormal opening.

hilus, hilum (hi′lus, hi′lum) depression where vessels enter an organ.

His (hiss) German anatomist of late nineteenth century.

histogenesis (his″to-jen′e-sis) formation of tissues from primary germ layers of embryo.

histology (his-tol′o-je) science of minute structure of tissues.

homeostasis (ho″me-o-sta′sis) relative uniformity of the normal body's internal environment.

hormone (hor′mōn) substance secreted by an endocrine gland.

hyaline (hi′ah-lin) glasslike.

hydrocephalus (hi-dro-sef′-ah-lus) abnormal accumulation of cerebrospinal fluid—"water on the brain."

hydrocortisone (hi″dro-kor′tĭ-sōn) a hormone secreted by the adrenal cortex; cortisol; compound F.

hydrolysis (hi-drol′ĭ-sis) literally "splitting by water"; chemical reaction in which a compound reacts with water to form simpler compounds.

hymen (hi′men) Greek for "membrane"; mucous membrane that may partially or entirely occlude the vaginal outlet.

hyoid (hi′oid) U-shaped; bone of this shape at the base of the tongue.

hyperemia (hi″per-e′me-ah) increased blood in a part of the body.

hyperkalemia (hi″per-kah-le′me-ah) higher than normal concentration of potassium in the blood.

hypernatremia (hi″per-na-tre′me-ah) higher than normal concentration of sodium in the blood.

hyperopia (hi″per-o′pe-ah) farsightedness.

hyperplasia (hi″per-pla′ze-ah) increase in the size of a part caused by an increase in the number of its cells.

hyperpnea (hi″-perp-ne′ah) abnormally rapid breathing; panting.

hypertension (hi″per-ten′shun) abnormally high blood pressure.

hyperthermia (hi″per-ther′me-ah) fever; body temperature above 37° C.

hypertonic (hi-per-ton′-ic) a solution containing a higher level of salt (NaCl) than is found in a living red blood cell (above 0.9% NaCl).

hypertrophy (hi-per′tro-fe) increased size of a part caused by an increase in the size of its cells.

hypoglycemia (hi″po-gli-se′me′ah) abnormally low blood glucose level.

hypokalemia (hi″po-kah-le′me-ah) lower than normal concentration of potassium in the blood.

hyponatremia (hi″po-na-tre′me-ah) lower than normal concentration of sodium in the blood.

hypophysis (hi-pof′ĭ-sis) Greek for "outgrowth"; hence the pituitary gland, which grows out from the undersurface of the brain.

hypothalamus (hi″po-thal′ah-mus) portion of the floor lateral wall of the third ventricle of the brain.

hypothermia (hi″po-ther′me-ah) subnormal body temperature below 37° C.

hypotonic (hi-po-ton′-ic) a solution containing a lower level of salt (NaCl) than is found in a living red blood cell (below 0.9% NaCl).

hypoxia (hi-pok′se-ah) oxygen deficiency.

hysterectomy (his″te-rek′to-me) surgical removal of the uterus.

incus (ing′kus) anvil; the middle ear bone that is shaped like an anvil.

inferior (in-fe′re-or) lower; opposite of superior.

inguinal (ing′gwĭ-nal) of the groin.

inhalation (in″hă-la′shun) inspiration or breathing in; opposite of exhalation or expiration.

inhibition (in″hĭ-bish′un) checking or restraining of action.

innominate (in-nom′ĭ-nāt) not named, anonymous; for example, ossa coxae (hipbones), formerly known as innominate bones.

insertion (in-ser′-shun) attachment of a muscle to the bone that it moves when contraction occurs (as distinguished from its origin).

insulin (in′su-lin) hormone secreted by islands of Langerhans in the pancreas.

integument (in-teg′-u-ment) refers to the skin.

intercellular (in″ter-sel′yu-lar) between cells; interstitial.

interferon (in″ter-fer′on) small proteins produced by immune system which inhibit virus multiplication.

internuncial (in″ter-nun′she-al) like a messenger between two parties; hence an internuncial neuron (or interneuron) is one that conducts impulses from one neuron to another.

interstitial (in″ter-stish′al) forming small spaces between things; intercellular.

intrinsic (in-trin′sik) not dependent on externals; located within something; opposite of extrinsic.

involuntary (in-vol′un-ter″e) not willed; opposite of voluntary.

involution (in″vo-lu′shun) return of an organ to its normal size after enlargement; also retrograde or degenerative change.

ion (i′on) electrically charged atom or group of atoms.

ipsilateral (ip″sĭ-lat′er-al) on the same side; opposite of contralateral.

IRDS infant respiratory distress syndrome.

irritability (ir″ĭ-tah-bil′ĭ-te) excitability; ability to react to a stimulus.

ischemia (is-ke′me-ah) local anemia; temporary lack of blood supply to an area.

ischium (is′ke-um) component part of the hip bone.

isometric (i″so-met′rik) type of muscle contraction in which muscle does not shorten.

isotonic (i″so-ton′ik) of the same tension or pressure.

jaundice (jawn′-dis) abnormal yellowing of skin, mucous membranes, and white of eyes.

keratin (ker′-ah-tin) protein substance found in hair, nails, outer skin cells, and horny tissues.

ketones (ke′tōnz) acids (acetoacetic, beta-hydroxybutyric, and acetone) produced during fat catabolism.

ketosis (ke-to′sis) excess amount of ketone bodies in the blood.

kilogram (kil′o-gram) 1,000 grams; approximately 2.2 pounds.

kinesthesia (kin″es-the′ze-ah) "muscle sense"; that is, sense of position and movement of body parts.

kyphosis (ki-fo′-sis) increased convexity in curvature of the thoracic spine—"hunchback."

labia (la′be-ah) lips.

lacrimal (lak′ri-mal) pertaining to tears.

lactation (lak-ta′shun) secretion of milk.

lactose (lak′tōs) milk sugar, a disaccharide.

lacuna (lah-kyu′nah) space or cavity; for example, lacunae in bone contain bone cells.

lamella (lah-mel′ah) thin layer, as of bone.

lateral (lat′er-al) of or toward the side; opposite of medial.

leukemia (lu-ke′me-ah) blood cancer characterized by an increase in white blood cells.

leukocyte (lu′ko-sīt) white blood cells.

leukocytosis (lu″ko-si-to′sis) abnormally high white blood cell numbers in the blood.

leukopenia (lu″ko-pe′ne′ah) abnormally low white blood cell numbers in the blood.

ligament (lig′ah-ment) bond or band connecting two objects; in anatomy a band of white fibrous tissue connecting bones.

lipid (lip′id) fats and fatlike compounds.

loin (loin) part of the back between the ribs and hipbones.

lordosis (lor-do′-sis) anterior concavity in curvature of lumbar spine—"swayback."

lumbar (lum′ber) of or near the loins.

lumen (loo′men), plural **lumina** (loo′mĭ-nah) passageway or space within a tubular structure.

lunula (lu′-nu-lah) crescent-shaped white area under proximal nail bed.

luteum (lu′te-um) golden yellow.

lymph (limf) watery fluid in the lymphatic vessels.

lymphocyte (lim′fo-sīt) one type of white blood cell.

lymphotoxin (lim-fo-tok′-sin) lethal substance released by T cells to kill "enemy" cells.

lyse (līz) disintegration of a cell.

lysosomes (li′so-sōmz) membranous organelles containing various enzymes that can dissolve most cellular compounds; hence called "digestive bags" or "suicide bags" of cells.

macrophage (mak′-ro-fage) phagocytic cells in the immune system.

malleolus (mal-le′o-lus) small hammer; projections at the distal ends of the tibia and fibula.

malleus (mal′e-us) hammer; the tiny middle ear bone that is shaped like a hammer.

Malpighi (mal-pig′e) seventeenth century Italian anatomist.

maltose (mawl′tōs) disaccharide or "double" sugar.

mammary (mam′er-e) pertaining to the breast.

manometer (mah-nom′e-ter) instrument used for measuring the pressure of fluids.

manubrium (mah-nu′bre-um) handle; upper part of the sternum.

mastication (mas″ti-ka′shun) chewing.

mastoiditis (mas″-toi-di′-tis) inflammation of air cells in mastoid bone of the skull.

matrix (ma′triks) ground substance in which cells are embedded.

meatus (me-a′tus) passageway.

medial (me′de-al) of or toward the middle; opposite of lateral.

mediastinum (me″de-as-ti′num) middle section of the thorax; that is, between the two lungs.

medulla (me-dul′lah) Latin for "marrow"; hence the inner portion of an organ in contrast to the outer portion, or cortex.

meiosis (mi-o′sis) nuclear division in which the number of chromosomes are reduced to half their original number before the cell divides in two.

melanin (mel′-ah-nin) brown skin pigment.

membrane (mem′brān) thin layer or sheet.

menarche (men-ar′kē) beginning of the menstrual function.

meninges (me-nin′-jez) fluid-containing membranes surrounding the brain and spinal cord.

menopause (men′o-pawz) termination of menstrual cycles.

menstruation (men″stroo-a′shun) monthly discharge of blood from the uterus.

mesentery (mes′en-ter″e) fold of peritoneum that attaches the intestine to the posterior abdominal wall.

mesial (me′ze-al) situated in the middle; median.

metabolism (mĕ-tab′o-lizm) complex process by which food is utilized by a living organism.

metacarpus (met″ah-kar′pus) "beyond" the wrist; hence the part of the hand between the wrist and fingers.

metaphase (met′-ah-faze) second stage of mitosis, during which nuclear membrane and nucleolus disappear.

metatarsus (met″ah-tar′sus) "beyond" the instep; hence the part of the foot between the tarsal bones and toes.

meter (mē'ter) about 39.5 inches.

microglia (mi-krog'le-ah) one type of connective tissue found in the brain and cord.

micron (mi'kron) 1/1,000 millimeter; 1/25,000 inch.

micturition (mik"tu-rish'un) urination, voiding.

midsagittal (mid-saj'-i-tal) a cut or plane that divides the body or any of its parts into two equal halves.

millimeter (mil'ĭme-ter) 1/1,000 meter; about 1/25 inch.

mineralocorticoids (min"er-al-o-kor'tĭ-koidz) hormones that influence mineral salt metabolism; secreted by adrenal cortex; aldosterone is the chief mineralocorticoid.

mitochondria (mi"to-kon'dre-ah) threadlike structures.

mitosis (mi-to'sis) indirect cell division involving complex changes in the nucleus.

mitral (mi'tral) shaped like a miter.

monocyte (mon-o-sit) nongranular white blood cell.

monosaccharide (mon"o-sak'ah-rīd) simple sugar; for example, glucose.

morula (mor'u-lah) a solid mass of cells formed by cleavage of a fertilized egg.

motor neurons (mo"to-nu'ronz) cells that transmit nerve impulses away from the brain or spinal cord; also called motor, or efferent, neurons.

muscular dystrophy (dis'-tro-fe) degenerative muscle disease characterized by weakness and atrophy.

myasthenia gravis (mi"as-the'ne-ah gra'vis) disease marked by progressive muscular weakness.

myelin (mi'ĕ-lin) lipoid substance found in the myelin sheath around some nerve fibers.

myeloid (mi'e-loid) pertaining to bone marrow.

myocardial infarction (mi"o-kar'de-al in-fark'shun) death of cardiac muscle cells resulting from inadequate blood supply as in coronary thrombosis.

myocarditis (mi"o-kar-di'tis) inflammation of the myocardium (heart muscle).

myocardium (mi"o-kar'de-um) muscle of the heart.

myofilament (mi"o-fil"ah-ment) ultramicroscopic threadlike structure found in myofibrils.

myopia (mi-o'pe-ah) nearsightedness.

myosin (my-o-sin) contractile protein found in the thick myofilaments of skeletal muscle.

myxedema (mik"se-de'mah) condition caused by deficiency of thyroid hormone in adult.

nares (na'rēz) nostrils.

neonatology (ne"o-na-tol'o-je) diagnosis and treatment of disorders of the newborn infant.

nephron (nef'ron) anatomical and functional unit of kidney, consisting of renal corpuscle and renal tubule.

nerve (nerv) collection of nerve fibers.

neurilemma, neurolemma (nu"rĭ-lem'mah) nerve sheath.

neuroglia (nu-rog'le-ah) fine-webbed supporting structures of nervous tissue.

neurohypophysis (nu"ro-hi-pof'ĭ-sis) posterior pituitary gland.

neuron (nu'ron) nerve cell, including its processes.

neutrophil (nu'tro-fil) white blood cell that stains readily with neutral dyes.

norepinephrine (nor-ep-e-nef'-rin) hormone secreted by adrenal medulla—released by sympathetic nervous system stimulation.

nucleus (nu'kle-us) spherical structure within a cell; a group of neuron cell bodies in the brain or cord.

occiput (ok'sĭ-put) back of the head.

olecranon (o-lek'rah-non) elbow.

olfactory (ol-fak'to-re) pertaining to the sense of smell.

oophorectomy (o'-of-o-rek'-to-me) surgical removal of the ovaries.

ophthalmic (of-thal'mik) pertaining to the eyes.

opsonin (op-so'-nin) antibody fractions that promote phagocytosis.

orchis (or'kis) the testis.

organ (or'gan) group of several tissue types that perform a special function.

organelle (or"gan-el') cell organ; one of the specialized parts of a single-celled organism (protozoon) serving for the performance of some individual function.

organogenesis (or"gah-no-jen'e-sis) formation of organs from primary germ layers of embryo.

origin (or'-i-jin) attachment of a muscle to the bone, which does not move when contraction occurs (as distinguished from its insertion).

os (ahs) Latin for "mouth" (plural **ora**) and for "bone" (plural **ossa**).

osmosis (oz-mo'sis) movement of a fluid through a semipermeable membrane.

ossicle (os'sĭ-k'l) little bone.

osteoclast (os'-te-o-klast) bone-absorbing cell.

osteocyte (os'te-o-sit) bone cell.

oxidation (ok"sĭ-da'shun) loss of hydrogen or electrons from a compound or element.

oxyhemoglobin (ok"se-he"mo-glo'bin) a compound formed by union of oxygen with hemoglobin.

palate (pal'ĭt) roof of the mouth.

palpebrae (pal'pe-bre) eyelids.

papilla (pah-pil'lah) small nipple-shaped elevation.

paralysis (pah-ral'ĭ-sis) loss of the power of motion or sensation, especially loss of voluntary motion.

parenchyma (par-eng'kĭ-mah) essential or functional tissue of an organ.

parietal (pah-ri'ĕtal) of the walls of an organ or cavity.

parotid (pah-rot'id) located near the ear.

parturition (par"tu-rish'un) act of giving birth.

patella (pah-tel'lah) small, shallow pan; the kneecap.

pectoral (pek'to-ral) pertaining to the chest or breast.

pelvis (pel'vis) basin or funnel-shaped structure.

pericardium (per"i-kar-de-um) membrane that surrounds the heart.

periosteum (per"-os'-te-um) tough, connective tissue covering bone.

peripheral (pĕ-rif'er-al) pertaining to an outside surface.

peristalsis (per-i-stal'-sis) contractions of stomach and intestines.

peritoneum (per"i-to-ne'um) saclike membrane lining the abdominopelvic walls.

pH (pe'āch') hydrogen ion concentration.

phagocytosis (fag"o-si-to'sis) ingestion and digestion of particles by a cell.

phalanges (fah-lan'jēz) finger or toe bones.

phrenic (fren'ik) pertaining to the diaphragm.

physiology (fiz"-e-ol'-o-je) study of body function.

pia mater (pi'ah ma'ter) Latin for "gentle mother"; the vascular innermost covering (meninges) of the brain and cord.

pineal (pin'e-al) shaped like a pinecone.

pinocytosis (pi-no-si-toe'-sis) active transport mechanism used to transfer fluids or dissolved substances into cells.

piriformis (pir"i-for'mis) pear-shaped.

pisiform (pi'sĭ-form) pea-shaped.

plantar (plan'tar) pertaining to the sole of the foot.

plasma (plaz'mah) liquid part of the blood.

plasmolysis (plaz-mol'ĭ-sis) shrinking of a cell caused by water loss by osmosis.

pleural (ploor'al) refers to the lungs.

pleurisy (ploor'i-se) inflammation of the pleura.

plexus (plek'sus) network.

pneumothorax (nu"mo-tho'raks) accumulation of air in the pleural space causing collapse of lung.

polycythemia (pol"e-si-the'me-ah) an excessive number of red blood cells.

polymorphonuclear (pol"e-mor'fo-nu'kle-ar) having many-shaped nuclei.

polysaccharide (pol"e-sak'ah-rīd) complex sugar.

pons (ponz) bridge.

popliteal (pop-lit'e-al) behind the knee.

posterior (pos-te're-or) located behind; opposite of anterior.

posture (pos'chur) position of the body.

presbyopia (pres"be-o'pe-ah) "oldsightedness"; farsightedness of old age.

pronate (pro'nāt) to turn palm downward.

prophase (pro'-faz) first stage of mitosis during which chromosomes become visible.

proprioceptors (pro"pri-o-sep'tors) receptors located in the muscles, tendons, and joints.

prostaglandins (pros"tah-glan'dins) a group of naturally occurring fatty acids that affect many body functions.

prothrombin (pro-throm'-bin) chemical required for blood clotting.

protoplasm (pro'to-plazm) living substance.

proximal (prok'sĭ-mal) next or nearest; located nearest the center of the body or the point of attachment of a structure.

psoas (so'iss) pertaining to the loin, the part of the back between the ribs and hipbones.

psychosomatic (si"ko-so-mat'ik) pertaining to the influence of the mind, notably the emotions, on body functions.

puberty (pyu'ber-te) age at which the reproductive organs become functional.

Purkinje system (pur'kin'jē) specialized fibers in the heart that conduct cardiac impulses into the walls of the ventricles.

quickening (kwik'en-ing) first recognizable movements of the fetus.

receptor (re-sep'tor) peripheral beginning of a sensory neuron's dendrite.

reflex (re'fleks) involuntary action.

refraction (re-frak'shun) bending of a ray of light as it passes from a medium of one density to one of a different density.

renal (re'nal) pertaining to the kidney.

residual volume (re-sid'u-al vol'um) air that remains in the lungs after the most forceful expiration.

reticular (re-tik'u-lar) netlike.

reticulum (re-tik'u-lum) a network.

ribosomes (ri'bo-sōmz) organelles in cytoplasm of cells; synthesize proteins, so nicknamed "protein factories."

rugae (roo'jee) wrinkles or folds.

sagittal (saj'ĭ-tal) like an arrow; longitudinal.

salpingitis (sal"pin-ji'tis) inflammation of the uterine (fallopian) tubes.

salpinx (sal'pinks) tube; oviduct.

sarcomere (sar'ko-mer) contractile unit of muscle; length of a myofibril between two Z bands.

sartorius (sar-to're-us) "tailor"; hence, one uses the thigh muscle to sit cross-legged like a tailor.

sciatic (si-at'ik) pertaining to the ischium.

sclera (skle'rah) from Greek for "hard"; white outer coat of eyeball.

scoliosis (sko"-le-o'-sis) abnormal lateral or side-to-side deviation or curvature of the spine.

scrotum (skro'tum) bag around testicles.

sebum (Latin for "tallow"); secretion of sebaceous glands.

sella turcica (sel'ah tur'sikah) Turkish saddle; saddle-shaped depression in the sphenoid bone.

semen (se'men) Latin for "seed"; male reproductive fluid.

semilunar (sem"e-lu'nar) half-moon–shaped.

senescence (se-nes'ens) old age.

serratus (ser-ra'tus) saw-toothed.

serum (se'rum) any watery animal fluid; clear, yellowish liquid that separates from a clot of blood.

sesamoid (ses'ah-moid) shaped like a sesame seed.

shingles (shing'-glz) an acute viral infection of the peripheral nervous system.

sigmoid (sig'moid) C-shaped; S-shaped.

sinus (si'nus) cavity.

soleus (so'le-us) pertaining to a sole; a muscle in the leg shaped like the sole of a shoe.

somatic (so-mat'ik) of the body framework or walls, as distinguished from the viscera or internal organs.

spermatogenesis (sper"mah-to-jen'e-sis) production of sperm cells.

spermatozoa (sper"mah-to-zo'ah) sperm cells.

spirometer (spi-rom'-e-ter) device used to measure the amount of air exchanged in breathing.

sphenoid (sfe'noid) wedge-shaped.

sphincter (sfingk'ter) ring-shaped muscle.

splanchnic (splank'nik) visceral.

sputum (spu'tum) respiratory mucus.

squamous (skwa'mus) scalelike.

stapes (sta'pēz) stirrup; tiny stirrup-shaped bone in the middle ear.

stimulus (stim'u-lus) agent that causes a change in the activity of a structure.

stress (stress) according to Selye, physiological stress is a condition in the body produced by all kinds of injurious factors that he calls "stressors" and manifested by a syndrome (a group of symptoms that occur together).

stressor (stres'sor) any injurious factor that produces biological stress; for example, emotional trauma, infections, severe exercise.

striated (stri'āt-id) marked with parallel lines.

sudoriferous (su"dor-if'er-us) secreting sweat.

sulcus (sul'kus) furrow or groove.

superficial (su"per-fish'al) near the body surface.

superior (su-pe're-or) higher, opposite of inferior.

supinate (su'pĭ-nāt) to turn the palm of the hand upward; opposite of pronate.

surfactant (ser-fak'tant) substance that lines alveolar sacs to reduce surface tension.

Sylvius (sil've-us) seventeenth century Dutch anatomist; also a sixteenth century French anatomist.

symphysis (sim'fĭ-sis) Greek for "growing together"; place where two bones fuse.

synapse (sin'aps) joining; point of contact between adjacent neurons.

synarthrosis (sin"ar-thro'sis) freely movable joint.

synergist (sin'er-jist) muscle that assists a prime mover.

synovia (sĭ-no've-ah) literally "with egg"; secretion of the synovial membrane; resembles egg white.

synthesis (sin'thĭ-sis) putting together of parts to form a more complex whole.

system (sis'tem) a group of differing organs arranged to perform a complex body function.

systole (sis'to-le) contraction of the heart muscle.

systolic pressure (sis'tol'ik pre'shur) blood pressure in arteries during systole (contraction of heart).

talus (ta'lus) ankle; one of the bones of the ankle.

target organ cell (tahrget or'gan sel) organ (cell) acted on by a particular hormone and responding to it.

tarsus (tahr'sus) instep.

telophase (tel'-o-faz) last stage of mitosis—cell divides.

tendon (ten'dun) band or cord of fibrous connective tissue that attaches a muscle to a bone or other structure.

tenosynovitis (ten-o-sin-o-vi'-tis) inflammation of a tendon sheath.

tetany (tet'ah-ne) continuous muscular contraction.

thoracentesis (tho-rah-sen-te'-sis) removal of excess pleural fluid using a tube inserted into the chest wall into the pleural space.

thorax (tho'raks) chest.

thrombin (throm'bin) protein important in blood clotting.

thrombocyte (throm'-bo-sit) also called platelets—play a role in blood clotting.

thrombosis (throm-bo'sis) formation of a clot in a blood vessel.

thrombus (throm'bus) stationary blood clot.

tibia (tib'e-ah) Latin for "shinbone."

tidal volume (tid'el vol'yum) amount of air breathed in and out with each breath.

tissue (tish'u) group of similar cells that perform a common function.

tonicity (to-nis'i-te) in fluid physiology is the effective osmotic pressure equivalent.

tonus (to'nus) continued, partial contraction of muscle.

tract (trakt) bundle of axons located within the central nervous system.

transverse (trans-vers') a horizontal or crosswise plane or cut.

trauma (traw'mah) injury.

treppe (trep'e) increase in extent of muscular contraction following rapidly repeated stimulation.

trigone (tri'gōn) triangular area on wall of urinary bladder.

trochlear (trok'li-ar) pertaining to a pulley.

trophic (trof'ik) having to do with nutrition.

turbinate (tur'bĭ-nāt) shaped like a cone or like a scroll or spiral.

tympanum (tim'pah-num) drum.

ulcer (ul'ser) a necrotic open sore or lesion.

umbilicus (um-bil'ĭ-kus) navel.

universal donor blood (yu-ne-versel do'nor blud) type O blood (contains neither A nor B antigens).

universal recipient blood (yu-ne-ver'sel re-sip'e-ent blud) type AB blood (contains neither anti-A nor anti-B antibodies).

uremia (u-re'me-ah) high levels of waste products in the blood.

urine (u'rin) fluid waste excreted by kidneys.

utricle (yu'tre-k'l) little sac.

uvula (yu'vyu-lah) Latin for "little grape"; a projection hanging from the soft palate.

vagina (vah-ji'nah) sheath; internal tube from uterus to vulva.

vagus (va'gus) Latin for "wandering."

valve (valv) structure that permits flow of a fluid in one direction only.

vas (vass) vessel or duct.

vasectomy (vah-sek'to-me) surgical removal of a portion of the vas deferens to induce sterility.

vastus (vas'tus) wide; of great size.

vein (vān) vessel carrying blood toward the heart.

ventral (ven'tral) of or near the belly; in humans, front or anterior; opposite of dorsal or posterior.

ventricle (ven'trĭk'l) small cavity.

vermiform (ver'mĭ-form) worm-shaped.

villus (vil'lus) hairlike projection.

viscera (vis'er-ah) internal organs (singular, viscus).

visceral (vis'-er-al) pertaining to an internal organ or viscus.

vital capacity (vi'tal kah-pas'i-te) largest amount of air that can be moved in and out of the lungs in one inspiration and expiration.

vomer (vo'mer) "plowshare"; unpaired flat bone of nasal septum.

vulva (vul'vah) external genitals of the female.

xiphoid (zif'oid) sword-shaped; pertaining to cartilaginous lower end of sternum.

zygoma (zi-go'mah) "bar," "bolt"; zygomatic bone in cheek.

zygote (zi-gōt) a fertilized ovum.

Index

t indicates table.

Joint(s)—cont'd
  elbow, 105, 106, 120
    bones of, 106
    bursae of, 130
  freely movable; *see* Diarthrosis
  functions of, 118
  gliding, 121
  hinge, 120, 121
  hip, 120
  kinds of, 118, 120-123
  knee, 120
    bones of, 112
    movement of, 134
  pivot, 120, 121, 123
  saddle, 121, 123
  shoulder, 120
  in skeletal system, 61
  sternoclavicular, 99
  synarthrotic, 118, 120
  between vertebrae, 123
Joint capsule, 118, 120
Jugular veins, 266
Juice(s)
  digestive, 37
  gastric, 312
  pancreatic, 318
Junction
  dermal-epidermal, 71
  neuromuscular, 137

K

Keratin, 69-70
  protective functions of, 77
Kidney
  acidification of urine and conserva-
      tion of base by, 451-452
  artificial, 367
  functions of, 62, 363-370
  structure of, 361-363
Kidney stones, 370
  removal of, using ultrasound, 371
Kinesthesia, 181
Knee cap; *see* Patella
Knee jerk; *see* Patellar reflex
Knee joint, 120
  bones of, 112
  movement of, 134
Knob, synaptic, 170, 171
Knuckles, 110
Krause's end-bulbs, 183t, 184
Kyphosis, 99, 102

L

Labia majora, 393, 394, 400
Labia minora, 393, 394, 400
Labyrinth, 187
Lacrimal bone, 100t, 102, 103
Lactase, 321, 322, 323
Lacteal, 317
Lactic acid
  buffering of, 449-450
  build-up of, 138
Lactiferous ducts, 399, 401
Lactogenic hormone; *see* Prolactin
Lactose, 322
Lacunae, 90, 91, 92

Lambdoidal suture, 102, 115t
Lamella, 91
Lamina cribrosa, 188
Langerhans, islands of, 223-224, 318
"Lantern jaw," 214
Lanugo, 72
Large intestine, 63, 318-320
Laryngeal mirror, 341
Laryngopharynx, 339
Larynx, 63, 333, 340-342
Lasix; *see* Furosemide
Latent period of twitch contraction,
      139
Lateral, 5
Lateral femoral cutaneous nerve, 158
Lateral longitudinal arch, 112, 118
Lateral malleolus, 117t
Lateral rectus muscle, 186
Latissimus dorsi, 136, 147t
  extension of, 143
  origin, insertion, and functions of,
      147t
Lecithin and IRDS, 334
Leg, 14t
  lower, flexion and extension of, 144
Length, metric units of, 21
Lens of eye, 182, 185, 186
Lesser curvature of stomach, 312, 313
Lesser pelvis, 117t
Lesser trochanter, 117t
Leukemia, 237-238
Leukocytes, 235, 236; *see also* White
      blood cells
  granular, 235, 236
  nongranular, 235, 236
Leukocytosis, 237
Leukopenia, 237
LH; *see* Luteinizing hormone
Lienal vein, 268
Life, cycle of, 378-429
Ligament(s), 120
  cruciate, 135
  in skeletal system, 61
  suspensory
    of Cooper, 399, 401
    of eye, 186
Light touch, receptors of, 74
Lines
  epiphyseal, 94
  Z, 131, 132, 133
Lingual tonsil, 341
Lipase
  gastric, 322, 323
  pancreatic, 322, 323
"Lipping" of bones, 420
Lithotriptor, 371
Liver, 63, 317-318
Liver glycogenolysis, 223
"Living matter"; *see* Cytoplasm
Lobe(s)
  of breast, 399
  ear, 187
  frontal, 177, 179
  occipital, 177, 179
  parietal, 177, 179

Lobe(s)—cont'd
  temporal, 177, 179
Lobule(s)
  of breast, 399, 401
  of tunica albuginea, 384
"Lockjaw"; *see* Tetanus
Loop of Henle, 363, 364, 366
  and urine formation, 370t
Long bone
  juvenile, 94
  structure of, 92-93, 94
Longitudinal arches, 117t
  medial and lateral, 112, 118
Longitudinal fissure, 178
Lordosis, 99, 102
Low blood pressure, 271
Lower abdominal regions, 9
Lower arm muscles, 147t
Lower extremity, 106, 110-112
  bones of, 101t
  muscles of, 147t
Lower respiratory tract, 334
Lumbar curvature in infancy, 417
Lumbar curve, 99, 104
Lumbar plexus, 158
Lumbar puncture, 172
Lumbar regions of abdomen, 9, 14t
Lumbar vertebrae, 98, 100t, 104
Lumbodorsal fascia, 136
Lumen of digestive tract, 303
Lunate bone, 108, 109
Lung(s), 63, 333, 335, 336, 338
  apex of, 344
  base of, 344
  cancer of, 351
  collapsed, 346
  exchange of gases in, 347, 348
  and pleura, 344-346
Lunula, 75
Lup-dup heart sound, 257
Luteinization, 209, 402
Luteinizing hormone, 209, 211
  functions of, 210
  and menstrual cycle, 402-404
Lymph, 62, 273
  drainage of, 274
    from breast, 276
  formation of, 273
Lymph ducts, 62
Lymph glands; *see* Lymph nodes
Lymph nodes, 62, 273-274, 277
  structure of, 275
  and white blood cell formation, 277
Lymph vessels, 274, 275
Lymphatic system, 273-277
Lymphatic subdivision of circulatory
      system, 58, 62, 253, 273-277
Lymphatic tissue, 235
Lymphatic vessels, 62, 273, 274
Lymphocytes, 235, 236, 240, 287-292
  functions of, 240
  types of, 288
Lymphokines, 291
Lymphotoxin, 291
Lysis of cells, 29
Lysosomes, 25

Illustration Credits

4-3; 4-7: Habif, T.P.: Clinical dermatology: a color guide to diagnosis and therapy, St. Louis, 1985, The C.V. Mosby Co.

5-3: Courtesy Dr. C.R. McMullen, Department of Biology, South Dakota State University.

5-7: Booher, J.M., and Thibodeau, G.A.: Athletic injury assessment, St. Louis, 1985, Times Mirror/Mosby College Publishing.

5-8, B; 5-9, B; 5-15; 5-16; p. 111: Vidić, B., and Suarez, F.R.: Photographic atlas of the human body, St. Louis, 1984, The C.V. Mosby Co.

5-22: Beck, E.W.: Mosby's atlas of functional human anatomy, St. Louis, 1982, The C.V. Mosby Co.

6-2: Malasanos, L., and others: Health assessment, ed. 3, St. Louis, 1986, The C.V. Mosby Co.

6-4: Courtesy Dr. H.E. Huxley; from Bevelander, G., and Ramaley, J.A.: Essentials of histology, ed. 8, St. Louis, 1979, The C.V. Mosby Co.

6-8, B: Raven, P.H., and Johnson, G.B.: Biology, St. Louis, 1986, Times Mirror/Mosby College Publishing.

6-9; 6-11; 6-12: Stacy, R.W., and Santolucito, J.A.: Modern college physiology, St. Louis, 1966, The C.V. Mosby Co.

7-1; 7-5; 7-9, A; 7-29: Beck, E.W.: Mosby's atlas of functional human anatomy, St. Louis, 1982, The C.V. Mosby Co.

7-9, B: Bevelander, G., and Ramaley, J.A.: Essentials of histology, ed. 8, St. Louis, 1979, The C.V. Mosby Co.

10-3: Bevelander, G., and Ramaley, J.A.: Essentials of histology, ed. 8, St. Louis, 1979, The C.V. Mosby Co.

11-2; 11-7, A: Beck, E.W.: Mosby's atlas of functional human anatomy, St. Louis, 1982, The C.V. Mosby Co.

11-4: Raven, P.H., and Johnson, G.B.: Biology, St. Louis, 1986, Times Mirror/Mosby College Publishing.

11-6, B: Kaye, D., and Rose, L.F.: Fundamentals of internal medicine, St. Louis, 1983, The C.V. Mosby Co.

12-3: Raven, P.H., and Johnson, G.B.: Biology, St. Louis, 1986, Times Mirror/Mosby College Publishing.

13-3; 13-9; 13-13; 13-14: Beck, E.W.: Mosby's atlas of functional human anatomy, St. Louis, 1982, The C.V. Mosby Co.

13-17: Daffner, R.H.: Introduction to clinical radiology, St. Louis, 1978, The C.V. Mosby Co.

14-2; 14-6; 14-11; 14-17: Beck, E.W.: Mosby's atlas of functional human anatomy, St. Louis, 1982, The C.V. Mosby Co.

14-8: Prior, J.A., Silberstein, J.S., and Stang, J.M.: Physical diagnosis: the history and examination of the patient, ed. 6, St. Louis, 1981, The C.V. Mosby Co.

14-10: Adapted from Lang and Wachsmith: Praktische anatomie, Berlin, 1955, Springer-Verlag.

14-13: Daffner, R.H.: Introduction to clinical radiology, St. Louis, 1978, The C.V. Mosby Co.

15-4: Courtesy Dr. H. Nakahara; from Bevelander, G., and Ramaley, J.A.: Essentials of histology, ed. 8, St. Louis, 1979, The C.V. Mosby Co.

15-5: Beck, E.W.: Mosby's atlas of functional human anatomy, St. Louis, 1982, The C.V. Mosby Co.

p. 367: Courtesy Travenol Laboratories, Inc., Deerfield, Illinois; From Brooks, S.M., and Paynton-Brooks, N.: The human body: structure and function in health and disease, ed. 2, St. Louis, 1980, The C.V. Mosby Co.

15-7: Bevelander, G. and Ramaley, J.A.: Essentials of histology, ed. 8, St. Louis, 1979, The C.V. Mosby Co.

16-2; 16-3: Beck, E.W.: Mosby's atlas of functional human anatomy, St. Louis, 1982, The C.V. Mosby Co.

18-6: Beck, E.W.: Mosby's atlas of functional human anatomy, St. Louis, 1982, The C.V. Mosby Co.

18-8; 18-9; 18-10; 18-13; 18-15; 18-16; 18-17: Bowers, A.C., and Thompson, J.M.: Clinical manual of health assessment, ed. 2, St. Louis, 1984, The C.V. Mosby Co.

18-11: Malasanos, L., and others: Health assessment, ed. 3, St. Louis, 1986, The C.V. Mosby Co.

18-18: Raven, P.H., and Johnson, G.B.: Biology, St. Louis, 1986, Times Mirror/Mosby College Publishing.

p. 454: Beck, E.W.: Mosby's atlas of functional human anatomy, St. Louis, 1982, The C.V. Mosby Co.